Population-Level Ecological Risk Assessment

Other titles from the Society of Environmental Toxicology and Chemistry (SETAC)

Ecosystem Responses to Mercury Contamination: Indicators of Change
Harris, Krabbenhoft, Mason, Murray, Reash, Saltman, editors
2007

Genomic Approaches for Cross-Species Extrapolation in Toxicology
Benson and Di Giulio, editors
2007

New Improvements in the Aquatic Ecological Risk Assessment of Fungicidal Pesticides and Biocides
Van den Brink, Maltby, Wendt-Rasch, Heimbach, Peeters, editors
2007

Freshwater Bivalve Ecotoxicology
Farris and Van Hassel, editors
2006

Estrogens and Xenoestrogens in the Aquatic Environment: An Integrated Approach for Field Monitoring and Effect Assessment
Vethaak, Schrap, de Voogt, editors
2006

Assessing the Hazard of Metals and Inorganic Metal Substances in Aquatic and Terrestrial Systems
Adams and Chapman, editors
2006

Perchlorate Ecotoxicology
Kendall and Smith, editors
2006

Natural Attenuation of Trace Element Availability in Soils
Hamon, McLaughlin, Stevens, editors
2006

For information about SETAC publications, including SETAC's international journals, *Environmental Toxicology and Chemistry* and *Integrated Environmental Assessment and Manage-ment*, contact the SETAC Administrative Office nearest you:

SETAC Office
1010 North 12th Avenue
Pensacola, FL 32501-3367 USA
T 850 469 1500 F 850 469 9778
E setac@setac.org

SETAC Office
Avenue de la Toison d'Or 67
B-1060 Brussels, Belgium
T 32 2 772 72 81 F 32 2 770 53 86
E setac@setaceu.org

www.setac.org
Environmental Quality Through Science®

Population-Level Ecological Risk Assessment

Lawrence W. Barnthouse
Wayne R. Munns, Jr.
Mary T. Sorensen

Coordinating Editor of SETAC Books
Joseph W. Gorsuch
Gorsuch Environmental Management Services, Inc.
Webster, New York, USA

CRC Press
Taylor & Francis Group
Boca Raton London New York

CRC Press is an imprint of the
Taylor & Francis Group, an **informa** business

First published 2008 by the Society of Environmental Toxicology and Chemistry (SETAC)

Published 2019 by CRC Press
Taylor & Francis Group
6000 Broken Sound Parkway NW, Suite 300
Boca Raton, FL 33487-2742

First issued in paperback 2019

ISBN-13: 978-0-367-45292-6 (pbk)
ISBN-13: 978-1-4200-5332-6 (hbk)

Visit the Taylor & Francis Web site at
http://www.taylorandfrancis.com

and the CRC Press Web site at
http://www.crcpress.com

Library of Congress Cataloging-in-Publication Data

Barnthouse, L. W. (Lawrence W.)
Population-level ecological risk assessment / Lawrence W. Barnthouse, Wayne R. Munns, Jr., Mary T. Sorensen.
p. cm.
Includes bibliographical references.
ISBN 978-1-4200-5332-6 (alk. paper)
1. Population biology. 2. Ecological risk assessment. I. Munns, Wayne R. II. Sorensen, Mary T. III. Title.

QH352.B374 2007
577.8'8--dc22 2007018699

SETAC Publications

Books published by the Society of Environmental Toxicology and Chemistry (SETAC) provide in-depth reviews and critical appraisals on scientific subjects relevant to understanding the impacts of chemicals and technology on the environment. The books explore topics reviewed and recommended by the Publications Advisory Council and approved by the SETAC North America, Latin America, or Asia/Pacific Board of Directors; the SETAC Europe Council; or the SETAC World Council for their importance, timeliness, and contribution to multidisciplinary approaches to solving environmental problems. The diversity and breadth of subjects covered in the series reflect the wide range of disciplines encompassed by environmental toxicology, environmental chemistry, and hazard and risk assessment, and life-cycle assessment. SETAC books attempt to present the reader with authoritative coverage of the literature, as well as paradigms, methodologies, and controversies; research needs; and new developments specific to the featured topics. The books are generally peer reviewed for SETAC by acknowledged experts.

SETAC publications, which include Technical Issue Papers (TIPs), workshop summaries, newsletter (*SETAC Globe*), and journals (*Environmental Toxicology and Chemistry* and *Integrated Environmental Assessment and Management*), are useful to environmental scientists in research, research management, chemical manufacturing and regulation, risk assessment, and education, as well as to students considering or preparing for careers in these areas. The publications provide information for keeping abreast of recent developments in familiar subject areas and for rapid introduction to principles and approaches in new subject areas.

SETAC recognizes and thanks the past coordinating editors of SETAC books:

A.S. Green, International Zinc Association, Durham, North Carolina, USA
C.G. Ingersoll, Columbia Environmental Research Center
US Geological Survey, Columbia, Missouri, USA
T.W. La Point, Institute of Applied Sciences
University of North Texas, Denton, Texas, USA
B.T. Walton, US Environmental Protection Agency
Research Triangle Park, North Carolina, USA
C.H. Ward, Department of Environmental Sciences and Engineering
Rice University, Houston, Texas, USA

Contents

APPROACHES TO POPULATION-LEVEL ECOLOGICAL RISK ASSESSMENT

List of Figures

List of Tables

Foreword

The workshop from which this book resulted, Population-Level Ecological Risk Assessment, held in Roskilde, Denmark, 23–27 August 2003, was part of the successful "Pellston Workshop Series," from the Society of Environmental Toxicology and Chemistry (SETAC). Since 1977, 44 Pellston Workshops have brought scientists together to evaluate current and prospective environmental issues. Each workshop has focused on a relevant environmental topic, and the proceedings of each have been published as peer-reviewed or informal reports. These documents have been widely distributed and are valued by environmental scientists, engineers, regulators, and managers for their technical basis and their comprehensive, state-of-the-science reviews. The other workshops in the Pellston series follow:

- Estimating the Hazard of Chemical Substances to Aquatic Life; Pellston, Michigan, 13–17 June 1977. Published by the American Society for Testing and Materials, STP 657, 1978.
- Analyzing the Hazard Evaluation Process; Waterville Valley, New Hampshire, 14–18 August 1978. Published by The American Fisheries Society, 1979.
- Biotransformation and Fate of Chemicals in the Aquatic Environment; Pellston, Michigan, 14–18 August 1979. Published by The American Society of Microbiology, 1980.
- Modeling the Fate of Chemicals in the Aquatic Environment; Pellston, Michigan, 16–21 August 1981. Published by Ann Arbor Science, 1982.
- Environmental Hazard Assessment of Effluents; Cody, Wyoming, 23–27 August 1982. Published as a SETAC Special Publication by Pergamon Press, 1985.
- Fate and Effects of Sediment-Bound in Aquatic Systems; Florissant, Colorado, 11–18 August 1984. Published as a SETAC Special Publication by Pergamon Press, 1987.
- Research Priorities in Environmental Risk Assessment; Breckenridge, Colorado, 16–21 August 1987. Published by SETAC, 1987.
- Biomarkers: Biochemical, Physiological, and Histological Markers of Anthropogenic Stress, Keystone; Colorado, 23–28 July 1989. Published as a SETAC Special Publication by Lewis Publishers, 1992.
- Population Ecology and Wildlife Toxicology of Agricultural Pesticide Use: A Modeling Initiative for Avian Species; Kiawah Island, South Carolina, 22–27 July 1990. Published as a SETAC Special Publication by Lewis Publishers, 1994.
- A Technical Framework for [Product] Life-Cycle Assessments; Smuggler's Notch, Vermont, 18–23 August 1990. Published by SETAC, January 1991; 2nd printing September 1991; 3rd printing March 1994.

- Aquatic Microcosms for Ecological Assessment of Pesticides; Wintergreen, Virginia, 7–11 October 1991. Published by SETAC, 1992.
- A Conceptual Framework for Life-Cycle Assessment Impact Assessment; Sandestin, Florida, 1–6 February 1992. Published by SETAC, 1993.
- A Mechanistic Understanding of Bioavailability: Physical-Chemical Interactions; Pellston, Michigan, 17–22 August 1992. Published as a SETAC Special Publication by Lewis Publishers, 1994.
- Life-Cycle Assessment Data Quality Workshop; Wintergreen, Virginia, 4–9 October 1992. Published by SETAC, 1994.
- Avian Radio Telemetry in Support of Pesticide Field Studies; Pacific Grove, California, 5–8 January 1993. Published by SETAC, 1998.
- Sustainability-Based Environmental Management; Pellston, Michigan, 25–31 August 1993. Cosponsored by the Ecological Society of America. Published by SETAC, 1998.
- Ecotoxicological Risk Assessment for Chlorinated Organic Chemicals; Alliston, Ontario, Canada, 25–29 July 1994. Published by SETAC, 1998.
- Application of Life-Cycle Assessment to Public Policy; Wintergreen, Virginia, 14–19 August 1994. Published by SETAC, 1997.
- Ecological Risk Assessment Decision Support System; Pellston, Michigan, 23–28 August 1994. Published by SETAC, 1998.
- Avian Toxicity Testing; Pensacola, Florida, 4–7 December 1994. Cosponsored by Organisation for Economic Co-operation and Development. Published by OECD, 1996.
- Chemical Ranking and Scoring (CRS): Guidelines for Developing and Implementing Tools for Relative Chemical Assessments; Sandestin, Florida, 12–16 February 1995. Published by SETAC, 1997.
- Ecological Risk Assessment of Contaminated Sediments; Pacific Grove, California, 23–28 April 1995. Published by SETAC, 1997.
- Ecotoxicology and Risk Assessment for Wetlands; Fairmont, Montana, 30 July–3 August 1995. Published by SETAC, 1999.
- Uncertainty in Ecological Risk Assessment; Pellston, Michigan, 23–28 August 1995. Published by SETAC, 1998.
- Whole-Effluent Toxicity Testing: An Evaluation of Methods and Prediction of Receiving System Impacts; Pellston, Michigan, 16–21 September 1995. Published by SETAC, 1996.
- Reproductive and Developmental Effects of Contaminants in Oviparous Vertebrates; Fairmont, Montana, 13–18 July 1997. Published by SETAC, 1999.
- Multiple Stressors in Ecological Risk Assessment; Pellston, Michigan, 13–18 September 1997. Published by SETAC, 1999.
- Re-evaluation of the State of the Science for Water Quality Criteria Development; Fairmont, Montana, 25–30 June 1998. Published by SETAC, 2003.
- Criteria for Persistence and Long-Range Transport of Chemicals in the Environment; Fairmont Hot Springs, British Columbia, Canada, 14–19 July 1998. Published by SETAC, 2000.

- Assessing Contaminated Soils: From Soil-Contaminant Interactions to Ecosystem Management; Pellston, Michigan, 23–27 September 1998. Published by SETAC, 2003.
- Endocrine Disruption in Invertebrates: Endocrinology, Testing, and Assessment (EDIETA); Amsterdam, The Netherlands, 12–15 December 1998. Published by SETAC, 1999.
- Assessing the Effects of Complex Stressors in Ecosystems; Pellston, Michigan, 11–16 September 1999. Published by SETAC, 2001.
- Environmental-Human Health Interconnections; Snowbird, Utah, 10–15 June 2000. Published by SETAC, 2002.
- Ecological Assessment of Aquatic Resources: Application, Implementation, and Communication; Pellston, Michigan, 16–21 September 2000. Published by SETAC, 2004.
- The Global Decline of Amphibian Populations: An Integrated Analysis of Multiple Stressors Effects; Wingspread, Racine, Wisconsin, 18–23 August 2001. Published by SETAC, 2003.
- Methods of Uncertainty Analysis for Pesticide Risks; Pensacola, Florida, 24 February–1 March 2002.
- The Role of Dietary Exposure in the Evaluation of Risk of Metals to Aquatic Organisms; Fairmont Hot Springs, British Columbia, Canada, 27 July–1 August 2002. Published by SETAC, 2005.
- Use of Sediment Quality Guidelines (SQGs) and Related Tools for the Assessment of Contaminated Sediments; Fairmont Hot Springs, Montana, 17–22 August 2002. Published by SETAC, 2005.
- Science for Assessment of the Impacts of Human Pharmaceuticals on Aquatic Ecosystem; Snowbird, Utah, 3–8 June 2003. Published by SETAC, 2005.
- Valuation of Ecological Resources: Integration of Ecological Risk Assessment and Socio-Economics to Support Environmental Decisions; Pensacola, Florida, 4–9 October 2003. To be published by SETAC, 2007.
- Emerging Molecular and Computational Approaches for Cross-Species Extrapolations; Portland, Oregon, 18–22 July 2004. Published by SETAC, 2006.
- Molecular Biology and Risk Assessment: Evaluation of the Potential Roles of Genomics in Regulatory Ecotoxicology; Pellston, Michigan, 18–22 September 2006. To be published by SETAC, 2007.
- Veterinary Pharmaceuticals; Pensacola, Florida, 12–16 February 2006. To be published by SETAC, 2007.

Preface

On 22 August 2003, the Pellston Workshop on Population-Level Ecological Risk Assessment convened at the Comwell Conference Center in Roskilde, Denmark. Over the next 5 days, 37 experts discussed a wide range of topics related to assessing risks of chemical and nonchemical stressors to higher levels of biological organization. The workshop had its origin in a very popular debate session on "Protection of Populations in Ecological Risk Assessment" held as part of the 22nd Annual Meeting of the Society of Environmental Toxicology and Chemistry (SETAC) North America (Baltimore, Maryland, November 2001). The content of the workshop was also shaped by a special session on ecological modeling held in conjunction with the 12th SETAC Europe meeting (Vienna, Austria, May 2002) and by an interactive poster session held at the 2002 SETAC North America meeting in Salt Lake City, Utah.

The Roskilde workshop continued a trend begun more than 10 years before, toward linking environmental toxicology and chemistry with ecology and other environmental disciplines to create new approaches for quantifying ecological risks. The workshop also followed the move toward globalization of SETAC, in that the steering committee included members of both SETAC North America and SETAC Europe, and the workshop itself included participants from North America, Europe, Japan, and Australia. Workshop planning included a deliberate effort to identify and support 3 highly qualified student participants.

The deliberations of the workshop made it clear that the intention, and often requirement, for protecting populations exist in the laws and regulatory processes of many jurisdictions. Further, the science supporting assessment of risks to population is rapidly developing in North America, Europe, and Asia and is already sufficient to support a diverse array of population-level assessments.

Knowledge bases, tools, and experiences available from the fields of conservation biology and resource management provide a solid foundation from which to learn and build. Greater interaction and collaboration among all of these groups is a necessary ingredient to advancing the science of population-level ecological risk assessment and its use in environmental decision-making.

Lawrence W. Barnthouse
Wayne R. Munns, Jr.
Mary T. Sorensen

Acknowledgments

The authors and editors of this book wish to acknowledge the following sponsors of a Society of Environmental Toxicology and Chemistry (SETAC) Pellston Workshop on Population-Level Ecological Risk Assessment in Roskilde, Denmark, 23–27 August 2003:

American Chemistry Council
American Petroleum Institute
ARCADIS
Bayer CropScience
Danish Environmental Protection Agency
Department of Energy
ENVIRON International Corp.
Health Canada
ExxonMobil
General Electric
Menzie-Cura & Associates, Inc.
NOAA/OR&R/CPRD
Roskilde University
Syngenta
US Army Center for Health Promotion and Preventive Medicine
US Environmental Protection Agency

Going back in time, interest and energy in the development of this book was sparked by a lively debate at the 2001 SETAC meeting in Baltimore. Many thanks go to Miranda Henning and Greg Biddinger for their assistance in the development of debate topics and chairing the infamous event. Thanks and acknowledgments also go to Ken Jenkins and John Emlen, 2 highly respected scientists who participated in the debate but for a variety of circumstances did not participate in the book itself.

This book is the product of the workshop and subsequent efforts by the authors to turn rough drafts of chapters into polished manuscripts. The book has also benefited greatly from the comments of the 2 anonymous reviewers.

We wish to thank SETAC for supporting the workshop from beginning to end. In particular, we are grateful to Greg Schiefer for his support in organizing the workshop and managing the funding associated with this effort and to Mimi Meredith for arranging the peer review, coordinating publication activities, and assisting with countless questions and issues.

We also would like to thank Valery Forbes for offering Roskilde University as the workshop site and for her enthusiastic and tireless effort as local meeting coordinator. The general support and participation of Roskilde University students Anne Jørgensen and Anders Elleby Engell-Kofoed also are appreciated.

We appreciate that Mike Thompson provided the research and efforts associated with the compilation of supplemental reading material provided in Appendix 3.

We want to acknowledge 3 individuals from Taylor and Francis who contributed greatly to the quality of this book. Susan Horwitz, our project editor, assisted with all aspects of our editorial needs from format to content, and she worked diligently to assist us on our final product (she will smile when we say thank/you and ask to keep just this one slash *please!*). In addition, Suzanne Lassandro provided support and wisdom regarding book production. The cover of this book was designed by Shayna Murry.

And finally, many thanks to Paige Leitman for the feature photograph that gives the cover its striking appearance (shark and rays). Additional photographs were provided by Wayne Munns (sea lions, common noddy, and flying foxes) and an anonymous photographer (meerkats and jungle).

Each of the contributions in this book has been peer reviewed. The opinions expressed in this book are those of the participants and may not reflect those of any of their agencies, the funding agencies, or SETAC.

About the Editors

Lawrence W. Barnthouse, PhD, is president and principal scientist of LWB Environmental Services, Inc., and adjunct associate professor of zoology at Miami University. After receiving his PhD in biology from the University of Chicago in 1976, he spent the next 19 years as a research staff member and group leader in the Oak Ridge National Laboratory's (ORNL) Environmental Sciences Division. During his years at ORNL, he was involved in dozens of environmental research and assessment projects involving the development of new methods for predicting and measuring environmental risks of energy technologies. Impacts of environmental stressors on populations were a major focus of many of these projects. In 1981, Dr Barnthouse became coprincipal investigator (with SETAC Founder's Award recipient Glenn Suter) on the US Environmental Protection Agency's (USEPA) first research project on ecologic risk assessment. Since that time, he has been active in the development and application of ecologic risk assessment methods for USEPA, other federal agencies, state agencies, and private industry. He was an advisor to USEPA's Risk Assessment Forum during the development of the Framework for Ecological Risk Assessment and Guidelines for Ecological Risk Assessment and served on the peer review panels for both documents. He is a Fellow of the American Association for the Advancement of Science, hazard and risk assessment editor of *Environmental Toxicology and Chemistry*, and founding editorial board member of *Integrated Environmental Assessment and Management.* He chaired a Pellston workshop on sustainable environmental management in 1993, and has participated in several other Pellston workshops. He currently chairs SETAC's Population-Level Ecological Risk Assessment Work Group.

Wayne R. Munns, Jr., PhD, is the associate director for science for the US Environmental Protection Agency's (USEPA) Atlantic Ecology Division (Office of Research and Development) in Narragansett, RI. A marine ecologist by training (University of Rhode Island, 1984), Dr Munns has expertise in developing and applying quantitative methods for ecological risk assessment, ecological modeling with particular emphasis on population dynamics, and large spatial scale environmental assessments. He has conducted research and managed programs addressing ocean disposal, hazardous waste sites, contaminated sediments, wildlife risk assessment, and environmental criteria development. His current interests include development of population-level ecological risk assessment methods to support regulatory decisions, ecological services and their valuation, and integration of assessment approaches to enhance the value of information supporting environmental protection decisions. Before joining USEPA, he was a senior scientist, division manager, and assistant vice president for Science Applications International Corporation. He has been a member of USEPA's Risk Assessment Forum and has advised the World Health Organization on the integration of human health and ecological risk assessment. Dr Munns has contributed to the development of several previous Pellston workshops as a steering committee member, editor, and participant. He is a past member of the editorial board for SETAC's *Environmental Toxicology and Chemistry* and currently serves as vice-chair of SETAC's Ecological Risk Assessment Advisory Group.

Mary T. Sorensen, CE, is a senior science advisor for ENVIRON International Corporation in Atlanta, GA. With a BS focused on animal behavior and environmental sciences, an MS in environmental biology (Georgia Institute of Technology 1992; 2000), and as a certified ecologist (Ecological Society of America 1999), Ms. Sorensen has focused on the integration of ecological principles into ecological risk assessment. She provides technical direction and senior management on projects related to hazardous waste sites, contaminated sediments, ecotoxicology, population-level and landscape-level wildlife risk assessments, net environmental benefit analysis, remedy alternatives analysis (including risk of remedy), and habitat restoration. Ms. Sorensen has participated in the development of the American Society of Testing and Materials' (ASTM) *Standard Guide for Risk-Based Corrective Action for Protection of Ecological Resources* and has served as a peer reviewer for the US Environmental Protection Agency publication, *Assessing Risks to Populations at Superfund and RCRA Sites: Characterizing Effects*. She currently cochairs SETAC's Population-Level Ecological Risk Assessment Work Group. As an ongoing pro bono effort, she serves as a biological monitoring instructor for the Georgia Environmental Protection Division Adopt-A-Stream (AAS) program; and, the upper Etowah River Alliance named her "Trainer of the Year" in 2006.

Workshop Participants

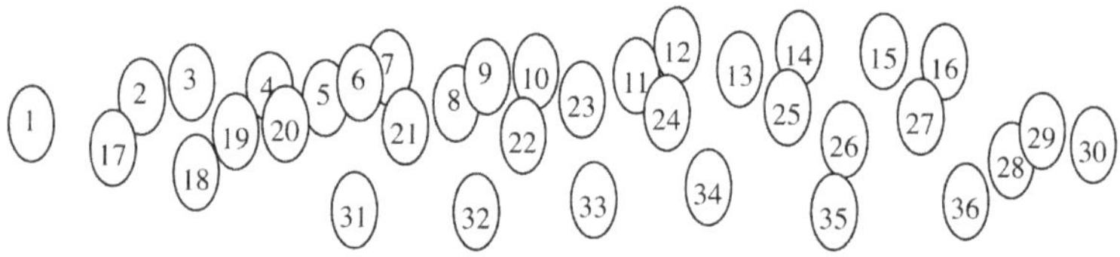

1. Gregory R. Biddinger
2. Anders Elleby Engell-Kofoed
3. Steve J. Maund
4. Mary T. Sorensen
5. Sandra M. Brasfield
6. Lawrence W. Barnthouse
7. Wayne R. Munns, Jr.
8. Nancy A. Bettinger
9. Niels Elmegaard
10. Helen Noel
11. Helen Regan
12. Wayne G. Landis
13. Randall S. Wentsel
14. Paul J. van den Brink
15. Udo Hommen
16. Peter D. Delorme
17. Peter Calow
18. Valery E. Forbes
19. Anne Jørgensen
20. Richard Sibly
21. Chris J. Topping
22. Jannicke Moe
23. W. Nelson Beyer
24. Peter F. Chapman
25. Glenn E. Harris
26. Charles A. Menzie
27. Robert A. Pastorok
28. Greg Schiefer
29. Lawrence A. Kapustka
30. Bruce K. Hope
31. Bin-Le Lin
32. Mayuko Nakamaru
33. Alyce T. Fritz
34. Julann A. Spromberg
35. Tina M. Carlsen
36. Jennifer A. Gervais

Missing from photo:

Ary A. Hoffmann
Vibeke Moller
Diane E. Nacci

Workshop Participants

Lawrence W. Barnthouse
(Workshop chair)
LWB Environmental Services, Inc
Hamilton, Ohio, USA

Nancy A. Bettinger
Massachusetts Department of Environmental Protection
Boston, Massachusetts, USA

W. Nelson Beyer
US Geological Survey
Patuxent Wildlife Research Center
Laurel, Maryland, USA

Gregory R. Biddinger
(Steering Committee member, Work Group leader)
ExxonMobil Refining & Supply Company
Fairfax, Virginia, USA

Sandra M. Brasfield
(Student participant)
University of New Brunswick
Saint John, New Brunswick, Canada

Peter Calow
University of Sheffield
Sheffield, England

Tina M. Carlsen
Lawrence Livermore National Laboratory
Livermore, California, USA

Peter F. Chapman
(Steering Committee member, Work Group leader)
Syngenta Ecological Science
Bracknell, Berkshire, England

Peter D. Delorme
PMRA, Health Canada
Environmental Assessment Division
Ottawa, Ontario, Canada

Niels Elmegaard
National Environmental Research Institute
Silkeborg, Denmark

Valery E. Forbes
(Steering Committee member)
Roskilde University
Roskilde, Denmark

Alyce T. Fritz
(Steering Committee member)
NOAA/OR&R/CPRD
Coastal Protection and Restoration Division
Seattle, Washington, USA

Jennifer A. Gervais
(Student participant)
Utah State University
Logan, Utah, USA

Glenn E. Harris
BC Ministry of Water, Land, and Air Protection
Victoria, British Columbia, Canada

Ary A. Hoffmann
La Trobe University
Centre for Environmental Stress & Adaptation Research
Victoria, Australia

Udo Hommen
Fraunhofer Institute for Molecular Biology and Applied Ecology
Schmallenberg, Germany

Bruce K. Hope
(Steering Committee member)
Oregon Dept. of Environmental Quality
Portland, Oregon, USA

Lawrence A. Kapustka
Ecological Planning & Toxicology, Inc
Corvallis, Oregon, USA

Wayne G. Landis
(Steering Committee member)
Western Washington University
Bellingham, Washington, USA

Bin-Le Lin
National Institute of Advanced Industrial Science
Research Center for Chemical Risk Management
Tukuba City, Japan

Steve J. Maund
Sygenta Crop Protection AG
Basel, Switzerland

Charles A. Menzie
(Steering Committee member, Work Group leader)
Exponent
Alexandria, Virginia, USA

Jannicke Moe
University of Oslo
Oslo, Norway

Vibeke Moller
Danish Environmental Protection Agency
Copenhagen, Denmark

Wayne R. Munns, Jr.
(Steering Committee member, Work Group leader)
US EPA, ORD NHEERL, Atlantic Ecology Division
Narragansett, Rhode Island, USA

Diane E. Nacci
US EPA, ORD NHEERL, Atlantic Ecology Division
Narragansett, Rhode Island, USA

Mayuko Nakamaru
Shizuka University
Department of Systems Engineering
Shizuoka, Japan

Helen Noel
(Student participant)
University of Reading
Reading, Berkshire, England

Robert A. Pastorok
Exponent, Inc
Bellevue, Washington
USA

Helen Regan
San Diego State University
San Diego, California, USA

Richard Sibly
University of Reading
Reading, England

Mary T. Sorensen
(Workshop cochair)
ENVIRON International Corporation
Atlanta, Georgia, USA

Julann A. Spromberg
NOAA /National Marine Fisheries Service
Northwest Fisheries Science Center
Seattle, Washington, USA

Chris J. Topping
NERI, Department of Wildlife Ecology & Biodiversity
Roende, Denmark

Paul J. van den Brink
(Steering Committee member)
Alterra Green World Research
The Netherlands

Randall S. Wentsel
(Steering Committee member, Work Group leader)
US EPA, ORD Office of Science Policy
Washington, DC, USA

Workshop participants had the opportunity to row and sail historic Viking ships on Roskilde Fjord.

Glossary of Key Terms

assessment endpoint an explicit expression of the environmental value that is to be protected, operationally defined by an ecological entity and its attributes

assessment population a group of conspecific organisms occupying a defined area that has been selected to serve as an assessment endpoint entity for an ecological risk assessment (see *population*)

attribute a quality or characteristic of an ecological entity

bottleneck effect a reduction in genetic heterogeneity within a population as a result of stressor-induced (or other) mortality

carrying capacity the maximum abundance of a biological population that is sustainable by a habitat or environment (see *equilibrium abundance*)

cohort a group of similarly aged members of the population

compensation a feedback between the density of a population and some biological property of that population (typically demographic rates) (synonymous with *density dependence*)

compensatory mechanism a biological mechanism, such as homeostatic acclimation of individuals, genetic adaptation, and density dependence in vital rates and migration, that can ameliorate adverse effects over the short or long term

demographic rates age- or stage-specific birth, death, and migration rates of individuals within the population (synonymous with *vital rates*)

density dependence a feedback between the density of a population and some biological property of that population (typically demographic rates) (synonymous with *compensation*) EC_{50} concentration causing a response in 50% of exposed organisms, EC_{100} concentration causing a response in 100% of exposed organisms

equilibrium abundance the abundance of a population at steady state (see *carrying capacity*)

error uncertainty resulting from the use of the wrong methods, models, and data in assessment activities (compare *ignorance*, *variability*)

ignorance uncertainty resulting from incertitude; a component of uncertainty resulting from lack of knowledge about the true value of a parameter that can result from inadequate or imperfect measurement (compare *error*, *variability*)

inbreeding depression detrimental changes in birth and death rates resulting from reduced genetic diversity (usually a problem of small effective population sizes) LD_{50} dose lethal to 50% of exposed organisms

life table age- or stage-specific schedules of fecundity and survivorship

metapopulation a biological population consisting of 2 or more subpopulations separated in space (see *subpopulation*) No Observed Effect Concentration

(NOEC), a concentration at which no statistically significant effect on exposed organisms is observed

parameterization quantification of the variables in a model

population variously: a collection of individuals of a single species that occupy some defined geographic space; a subset of all individuals of a given species that share a common area and that interbreed (see *assessment population*)

prediction a quantitative description of the future abundances or behavior of a population (compare *projection*)

projection a description of the future abundance or behavior of a population assuming constant environmental conditions (compare *prediction*)

sensitivity analysis an evaluation of the influences of model variables on model outputs

state variable a component or property of the system being modeled that, when aggregated with other state variables, determines what the system looks like

subpopulation an internally coherent subdivision of the larger population (see *metapopulation*)

variability uncertainty resulting from actual differences in the value of a parameter or attribute among units in a statistical population (compare *error*, *ignorance*)

vital rates age- or stage-specific birth, death, and migration rates of individuals within the population (synonymous with *demographic rates*)

1 Introduction

Lawrence W. Barnthouse, Wayne R. Munns, Jr., and Mary T. Sorensen

Regulations, policies, directives, and guidance documents frequently discuss the need for ecological risk assessments to consider risks to populations, not simply to individual organisms or organism-level attributes. The reason for this is that, from a management perspective, the population-level attributes such as abundance, persistence, age composition, and genetic diversity are usually more relevant than are the health or persistence of individual organisms. Despite the many published calls for assessing risks to populations, the overwhelming majority of assessments of ecological risks of environmental chemicals are still based on an organism-level approach. Although populations are by no means the only entities that can or should be addressed in ecological risk assessments, the general rarity of assessments that focus on population attributes is quite remarkable.

The relative rarity of population-level ecological risk assessments is not from a lack of scientific understanding and technical tools. The field of population ecology has a long history, building a solid foundation of theory and empirical demonstration. Consideration of population-level endpoints is fundamental to resource management, and the sophistication of approaches used in conservation biology and fisheries management has increased steadily. Moreover, applications in environmental toxicology have become increasingly common. Whereas only a handful of articles in *Environmental Toxicology and Chemistry* before 1990 addressed population-level effects, more than 30 such articles appeared between 1995 and 2002.[1] Although research activity relating to population-level ecological risk assessment is at an all-time high, interaction between workers on different continents, and even among workers on the same continent, has been limited. Risk assessors worldwide face similar problems in determining how population-level considerations can be integrated into environmental decision-making.

The Pellston Workshop on Population-Level Ecological Risk Assessment had its origin in a debate session on "Protection of Populations in Ecological Risk Assessment" held as part of the 22nd Annual SETAC North America meeting (Baltimore, November 2001). The participants in the debate touched on a wide variety of topics of importance to risk assessment, including ecological protection goals, *de minimis* criteria for population impacts, mathematical modeling, and risk management decision-making. More than 200 SETAC members attended the debate, and discussion between the panel and the audience continued for more than an hour

[1] *Environmental Toxicology and Chemistry* was searched online for all articles in which the words "population" or "populations" were included in the title. The abstract of each "hit" was examined to ensure that the article discussed effects of chemicals on populations.

after the official ending of the session. After the meeting, the participants continued their debate via e-mail and established a Population-Level Ecological Risk Assessment Work Group under the sponsorship of the SETAC Ecological Risk Assessment Advisory Group. Planning for this Pellston workshop began early in 2002. The content of the workshop was shaped by a special session on ecological modeling held in conjunction with the 12th SETAC Europe meeting (Vienna, May 2002) and by an interactive poster session held at the 2002 SETAC North America meeting in Salt Lake City. The workshop itself was held 22–27 August 2003, in Roskilde, Denmark.

The workshop built on several recent Pellston workshops, especially the 1999 workshop on "Ecological Variability: Separating Natural from Anthropogenic Causes of Ecosystem Impairment" (Baird and Burton 2001) and the earlier workshop on "Wildlife Toxicology and Population Modeling" (Kendall and Lacher 1994). It continued a trend begun more than 10 years ago toward linking environmental toxicology and chemistry with ecology and other environmental disciplines to create new approaches for quantifying risks of chemical and nonchemical stressors to higher levels of ecological organization. The workshop also followed the trend toward globalization of SETAC, in that the steering committee included members of both SETAC North America and SETAC Europe, and the workshop itself included participants from North America, Europe, Japan, and Australia.

The objective of this workshop was to advance the practice of population-level ecological risk assessment by establishing a framework that included definition of goals, identification of appropriate assessment methods, and specification of data needs for different types of assessment applications, all within the context of providing information supportive of risk management decisions. Most of the work was conducted by 5 breakout groups, which dealt with the themes of "Defining Ecological Protection Goals," "Risk Management Decision-Making," "Empirical Approaches to Population-Level Ecological Risk Assessment," "Modeling Approaches to Population-Level Ecological Risk Assessment," and "A Framework for Population-Level Ecological Risk Assessment." Preworkshop white papers were commissioned on 2 special topics of interest to all of the breakout groups — density-dependence in populations and population genetics. Additional articles discussing spatial considerations for population-level risk assessment and transferable experiences from conservation biology and resource management were developed during the course of the workshop. All 4 of these articles are included as stand-alone chapters in this book, along with the products of the 5 breakout groups.

Ecological risk assessment is an internationally recognized, science-based tool used to inform environmental decision-making. To a large degree, the management contexts within which decisions are made determine the form of a risk assessment and define the nature of the information needed. Reflecting this, Chapter 2 summarizes a number of legal, regulatory, business, and other decision contexts relative to protecting populations. Specific contexts (termed "management scenarios") are elaborated in Appendix 1 and are used throughout Chapter 2 (and subsequent chapters) to illustrate important concepts, issues, and approaches associated with population-level risk assessment. Considerations are offered in Chapter 2 regarding the availability and adequacy of current ecological risk assessment approaches in accommodating population-level assessment endpoints to evaluate population-level risk. The chapter concludes with a series of specific recommendations for improving the use of population-level risk assessment to support environmental management decisions.

The decision context and management goals also influence how populations should be defined in the assessment, and what it means to protect a population from adverse effects. Chapter 3 explores these issues, recommending that "Assessment Populations" be defined in the manner that best reflects the informational needs of the decision. Clear, explicit, operational definitions of the assessment population of interest should be included in an ecological risk assessment. Assessment populations may encompass the biological population (or a component of a relevant metapopulation) or they may be designated as a component of the biological population (e.g., the exposed individuals). For assessments that involve 2 or more species, there may be 2 or more definitions of assessment populations so that the relevant ecological relationships and management objectives can be satisfied. Attributes of populations, which differ uniquely from those of organisms in that they represent structures and processes of groups of interacting organisms, are described relative to their potential value in population-level risk assessments. Together with specification of the assessment population, these form the assessment endpoints for the risk assessment. Because risks to assessment endpoints cannot always be assessed directly, measures are described that can be related directly to population-level assessment endpoints. Issues important to protection of populations are considered at the end of this chapter as they pertain to management goals.

The next 4 chapters discuss issues in population ecology that are germane to improving the quality of population-level risk assessments. Chapter 4 introduces the concept of density dependence in population regulation, reviewing the state of science with respect to theory and its practical application. Density-dependent processes can both compensate for stressor effects on populations and aggravate effects under certain circumstances. Without adequate acknowledgment of such processes and feedbacks, risk to populations can be overestimated or underestimated. Chapter 5 focuses on the genetics of populations, and how genetic structure can both affect how populations respond to stressor exposure, and be affected by that exposure. Population genetic attributes are not typically considered in ecological risk assessment, but because of the central importance genetics plays in population dynamics, candidate measures are offered that can inform environmental decisions. Chapter 6 argues for the importance of spatial considerations in population-level risk assessment. Many populations of plants and animals are subdivided into spatially distinct subpopulations linked through immigration and emigration. This chapter synthesizes studies that have investigated the ways in which spatial structure affects the responses of populations to chemical exposures. Chapter 7 describes approaches used to support decision-making in the fields of conservation biology and resource management. Central to this presentation are considerations of the technology used and lessons learned in those fields that may have relevance to population-level risk assessment.

Approaches that accommodate these and others issues are presented in the next two chapters, which describe the "tool box" of empirical and modeling methods supporting characterization of ecological effects in population-level risk assessment. Chapter 8 summarizes available approaches for using laboratory and field data directly to quantify effects of chemical and other stressors on populations. This chapter also discusses approaches for estimating parameters used by models that extrapolate responses measured at the organism-level of biological organization to population-level responses. Modeling approaches are described in Chapter 9. Building from an earlier Pellston workshop (Baird and Burton 2001), this chapter develops a taxonomy of existing mechanistic population modeling approaches, and describes the general approaches, assumptions, data requirements, and strengths and limitations of each category relative to their use in risk assessment. It also considers the types of population-level attributes that can be evaluated by different modeling approaches. Both chapters consider the uses of tools within tiered assessment formats, and the nuances that derive from the nature of the assessment (prospective versus retrospective, site-specific versus site-independent).

The concepts and approaches communicated in the preceding chapters are synthesized in Chapter 10 into broad guidance for population-level ecological risk assessment following the general structure of the ecological risk assessment framework commonly used in North America. Focusing particularly on the problem formulation and risk characterization stages of an assessment, the chapter describes a process for framing questions and considerations that support selection of assessment endpoints and development of an analysis plan. The chapter then discusses how empirical data and models can be used to quantify and characterize risks to population-level assessment endpoints.

This volume concludes in Chapter 11 by summarizing the issues identified and recommendations generated over the course of the workshop. This chapter highlights key considerations in formulating and conducting population-level ecological risk assessments, describes guidance and communication activities that will foster broader acceptance of population-level ecological risk assessments in environmental management decisions, and describes the research and development needed to improve those assessments and their interpretation.

Appendix 1 presents a variety of scenarios that illustrate important concepts, issues and approaches associated population-level ecological risk assessment. Scenarios identify the regulatory context, role of risk assessment in support of the management decision, current risk assessment approach, and the rationale for population-level risk assessment methods. Scenarios include the European Water Framework Directive, estimating consequences of exceeding water and sediment quality thresholds, agricultural pesticide registration, and hazardous waste site management. Appendix 2 describes an exercise undertaken during the workshop to apply two population modeling approaches to assessments that might support one of these regulatory contexts, namely agricultural pesticide registration. This exercise illustrates how the modeling approach selected can affect population-level ecological risk assessment outcomes. Appendix 3 of this book presents a broad range of supplemental reading material in the fields of risk assessment, ecotoxicology, and population ecology. These references were not cited in the book (i.e., Chapter 12), but are considered relevant for a further understanding of this topic.

The unique insights and enthusiastic participation of workshop members were the key ingredients contributing to the success of the workshop and attainment of workshop objectives. In addition, and a novel experiment for the Pellston process, there was continued open dialog and interactions with scientists around the world before and after the workshop. Open dialog occurred via multiple direct solicitations using email, SETAC interactive poster sessions, the establishment of a SETAC Population-Level Ecological Risk Assessment Workgroup (a part of the Ecological Risk Assessment Advisory Group [ERAAG]), development and maintenance of an associated Internet Web site (under ERAAG). We hope the book motivates the many interested risk assessors and managers who were unable to participate directly in the workshop to become involved in the further development of and application of population-level ecological risk assessment.

2 Managing Risk to Ecological Populations

Gregory R. Biddinger, Peter Calow, Peter Delorme, Glenn Harris, Bruce Hope, Bin-Le Lin, Mary T. Sorensen, and Paul van den Brink

INTRODUCTION

With the exception of legally protected species (e.g., threatened or endangered species in the United States) individual organisms generally have little legal, societal, or ecological relevance. Thus ecological risk management decisions are (whether stated or implied) essentially aimed at ensuring protection of a population of individuals and not necessarily each individual in that population. Risk management and ecological risk-based decision-making has been the topic of numerous books, articles, and guidance documents in recent years (Kolluru et al. 1996; Barnthouse et al. 1998; Koller 1999; USEPA 1999; Swindoll et al. 2000; Stahl 2001). The purpose of this chapter is to discuss the current context for risk management decision-making that is considered protective of ecological populations.

RISK MANAGEMENT AND RISK ASSESSMENT CONTEXT FOR POPULATION-LEVEL ECOLOGICAL RISK ASSESSMENT

Risk management and risk assessment are separate, yet interrelated, activities. Risk management refers to the activities of identifying, evaluating and selecting among alternative regulatory actions. Risk managers typically deal with broad social, economic, ethical, and political issues when selecting among management options. Evaluating risk-benefit, cost-benefit, or risk-risk tradeoffs among options is part of risk management. Risk assessment is typically defined as the scientific analysis and characterization of adverse effects of environmental hazards. It may include both qualitative and quantitative descriptors, but it often excludes perceived risk, risk comparisons, and analysis of the social and economic effects of management or regulatory decisions. A clear distinction between risk management and risk assessment is useful for various important purposes, such as insulating scientific activity from political pressure and maintaining the analytical distinction between the magnitude of a risk and the cost of coping with it. However, maintaining too strict a separation between management and assessment can impede the decision-making process. A risk assessment must provide information that is useful for risk management decision-making (i.e., it is not sufficient to merely do a scientific study); thus, interaction between the assessment and management functions is warranted and necessary (NRC 1996; Suter et al. 2000).

At a practical level, a risk manager, unlike a risk assessor, has the responsibility and the authority to make decisions that lead to actions that reduce risk to, or maintain it at, an acceptable level for specific, ecological receptors. A risk manager must thus 1) decide, based in part on the results of a risk assessment, whether a chemical or other stressor in the environment presents an unacceptable risk and 2) if unacceptable risk is identified (or predicted), decide on appropriate actions that will protect the environment over time, and minimize or mitigate collateral injuries to natural resources. The risk manager may also have to determine the levels of risk that must be addressed through active measures versus those that may be addressed through passive means. The risk manager also must consider the level of adverse impacts to ecological populations that could occur from the very actions intended to protect ecological populations (e.g., excavation of a wetland to prevent chemical exposures). The ability to make sound risk management decisions rests heavily on information provided through the risk assessment process. Information available to risk managers can range from the qualitative to the quantitative and can be developed using a variety of sources, methods, and tools. An ecological risk assessment (ERA) is one tool available to provide risk managers with information about the nature, magnitude, and distribution of risk to ecological receptors. ERAs that are focused toward ecological entities (e.g., populations) that risk managers are interested in protecting provide useful information for the management decision.

POPULATION-LEVEL ERA APPLICATIONS

The use of population-level ERAs versus organism-level ERAs depends on the risk management context. The release of a chemical or nonchemical stressor into the environment, either intentionally or unintentionally, can result in risk to various ecological receptors, organized on a scale ranging from organisms to ecosystems. For the great majority of species (those that are not legally protected or aesthetically valued), losses of individual organisms are to be expected and management action is only necessary in the face of adverse impacts to populations, communities, and ecosystems. However, many ERAs focus on organism-level attributes. Although this focus may be appropriate under certain circumstances (e.g., when protecting threatened or endangered species), it is unlikely to be ecologically relevant for the great majority of species. Protecting all individuals at all times, when there is no clear ecological or legal reason to do so, can lead to disproportionate risk management actions. Populations and their attributes are responsive assessment endpoints for ERAs that can support most management decisions. Using population-level ERA can be expected to increase 1) the ecological relevance of decision-making, 2) better address the intent of laws and regulations, 3) allow greater recognition of societal and stakeholder values, and 4) provide managers with the breadth of information necessary to support practical, balanced, and implementable decisions.

Using population-level ERA can be expected to increase the ecological relevance of decision-making, better address the intent of laws and regulations, allow greater recognition of societal and stakeholder values, and provide managers with the breadth of information necessary to support practical, balanced, and implementable decisions.

There are several broad sectors where information about risks to populations could be used by public or private sector risk managers who must decide whether there is a problem for which a response action is required, and, if so, must then decide what form any such action should take. Some of these sectors are identified generally in the following section, and examples are provided in greater detail in subsequent chapters of this book:

1) **New products.** Population-level ERA tools may be useful in the design, registration, and use of new and existing chemicals (e.g., commodities, pesticides, pharmaceuticals), including product substitution, reformulation, reregistration, unintentional consequences, or changes in amount or extent of use. These activities may be directed by the need to comply with various local, national, and international laws, regulations, policies, or directives that require or allow use of risk assessment information. The use of population-level ERA tools to evaluate a hypothetical pesticide application scenario is provided in Chapter 10 as a case study.
2) **Controlled hazardous waste management.** Population-level ERA tools could be used in the management of controlled (permitted) releases of chemicals to the environment, as from industrial discharges, publicly owned treatment work (sewage) outfalls, stormwater runoff, and others, including both past and present such releases. As with product design, there are likely to be regulatory drivers and constraints on these management actions. A hypothetical scenario is outlined in Appendix 1.
3) **Uncontrolled/accidental hazardous waste releases.** Population-level ERA tools could be used in the management of uncontrolled continuous or single-pulse releases to the environment from past (historical or legacy) contamination or from current spills. Again, there are likely to be regulatory drivers and constraints on these management actions. The use of population-level ERA to evaluate a hypothetical hazardous waste site is outlined in Appendix 1.
4) **Biological monitoring.** Population-level ERA tools could be used in the interpretation of monitoring results. Observed adverse effects on organism-level attributes, or examination of raw monitoring data may identify a potential problem. Advancing the assessment to the population-level may help risk managers place these results in a broader, and perhaps more appropriate, ecological context.

5) **Strategic corporate environmental management.** Population-level ERA tools could be used in strategic management. For example, absent specific legal or regulatory drivers, commercial interests may still find information from a population-level ERAs useful during design of a chemical or other product. Such information could allow them to anticipate, and ideally avoid, adverse effects associated with use of the product. Conversely, a competing business could use such analyses to suggest that their competitor's product could be or is harmful to the environment. Similarly, a nongovernmental organization could use risk analysis to suggest that an industrial product destined for release into commerce could be harmful to the environment or to show that one already commercially available is potentially harmful. Product stewardship would not be the only useful application of population-level ERA. Obviously such approaches could be applied in design of controls for most environmental aspects of manufacturing and distribution.
6) **Comparative risk analysis for resource allocation.** Population-level ERA tools could be used as an input to a comparative risk analysis (CRA). This involves doing risk assessment for two or more risks at the same time to illuminate whether or not resources are sensibly allocated to reducing one set of risks rather than another. As long as cost is quite explicitly introduced, or risks are normalized for a given benefit, then comparative risk analysis provides a powerful methodology for improving resource allocation. The analysis may relate to health risks alone (health risk assessment) or to ecological risks alone (ERA). When applied to both, problems arise with respect to the aggregation of the risks since it is then important to know the relative weights to give to risk reduction to health as compared to ecosystems (but see Suter et al. 2007).
7) **Nonchemical stressor management.** Population-level ERA tools could be used in the management of nonchemical stressors. Population-level methods can be used to assess or anticipate risks associated with the intentional or unintentional release of genetically modified organisms into the environment or with the introduction of invasive species. Other risk management problems resulting from such nonchemical stressors as habitat destruction, harvesting pressure, climate change, or natural disasters can be similarly informed by population-level information. A hypothetical example is outlined in Appendix 1.
8) **Resource management and conservation.** Conservation and resource management is concerned primarily with the protection and management of free-living populations. Although use of population-level methods and models (e.g., POP2 for deer and elk; Table 2.1) is a common occurrence, the risk assessment paradigm has so far found only a limited role in resource and conservation management. However, even if this condition does not change in the foreseeable future, the experiences gained and tools developed over the years by resource managers have a great deal of relevance to population-level ERA. As population-level ERA moves into the complex and uncertain realities of ecological systems, there is much to be gained from examining the experiences and tools of the fields of

conservation biology and natural resource management. See Chapter 7 for more information on this topic.

LAWS, REGULATIONS, POLICIES, NARRATIVE GOALS, AND DIRECTIVES THAT PROTECT THE ENVIRONMENT

The risk management context is influenced by various local, national, and international laws, regulations, policies, narrative goals, and directives. The risk management context is also informed by the characteristics of the impacted environment, the stakeholders involved or concerned about protecting the impacted environment, and the ecological entities identified for protection. Despite this array of influencing factors, very often, ERAs that support risk management conditions are performed at the organism-level. However, the laws, regulations, policies, narrative goals, and directives do not preclude and in some cases support (ORDEQ 2001) developing and using population-level ERA information. In some of these, "populations" are explicitly identified as entities to be protected, whereas in others, the intention to protect populations can be implied within a broad mandate for environmental protection.

The risk management context is influenced by various local, national, and international laws, regulations, policies, narrative goals, and directives.

The purpose of this section is to provide examples of the laws, regulations, policies, goals, and directives that either explicitly or implicitly identify populations as an ecological entity for protection. In addition, an overview of how current ERA approaches are used as tools for managing risks at the population-level is discussed. This section is divided in geographic regions of the United States, Canada, the European Union, and Japan. Categories include new chemical approval and registration, chemical releases, fisheries management, wildlife management, and chemical substances management. Table 2.1 provides additional detail on this topic for each geographic region.

UNITED STATES

Environmental laws, regulations, and statutes authorizing actions by US federal and state agencies call for protection of a diverse array of organisms and their habitats. The terminology in the laws, regulations, and statutes often make general reference to protection of "the environment." The US Environmental Protection Agency (USEPA), a federal governing agency, has acknowledged that risk management needs vary considerably between situations, but the overall management goals generally involve protection of populations, communities, and ecosystems (USEPA 1997b, 1999, 2001). For example, the USEPA states "Superfund's number 1 principle is to reduce ecological risks to levels that will result in the recovery and maintenance of healthy local populations and communities of biota" (USEPA 1999). The US Fish

TABLE 2.1
Summary of population-level–related regulatory, legal, and other drivers — United States

Area	New chemical approval and registration
Jurisdiction	Federal
Activity	Toxic Substances Control Act (TSCA).
Population-level focus	Authorizes the USEPA Office of Prevention, Pesticides and Toxic Substances (OPPTS) to obtain data from industry on health and environmental effects of chemical substances and mixtures. If unreasonable risks to populations occur, USEPA may regulate, limit, or prohibit the manufacture, processing, commercial distribution, use, and disposal of such chemicals.
References	http://laws.fws.gov/lawsdigest/toxic.html
Area	New chemical approval and registration
Jurisdiction	Federal
Activity	Pesticide registration/re-registration under the US Federal Insecticide, Fungicide and Rodenticide Act of (FIFRA) 7 U.S.C. s/s 136 *et seq.* (1996).
Population-level focus	Follows USEPA Guidelines for Ecological Risk Assessment. USEPA, Office of Prevention, Pesticides and Toxic Substances (OPPTS), Office of Pesticide Programs (OPP). OPP is developing a tiered risk assessment process with initial tiers that focus on organismal-level effects. Tier 1 is a deterministic (conservative) screen of acute toxicity; Tier 2 is a preliminary probabilistic assessment of likelihood and magnitude of acute effects; Tiers 3 and 4 are refined probabilistic assessments addressing populations. Methods are still under development (demonstration projects under way); population models will be used to identify refinements to registrant data packages. Probabilistic methodology under development following recommendations of the Ecological Committee on FIFRA Risk Assessment Methods (ECOFRAM).
References	http://www.epa.gov/oppefed1/ecorisk http://cfpub.epa.gov/ncea/cfm/recordisplay.cfm?deid=12460
Area	Accidental releases and historic contamination
Jurisdiction	Federal
Activity	US Comprehensive Environmental Response, Compensation, and Liability Act (CERCLA - Superfund) 42 U.S.C. s/s 9601 *et seq.* (1980) and Superfund Amendments and Reauthorization Act (SARA) 42 U.S.C. 9601 *et seq.* (1986).
Population-level focus	USEPA, Office of Solid Waste and Emergency Response (OSWER), Office of Emergency Response and Remediation (OERR) states that Superfund's number 1 principle is to "reduce ecological risks to levels that will result in the recovery and maintenance of healthy local populations and communities of biota." No formal population-level risk assessment guidance offered, but "ecological effects of most concern are those that can impact populations…." Superfund follows an 8-step process for baseline risk assessments; Steps 1 and 2 are screening steps based on chemical exposure pathway analysis. Population-level effects have been evaluated as part of both terrestrial and aquatic site risk assessments in ad hoc fashion; information used to support Records of Decision for remediation and clean-up.

TABLE 2.1 (continued)
Summary of population-level–related regulatory, legal, and other drivers — United States

References	Ecological Risk Assessment for Superfund: Process For Designing and Conducting Ecological Risk Assessments (Interim Final), 1997b (http://www.epa.gov/oerrpage/superfund/programs/risk/ecorisk/intro.pdf), consistent with USEPA's Guidelines for Ecological Risk Assessment (http://cfpub.epa.gov/ncea/cfm/recordisplay.cfm?deid=12460).
Area	Accidental releases and historic contamination
Jurisdiction	State (Oregon)
Activity	Oregon Department of Environmental Quality (ORDEQ); Land Quality Division. Required by Oregon Revised Statute 465.315(1)(b)(A), 465.315(2)(a)(H) Implemented by Oregon Administrative Rules 340-122-115(6), -115(21).
Population-level focus	ORDEQ developed a three-tiered (scoping, screening, baseline) framework for ecologic risk, based loosely on USEPA guidance, but emphasizing simplicity and easy of implementation. Population-level risk can be assessed only at the baseline level. Acceptable population risk is defined in Oregon Administrative Rules 340-122-115(21). Effects in <20% of the population is a nominal de minimis level. However, there is a provision for action if adverse effects are observed, regardless of what the risk assessment indicates.
References	www.deq.state.or.us/wmc/cleanup/ecocover.htm www.epa.gov/superfund/programs/risk/tooltrad.htm#gdec
Area	Accidental releases and historic contamination
Jurisdiction	State (Texas)
Activity	Texas Commission on Environmental Quality (TCEQ); Remediation Division; Technical Support Section. Required by Chapters 350.77, 350.91(b)(7), 335.556(b), 335.559(d)(4), 335.562(c)(3), 335.563(i)(2)(B), and 335.563(j)(3) of the Texas Administrative Code.
Population-level focus	TCEQ developed our own three-tiered (exclusion criteria checklist, screening-level ERA, site-specific ERA) framework for ecologic risk, based loosely on USEPA 1997 Ecorisk Guidance for Superfund, but allowing more realistic inputs and more exit ramps. TCEQ is not currently using probabilistic or any other strictly population-based ecorisk assessment method. However, by relying on a process that develops clean-up levels that may approach a reasonable LOAEL and by focusing on the feeding guild, TCEQ and stakeholders that assisted in the development of this approach consider that the ERA program incorporates population-based thinking. The entire Tier 1 checklist is based on exclusion criteria and is designed to determine if ecological pathways are complete/significant. The last of these criteria is independent of community structure and excludes a de minimis size of 1 acre or less of contamination, provided other qualifying conditions are met.
References	http://www.tnrcc.state.tx.us/permitting/trrp.htm#eco
Area	Accidental releases and historic contamination
Jurisdiction	State (Massachusetts)

TABLE 2.1 (continued)
Summary of population-level–related regulatory, legal, and other drivers — United States

Activity	Massachusetts Department of Environmental Protection (MADEP), Bureau of Waste Site Cleanup. The statute (MGL Ch. 21E) and the Massachusetts Contingency Plan (MCP) regulations require a risk-based decision process at waste sites.
Population-level focus	Guidance states that risk characterization should focus on the groups of organisms actually exposed to hazardous materials at the site and that assessment objectives should not be limited to measurable ecosystem level effects. For this reason, population-level effects have not often been the assessment focus. No methodology has been adopted. For sites such as lakes and ponds, where true populations are affected and population-level assessment is applicable, measures of abundance have been used. Guidance suggests de minimis screening criteria for terrestrial sites of 2 acres, providing other criteria are also met.
References	MCP (www.state.ma.us/dep/bwsc/reg.htm). Guidance for Disposal Site Risk Characterization suggests focusing on harm to subpopulations or communities, not on individual organisms. www.state.ma.us/dep/ors/orspubs.htm
Area	Accidental releases and historic contamination
Jurisdiction	State (Pennsylvania)
Activity	Ecological screening process. Pennsylvania Department of Environmental Protection (PADEP).
Population-level focus	The site screening procedure defines substantial impact as the potential for constituents detected onsite to cause > 20% change in abundance of species of concern compared with an appropriate reference area, or > 50% change in the composition or diversity of a habitat of concern compared to an appropriate reference area. This goal acknowledges the "substantial acclimation capacity of natural populations to exposure to low or moderate concentrations of chemical residuals."
References	http://www.dep.state.pa.us/dep/subject/advcoun/cleanup/attachmentve3.doc
Area	Wildlife and fisheries management
Jurisdiction	State (Oregon)
Activity	Oregon Department of Fish and Game, Wildlife Management Unit. Oregon Revised Statutes 183 and 496 (Mule Deer) Oregon Revised Statutes 496.012, 138, 146, 162, 164 (Elk).
Population-level focus	POP2 is used for mule deer and elk. Models rely on empiric data for herd ratios, ages and productivity rates, harvest rates, natural mortality, weather severity, and total amount of various habitat types. Sightability indexes determined with over-flights. Use of models has been curtailed as budget constraints have limited ability to gather necessary empirical data.
References	Oregon Administrative Rules 635-190-0000 (Mule Deer Management Plan) www.dfw.state.or.us/ODFWhtml/InfoCntrWild/PDFs/MuleDeerPlanFinal.PDF Oregon Administrative Rules 635-160-0000 (Elk Management Plan) www.dfw.state.or.us/ODFWhtml/InfoCntrWild/PDFs/Elk%20Planfinal.PDF
Area	Wildlife and fisheries management
Jurisdiction	Federal

TABLE 2.1 (continued)
Summary of population-level–related regulatory, legal, and other drivers — United States

Activity	Endangered Species Act (ESA) and Migratory Bird Treaty Act.
Population-level focus	The ESA provided for the conservation of ecosystems on which threatened and endangered species of fish, wildlife, and plants depend. The ESA focuses on the protection of populations by also focusing on the protection of individuals within populations. The US Fish and Wildlife Service expresses risk mainly as predicted mean times to extinction. Many of the models address single stressors and fail to provide probability functions for the shorter time frames that might prove to be useful in ecological risk assessments. The National Research Council has recommended research to produce models that predict distributions of extinction time based on the integration of genetic, demographic, and environmental stochasticity into spatially explicit frameworks. Population viability analysis is also an approach used to evaluate risks for endangered species.
References	http://laws.fws.gov/lawsdigest/esact.html http://laws.fws.gov/lawsdigest/migtrea.html
Area	Wildlife and fisheries management
Jurisdiction	Federal
Activity	Anadromous Fish Conservation Act, Atlantic Striped Bass Conservation Act and Atlantic States Marine Fisheries Commission. Implementation of Magnuson–Stevens Sustainable Fishery Act.
Population-level focus	Fisheries science is a source of models applicable to other kinds of population-level assessments; the models used in fisheries management provide a useful upper bound on the precision, accuracy, and utility of population models used in other kinds of risk assessments. Age-structured matrix projection models are used, some of which include density-dependence. Documentation of models is in theory available from technical committees (e.g., the striped bass and weakfish technical committees), but can be difficult to obtain. Empirical data used include catch statistics; fishery-independent estimates of annual variations in abundance of young fish.
References	http://laws.fws.gov/lawsdigest/anadrom.html http://laws.fws.gov/lawsdigest/atlstr.html Descriptions of goals, activities, and publications from the ASMF are available from the ASMFC Web site (www.asmfc.org).
Area	Permitted water releases
Jurisdiction	Federal
Activity	US Federal Water Pollution Control Act, as amended through P.L. 107-303, (33 U.S.C. 1251 *et seq.*), commonly known as the Clean Water Act (2002).
Population-level focus	Models not yet used; modeling approaches being explored to address population-level and "community assemblage" impacts and recovery to establish risk-based national and site-specific criteria. Approaches are consistent with USEPA Guidelines for Ecological Risk Assessment. (http://cfpub.epa.gov/ncea/cfm/recordisplay.cfm?deid =12460).
References	No authoritative documentation; planning under way to incorporate population-level risks in water quality criteria and wildlife criteria development.

and Wildlife Service, an agency with responsibilities for protecting threatened and endangered species, also focuses on preservation of populations and their habitats. A difference in the US Fish and Wildlife Service approach, however, is that actions related to the preservation of threatened and endangered species involve the protection of individual organisms.

USEPA states "Superfund's number 1 principle is to reduce ecological risks to levels that will result in the recovery and maintenance of healthy local populations and communities of biota" (USEPA 1999).

Table 2.1 provides a range of examples for a variety of federal and state environmental protection programs, including the most common and overarching federal regulations in the United States, such as the following:

- **New Chemical Approval and Registration**
  - Federal Insecticide, Fungicide, and Rodenticide Act (FIFRA)
  - Toxic Substances Control Act (TSCA)
- **Accidental Releases and Historic Contamination**
  - Comprehensive Environmental Response, Compensation, and Liability Act (CERCLA - Superfund) and Superfund Amendments and Reauthorization Act (SARA)
  - Resource Conservation and Recover Act (RCRA)
- **Fisheries and Wildlife Management**
  - Endangered Species Act and Migratory Bird Treaty Act
  - Anadromous Fish Conservation Act; Atlantic Striped Bass Conservation Act
  - Magnuson–Stevens Sustainable Fishery Act
- **Permitted Water Releases**
  - US Federal Water Pollution Control Act, as amended through PL 107-303, (33 USC 1251 *et seq.*) commonly known as the Clean Water Act (2002)
- **Permitted Air Releases**
  - Clean Air Act amendments regarding Residual Risk (§112)

Examples of state level regulations and statutes summarized in Table 2.1 include the following:

- Accidental releases and historic contamination
  - Oregon Department of Environmental Quality (ORDEQ); Land Quality Division. Required by Oregon Revised Statute 465.315(1)(b)(A), 465.315(2)(a)(H). Implemented by Oregon Administrative Rules 340-122-115(6), -115(21)
  - Oregon Department of Environmental Quality (ORDEQ); Land Quality Division. Required by Oregon Revised Statute 465.315(1)(b)(A),

465.315(2)(a)(H) Implemented by Oregon Administrative Rules 340-122-115(6), -115(21)
- Massachusetts Department of Environmental Protection (MADEP), Bureau of Waste Site Cleanup, Statute (MGL Ch. 21E) and the Massachusetts Contingency Plan (MCP)

Examples of decision-making or guidance that addresses the population-level focus of these laws, regulations, or statutes include the following:

- USEPA states "Superfund's number 1 principle is to reduce ecological risks to levels that will result in the recovery and maintenance of healthy local populations and communities of biota" (USEPA 1999).
- USEPA has used Section 404 of the Clean Water Act to veto a permit for a dam and reservoir project after modeling projected that the project could lead to the extirpation of populations of birds of special interest (USEPA 1994).
- USEPA has used projection matrix population models to assess effects on rainbow trout population abundance by chloroparaffins regulated under TSCA (USEPA 1993b).
- Populations of piscivorous birds and mammals were the ecological assessment endpoint entities selected by USEPA in its Mercury Study Report to Congress (USEPA 1997a).
- Prevention of adverse effects to public welfare (that includes, but is not limited to, effects on soils, water, crops, vegetation, animals, and wildlife [§190]), is mandated under Section 108 of the Clean Air Act. USEPA has used this language to revise the secondary ozone standard to provide increased protection against ozone-induced effects on vegetation, such as agricultural crop loss of forest damage (USEPA 1997b).
- USEPA is authorized (by FIFRA [7 USC 135 *et seq.*]) to require pesticide registrants to submit tests on the effect of pesticides on plant and animal populations. Results are used, in conjunction with other information, in making pesticide registration decisions.
- The effects of a nonchemical stressor (recombinant rhizobia) on changes in production of specific legume species were addressed in an assessment authorized by TSCA (Orr et al. 1999).

In addition to those under the jurisdiction of USEPA, protection of populations is implied by the numerous federal laws and treaties that have the purpose of maintaining or increasing the production of game birds and mammals, commercially valuable fish, and timber species. Examples include the Migratory Bird Hunting Act Stamp (48 Stat. 451), Wildlife Restoration Act (50 Stat. 971), Fish Restoration and Management Act (64 Stat. 430), Convention on Great Lakes Fisheries (6 UST 2836), and the Fish and Wildlife Act of 1956 (70 Stat. 1119). In addition, the Clean Water Act sets a national goal of "protection and propagation of fish, shellfish, and wildlife," which implies protecting both the abundance and production of populations.

Risk Management Tools: US ERA Approaches Often Used to Address Population-Level Goals

ERAs are risk management tools used by risk managers in achieving protection goals established by laws, regulations, statutes, and the individual circumstances of the risk management decision. The ERA approaches that are available (whether using terminology of "tiers," or "screening-level, or baseline) are flexible enough to allow risk management at the population-level (e.g., USEPA 1997b, 1998; ASTM 2002). However, population-level models and studies are often avoided for reasons such as the perception of high costs, difficulty of measuring many wildlife species, and issues related with natural variability in wildlife populations. Only the state of Oregon's ERA guidance (ORDEQ 2001) explicitly requires consideration of population-level risks at the later stages of the ERA process. Although population models have been used routinely for fisheries and wildlife management, these approaches are often unknown or poorly understood by ecological risk assessors and risk managers in other regulatory programs. Risk managers and ERA practitioners acknowledge the general lack of accepted population-level ERA methodologic approaches and the general lack of understanding of available population-level ERA tools. Acknowledgement of this deficit was a driving force behind the development of this book.

ERA approaches that are available are flexible enough to allow risk management at the population-level.

Risk managers currently have a broad array of ERA approaches to use for managing risks at the population-level, though only one (ORDEQ) explicitly identifies population-level directive for decision-making. More than 100 ERA guidance documents, procedural guidelines, or other resources — spanning just the last decade — have been published for use in the evaluation of ecological risks (see annotated compilations by Sorensen 1994; Sorensen and Margolin 1998). The vast majority of these follow the USEPA Framework for ERA (1992), the USEPA ERA Guidance for Superfund (1997b), and the USEPA Guidelines for ERA (1998). Many of the ERA approaches currently used are based on chemical-specific estimates of risks to organism-level attributes through the use of a hazard quotient (e.g., USEPA 1997b, 2000). Some of these are based on "tiers" (e.g., ORDEQ 2001; ASTM 2002), with early tiers representing conservative, simplistic estimation of risks to organisms. Other approaches are based on screening level and baseline level ERAs, in which the screening-level ERA is based on simplistic estimates of risk (e.g., USEPA 1997b, 2000). Although later tiers in the ERA process or baseline ERAs could include a broad array of assessment endpoints, including those relevant to populations, often, ERA practitioners instead use the later tiers to refine the organism-level, chemical-specific approach (i.e., hazard quotients with less uncertainty are calculated). However, the ERA approaches (tiered or screening-level and baseline) offer little detailed information about how to extrapolate chemical-specific hazard quotients for organisms (often laboratory species) to risks that may exist for ecological populations exposed to chemical mixtures and nonchemical stressors. Therefore, risk managers often have limited information with which to make population-level management decisions. The greater use of population-level ERA approaches could provide greater insight for risk management decisions than the chemical-specific hazard quotient method.

The greater use of population-level ERA approaches could provide greater insight for risk management decisions than the chemical-specific hazard quotient method.

Contrary to the organism-level based chemical-specific hazard quotient approach, several US states have ERA approaches that use explicit *de minimis* ecological significance (or exclusion) criteria by considering concepts such as fate and transport, scale of release, and habitat issues that are inherently related to population-level wildlife exposures (e.g., MADEP 1996; PADEP 1998; TCEQ 2001). For example, in PA, MA, and TX, sites up to 2 acres could be excluded from the ERA process, providing concerns such as threatened and endangered wildlife exposures and biotic transfer via the food web have been addressed. The TCEQ Tier 1 Exclusion Criteria Checklist provides the most detailed checklist available for the use of such criteria (an analysis of criteria in a broad range of ERA and regulatory documents is provided in GRI 2002). A diverse stakeholder group, including representatives from the USEPA, the US Fish and Wildlife Service, the National Oceanic and Atmospheric Administration, the Texas Natural Resource Conservation Commission, and a variety of industrial sectors (TCEQ 2001), developed the criteria used by TCEQ in the Tier 1 Checklist. Risk management decision-making using these approaches is based on the assumption that small spatial scale impacts to organisms or their habitats are not likely to impact populations, communities, or ecosystems. An important exception to this assumption is when the acreage in question is considered a keystone habitat for a species of special interest or protective status. In that case, then spatial attributes would not be useful screening criteria to limit further analysis.

The State of Oregon provides the only US ERA guidance document (ORDEQ 2001) with an explicitly stated population-level directive. In 1997, the State of Oregon enacted amendments to the state hazardous waste site cleanup law that emphasize risk-based remedial action decisions. The amended statute and associated rules require that protection of ecological receptors occur at the population level for all plants and animals not listed as threatened or endangered. The clear legislative intent was to protect only the viability of the population and not that of any specific individual in the population. Risk management decisions thus focus on the policy goal of preventing significant adverse effects rather than preventing any adverse effects.

Oregon's specific regulatory language states that an unacceptable risk occurs when there is a >10% chance of ≥20% of the total local population receiving an exposure greater than the toxicity reference value (TRV) for a hazardous substance.

Conversely, risk is acceptable when either the chance of exposure exceeding the TRV is <10% regardless of the fraction of the population exposed or the chance of exposure exceeding the TRV is >10% for an individual organism but <20% of the local population is exposed. Exposure may exceed the TRV only if this does not occur in more than 20% of exposed individuals. The 20% threshold relates to the ability to discern and differentiate changes in population levels resulting from either natural or anthropogenic stressors. In extreme cases, this regulatory model would permit significant adverse effects in fully 20% of the population. For small local populations with few breeding individuals and restricted immigration, 20% could be a significant number of individuals. In these instances, the regulation does provide for an alternative, and preemptive, qualitative assessment of the health and viability of the local population.

Canada

Risk management in Canada is also generally focused toward the protection of ecological populations, communities, and ecosystems (Environment Canada 1997). In Canada, as in other jurisdictions, federal statutes providing protection to the environment do not explicitly require ERAs to use any specific assessment endpoint, such as a population of organisms. There are several federal and provincial statutes that provide environmental protection.

Examples of provincial level regulations and statutes summarized in Table 2.2 include the following:

- New chemical approval and registration
  - Canadian Environmental Protection Act (CEPA)
  - Pest Control Products Act (PCPA)
- Accidental releases and historic contamination
  - Canadian Environmental Protection Act (CEPA)
  - Province of British Columbia, Waste Management Act
  - Fisheries Act (FA)
- Fisheries and wildlife management
  - Canada Wildlife Act
  - The Fisheries Act (FA)
  - Species at Risk Act (SARA)

Of these statutes, CEPA, the PCPA, and FA address toxic chemicals and contaminated sites, whereas CEAA covers environmental impact assessments of major development projects. As with other jurisdictions, although protection of populations is not specifically identified in these acts, it maybe inferred by both risk assessors and risk managers.

For example, the assessment of priority substances[1] in Canada under CEPA considers effects observed at the population, community, or ecosystem level as more

[1] Priority substances are those substances which are identified to be given priority for assessment of environmental and human health effects as determined by an expert panel.

TABLE 2.2
Summary of population-level related regulatory, legal, and other drivers — Canada

Area	Accidental releases and historic contamination
Jurisdiction	Provincial (British Columbia)
Activity	British Columbia Ministry of Water, Land and Air Protection; Contaminated Sites Program. Waste Management Act (RSBC 1996) Chapter 482 and Contaminated Sites Regulation.
Population-level focus	The Ministry recognizes 3 tiers of ERA, but has only developed guidance for Tier 1 assessments. Sites requiring assessments at the Tier 3 level (which would consider population dynamics) typically constitute < 1% of all contaminated sites in British Columbia. These types of sites require specialized expertise and more consultation with ministry staff. ERA guidance focuses on organism-level attributes. Population-level effects have been evaluated for specific sites, but this was and is done on an ad hoc, site-specific basis in consultation with the ministry.
References	No formal population-level risk assessment guidance offered. Tier 1 ERA guidance is available. http://wlapwww.gov.bc.ca/epd/epdpa/contam_sites/policy_procedure_protocol/protocols/tier_1
Area	Accidental releases and historic contamination
Jurisdiction	Federal
Activity	Environment Canada (EC). Chemicals Evaluation Division.
Population-level focus	The Canadian Environmental Protection Act (CEPA) provides the legal tool for addressing toxic substances in the environment. Though population-level ERA approach is not specified, the use of statistical population protection is implied in the criteria development.
References	Environmental Assessments of Priority Substances under the Canadian Environmental Protection Act. Technical Guidance Manual Version 1.0. EC Chemicals Evaluation Division. http://www.ec.gc.ca/substances/ese/eng/psap/psl2manual.cfm

environmentally harmful and are of more concern than those effects observed only at lower levels of biological organization (Environment Canada 1997). Therefore, although not specified in the act, protection at the population level is inferred by those conducting assessments and undertaking risk management decisions.

Similarly, the PCPA identifies "protection of the environment and biodiversity" as a management goal for the registration of pest control products. The agency responsible recently held a workshop to identify assessment endpoints, from a scientific perspective, for conducting pesticide risk assessments. Participants in the workshop identified population level assessment endpoints as being appropriate for all the major taxonomic groups, and were the preferred assessment endpoint for 4 of the 7 taxonomic groups discussed (Kriz et al. 2003). In all cases, organism-level effects were identified as the assessment endpoint for species at risk. The results of the workshop are consistent with recent risk management decisions to remove certain uses of pesticides in Canada because of concerns about effects on populations.

Risk Management Tools: A Canadian ERA Example

The province of British Columbia provides for the use of risk-based standards and risk assessment in evaluating contaminated sites under the Contaminated Sites Regulation of the Waste Management Act. However, the ministry has only developed guidance for a screening level (i.e., Tier 1) ERA. The guidance places particular emphasis on situations typical in the province, including land uses and characteristic types of ecological structures. At the screening level, the focus is predominantly on the organism level of organization and ecological properties such as population dynamics are not considered.

In British Columbia, the Ministry of Water, Land and Air Protection (Contaminated Sites Program) expects that 90% of all contaminated sites within its jurisdiction can be successfully evaluated using a Tier 1 ERA. Although the focus in the Tier 1 ERA is on attributes of organisms, the development of the TRVs for the effect assessment component of the ERA is based on conservative, nonlethal endpoints. The ministry has adopted an effects concentration (EC_X) approach to set TRVs. An EC_X approach, based on endpoints such as reproduction and growth, allows the assessment to focus on important factors in maintaining viable populations. A lethality endpoint, such as the lethal concentration causing 50% mortality (LC_{50}), does not focus on this level of protection. Unlike the no-observable-adverse-effects level and lowest-observable-adverse-effects level (NOAEL, LOAEL) or no-observable-effects concentration and lowest-observable-effects concentration (NOEC, LOEC) approach, the EC_X approach is based on the dose response curve and allows the ministry to set different protection goals for different lands uses (e.g., agricultural versus industrial lands) and to use different endpoints. The aquatic EC_X for the protection of aquatic life is set at 20% (EC_{20}) to be conservative and allow for the protection of commercially important species such as salmonids. In contrast, the EC_X for industrial lands is set at 50% (EC_{50}); these lands typically offer limited habitat and are not considered to be critical habitat, habitat for valued species, or as breeding grounds. Of note, a NOAEL, NOEC approach is adopted for threatened or endangered species where explicit protection of individual organisms is required.

A second tier (Tier 2) ERA would be applicable to approximately 9% of all potential contaminated sites in British Columbia and involves more detailed analytical techniques (e.g., Monte Carlo analysis) and extensive sampling of the site and its resident organisms. The ministry expects that less than 1% of all sites would be dealt with at a Tier 3 assessment. This assessment would be a detailed quantitative assessment for sites with multiple, unrelated chemical stressors, a wide variety of habitats and terrain types, and encompassing a wide geographic area. For these sites, the ministry would expect extensive consultations between the responsible parties and the ministry, and that ecological properties such as population dynamics, would be addressed on a site-specific basis. The ministry has not established benchmarks for protection of populations, as has Oregon in the United States, because of the limited number of sites where population-level impacts would be expected to be a concern.

TABLE 2.3
Summary of population-level related regulatory, legal, and other drivers — European Union

Area	New chemical approval and registration
Jurisdiction	European Union
Activity	European Union (EU). 1997. Council Directive 97/57/EC of 21 September 1997; Establishing annex VI to Directive 91/414/EEC Concerning the placing of plant protection products on the market. Official Journal of the European Communities L265: 87-109.
Population-level focus	The EU tiered approach starts with single species studies and safety factors and ends with semifield tests. Population empirical approaches are somewhat in the middle of this process to study recovery.
References	Campbell PJ, Arnold DSJ, Brock TCM, Grandy NJ, Heger W, Heimbach F, Maund SJ, Streloke M, editors. 1999. Guidance document on Higher-tier Aquatic Risk Assessment for Pesticides (HARAP). SETAC-Europe publication, ISBN 90-5607-011-8.

European Union

There are 2 main types of environmental protection legislation in the European Union — those focusing on the causes of harm (contaminating agents) and those focusing on effects to particular ecological systems (ecological receptors; species and ecosystems) (the later being identified in Table 2.3 and former is described in Appendix 1 [A1.3]).

Not surprisingly, the causally orientated instruments refer only in very general terms to the protection targets. Thus, the legislation associated with plant protection products (Directive 91/414) specifies (Article 3) that a plant protection product shall not be authorized unless it has no unacceptable influence on the environment having particular regard to its fate and distribution and its impact on nontarget species. "Environment" is defined (Article 2) as water, air, land, and wild species of flora and fauna. Similarly, legislation relating to risks from industrial chemicals (Regulation 793/93) aims, under Article 1, to evaluate risks of existing substances to humans and the environment to ensure better management of these risks. Similar legislation exists for new chemicals (Directive 93/67). Protection targets have been defined more precisely in Annexes and Guidance notes. Thus Annex III of Directive 91/414, as modified by Commission Directive 96/12, specifies that the notifications contain sufficient information to permit an evaluation of short- and long-term risks for nontarget species including populations, communities, and processes as appropriate. Similarly the Technical Guidance Document (European Chemicals Bureau 2003) associated with the assessment of industrial chemicals refers to the protection of aquatic and terrestrial ecosystems and top predators within them. In practice, this has led to consideration of population-level effects in interpreting the results from both single-species (explicitly from microbial and algal systems) and from multispecies (micro- and macrocosms) ecotoxicity tests. Proposed new legislation for industrial chemicals (European

Commission 2001) is likely to put more emphasis on a hazard- rather than risk-based approach and may well lead to less consideration of population level effects

The target-orientated instruments are somewhat more precise in terms of what they aim to protect. Thus Directive 92/43 on the conservation of natural habitats and of wild fauna and flora aims (Article 2) to contribute toward "ensuring bio-diversity" through the conservation of natural habitats and of wild fauna and flora. It defines "conservation" (Article 1) as a series of measures required to maintain (or restore) natural habitats and the populations of wild fauna and flora at (to) a favorable status. On the other hand, the Water Framework Directive (2000/60) under Article 4 requires that member states protect, enhance, and restore all controlled waters with the aim of achieving good ecological status within a specified time frame. Ecological status is defined (Article 2) as an expression of the quality of the structure and functioning of ecosystems classified according to general criteria laid down in an Annex V and including population level measurements (e.g., of abundance, biomass and, at least for the fish fauna, age structure).

"Impacts legislation" refers to specific sites and thus to both specific causal agents and receptors. Directive 85/337 on impact assessment of certain public and private projects (amended by Directive 97/11) requires that for specified projects an impact assessment shall be carried out to identify, describe, and assess the direct and indirect effects of the project on human beings; fauna and flora; soil, water, air, climate, and landscape; and interaction between these. Depending on circumstances therefore, this is likely to lead to population level considerations. Interestingly, in Europe, this legislation does not specify what should be done in the event of an impact being identified but it does require that the information be put into the public domain.

Risk Management Tools: An ERA Approach for the European Commission

The Scientific Steering Committee of the European Commission's Health and Consumer Protection Directorate-General recently published a report on risk assessment for animal populations with emphasis on wildlife (SSC 2003). The main aim of the report was to compare the premises and methodologies employed for assessing animal health risks (i.e., largely for farm animals) and those employed for assessing population risks focusing on mammals and birds. The report considers approaches used for assessing infectious disease risk, immunological medicines, microorganisms used as active substances for plant protection products, industrial chemicals, pesticides, pharmaceuticals, and feed additives. The report describes the extent to which population risk assessment for wildlife is incorporated into the different risk assessment frameworks. It concludes that risk for animal populations "is assumed as an essential part of several risk assessment processes." With respect to industrial chemicals, it is noted that no specific protocols or guidance for assessing the risks to animal populations are defined in the Technical Guidance Document, and this aspect is not currently a key issue in the chemicals registration area. With regard to pesticides, the report notes that the environmental risk assessment for pesticides focuses on the protection of nontarget species, which in most cases is concerned with effects at the population level. Whereas the risks of pesticides for wild populations are included as part of the ecotoxicological assessment, the risks to domestic animal health are covered, simultaneously to human health impacts, in the toxicological and residue assessments. For the most part, effects of pesticides on populations of wildlife are estimated from toxicological information gathered from standard test species, extrapolated to wildlife species using fixed factors, or in some cases, statistical distributions of toxicity data. Exposure may be modeled from information on chemical persistence, bioaccumulation potential, and usage, usually starting with worst-case assumptions followed by more realistic refinements (e.g., that an animal does not spend 100% of its time feeding on treated fields) should the initial estimated risks fail accepted trigger values (Annex VI of Directive 91/414/EEC). Thus population-level risks are neither measured nor modeled, but are assumed on the basis of risks to organisms.

Japan

Movements toward expending the scope of regulation to include ecosystem protection in Japan have been concentrated in recent years. Several national laws, policies, regulations, or directives have been amended or established for the goal of prevention of adverse effects to ecosystems, and generally are inferred to protect populations, even if populations are not specifically cited by them.

- The Basic Environmental Law, which set out basic principles and directions for formulating environmental policies, was enacted in November 1993. Section 2 (Article 14) of Chapter 2 describes three objectives for policy formulation, one of them with respect to ecosystem protection described as protecting biodiversity, such as the diversity of ecosystems and wildlife species.
- The Basic Environmental Plan, established in accordance with the provision of Article 15 of the Basic Environmental Law (Law No. 91, 1993) in December 1994 and amended in December 2000, clearly states that not only human health but also ecosystem health should take into account the adverse effects of chemical substances. The Basic Environment Plan is designed to engage all sectors of the society in a concerted effort to protect the environment. Part II, principles of environment policy, states that the ecosystem is formed by many delicate balances and that precautionary measures employing scientific knowledge should be applied to avoid serious, irreversible negative impacts on the environment.
- The National Strategy on Biological Diversity finalized on October 1995 was amended in March 2002 to promote policy to deal with the adverse effects of chemical substances on ecosystems. In its second part, it describes Japan's basic polices for conservation and sustainable use of biological diversity and its long-term goals, clearly stating that one of the urgent policy objectives for achieving the long-term goals is "no plants or animals must be threatened with extinction."
- The Chemical Substances Control Law, first issued in 1973 then amended in 1986, was amended again based on the Basic Environmental Law and

TABLE 2.4
Summary of population-level related regulatory, legal, and other drivers — Japan

Area	Chemical substances management
Jurisdiction	Federal
Activity	Research Center for Chemical Risk Management (CRM), National Institute of Advanced Industrial Science and Technology (AIST). Related to Basic Environment Law and the Amended Chemical Substances Control Law. Research funded by Japan Ministry of Economy, Trade and Industry (METI).
Population-level Focus	This is the first population-level risk assessment example for decision making in Japan. Two approaches have been applied to the risk assessments of some high-priority chemicals such as nonylphenol. Approaches included the derivation of concentration and the predicted-no-effect concentration (PNEC) for medaka population-level impact based on the threshold concentration (defined as the chemical concentration at lamda (population growth rate =1), as well as the lowest-observed-effect concentration (LOEC), the no-observed-effect concentration (NOEC), and the maximum-acceptable-toxic concentration (MATC).
References	(1) CRM of AIST: http://unit.aist.go.jp/crm/crm_e/index.html (2) Basic environment law: http://www.env.go.jp/en/lar/blaw/index.html (3) Amended Chemical Substances Control Law: http://www.safe.nite.go.jp/kasinn/pdf/kaiseieng.pdf

the Basic Environmental Plan in May 2003. The amended law extends the aim and scope of regulation to include ecosystem protection, and further improvements in the effectiveness and efficiency of chemical management.

- The Environmental Quality Standards (EQS) for Water Pollutants was issued in accordance with the provision of Article 16 of the Basic Environmental Law in 1993. Recent movement toward establishing the EQS for protecting aquatic ecosystems from adverse effects of chemical substances has been seen and the law is expected to be amended. Establishing EQS values for several chemicals based on the assessment endpoint that related to population viability are clearly recognized. Table 2.4 illustrates a case study of population-level ERA for nonylphenol that proposes approaches for setting the EQS values.
- The Water Pollution Control Law issued in 1970 and amended in 1995 aims at preventing the pollution of water in public water areas by regulating effluent discharged by factories or establishments into the areas. Movement toward regulation of discharge of effluent in the context of chemical substances is expected. The case study of population-level ERA for nonylphenol will be proposed and applied for the amendment.
- The Agricultural Chemicals Regulation Law issued in 1948 and finally amended in 1984 aims at protection for health of people. Recent movement toward extending the regulatory scope to protection of ecosystems has been seen and the law is expected to be amended. The case study of population-level ERA for nonylphenol will be proposed and applied for the amendment.

Risk Management Tools: A Case Study for Japan

To date, population-level decision-making has been proposed in Japan for the risk management of nonylphenol, coplanar PCB, and tributyltin, in expectation that the case study for nonylphenol can be used in the future for other chemicals, as appropriate. The case study of 4-nonylphenol used full lifecycle toxicity data for medaka (*Oryzias latipes*). It proposed to set population growth rate (λ) as the assessment endpoint for the population-level ERA. Although calculating from laboratory toxicity test is not always possible because of the scarcity of full lifecycle toxicity test, two approaches were used in the case study.

1) Using the theory of population growth — that is, $\lambda > 1$ indicates a growing population; $\lambda < 1$ indicates a population in decline and possibly headed toward extinction; and $\lambda = 1$ indicates a stable population in which the number of individuals does not fluctuate. It was therefore proposed to set $\lambda = 1$ as the threshold concentration for population growth impacts.
2) Lowest observable effects concentrations (LOEC), no observed effects concentrations (NOEC), and maximum-acceptable-toxic concentration (MATC) were derived for population-level based on the calculated λs. Statistically significant differences in the calculated s at different exposure concentrations were conducted to establish the LOEC, the NOEC, and the MACT.

BENEFITS AND CHALLENGES OF RISK MANAGEMENT WITH POPULATION-LEVEL ERA

Ecological Relevance

With the exception of legally protected species (e.g., threatened or endangered species in the United States) individual organisms generally have little legal, societal, or ecological relevance. Thus ecological risk management decisions are essentially aimed at ensuring protection of a population of organisms and not necessarily each individual organism in that population. It is now better recognized that populations have unique characteristics that are not captured or taken into account when an ERA deals only with effects on organisms and then simply ascribes these to effects on the population (see Chapter 3). A focus on organisms can also easily miss important spatial and temporal factors that may be important for the decision. But, from a purely technical perspective, it has been easier, for a number of reasons, to perform ERAs at the level of the organism. This disjuncture between management aims and technical execution too often results in risk management false positives — in which a risk manager undertakes a significant (in terms of money, time, and political capital) response to potential adverse effects on organisms that are unlikely to actually translate into risk to the population. This is often the case with conservative, worst-case screening level methods that compel action when the exposure of a single individual exceeds some regulatory threshold. Given this current state of affairs, performing ERAs at the population level (when appropriate) would be expected to benefit the decision-making process by 1) shifting the technical focus of an assessment to a level that is both more ecologically meaningful and that allows for consideration of the unique attributes of populations and 2) supporting of risk management decisions that lead to more reasonable, pragmatic, and implementable outcomes and which do not divert limited resources toward addressing inconsequential risk.

Ecological risk management decisions are (whether stated or implied) essentially aimed at ensuring protection of a population of organisms and not necessarily each individual organism in that population, with exception for threatened or endangered species.

Regulatory Value

Greater implicit and explicit acceptance of population-level ERA approaches and results by risk managers, when it is appropriate to do so, would be an improvement over decision-making based on assessment of risk to organisms. As was noted previously, there are several instances in which regulations implicitly allow consideration of population-level endpoints, but currently only one instance in which they are explicitly required by legislation. This reflects a long-standing practice within the regulatory community of focusing on organism-level effects when assessing chemical stressors. This practice is, in large part, a response by risk assessors to the types of data and methods available to them. For example, toxicity data, although they may have population-level implications (e.g., through the birth and death of organisms), are usually used to assess organism-level effects and most exposure estimates, whether based on point estimates or Monte Carlo simulation, generally represent exposure of organisms to contaminants. However, for the great majority of species not protected as individuals (i.e., other than rare species in the United States under the Endangered Species Act; passerine birds in the United States, Canada, and Mexico under the Migratory Bird Treaty Act), losses of individual organisms because of natural causes or stressors are accepted and, for commercial and game species, an important factor in their management. As a result, many resource management decisions for ecological effects are based on population-level effects. Decisions based on organism-level response are only implemented in resource management when there is perceived risk of extinction or greatly reduced abundance.

Greater implicit and explicit acceptance of population-level ERA approaches and results by risk managers, when it is appropriate to do so, would be an improvement over decision-making based on assessment of risk to organism-level attributes.

A risk manager must, of course, balance the perception that loss of a few individual organisms is a significant event against the science-based understanding that small losses are unlikely to affect the sustainability of most populations. Nonetheless, we would argue that regulating to protect every organism in a population, when some level of naturally occurring loss is almost assured and the species in question is not valued as an individual, is essentially a worst-case analysis. Although worst-case analyses are appropriate under certain circumstances, their routine use is unlikely to support reasonable, flexible, or cost-effective risk management actions. Greater implicit and explicit acceptance of population-level ERA by risk managers, when it is appropriate to do so, would be an improvement with respect to such worst-case analysis. Use of population-level assessment endpoints should reduce false positives (i.e., instances in which a risk assessment at the organism-level estimates significant effects when, in fact, the sustainability of that population is not at risk). Reducing this clutter of false positives would allow limited regulatory resources (time, money, and political or social capital) that could then be directed on instances where a population is truly jeopardized. It would also signal to the regulated community that regulators are making significant efforts to gain a clearer view of the ecological risks posed by stressors to populations.

Challenges to Risk Management with Population-Level ERA

The use of population-level endpoints and population-level ERA methods, when it is appropriate to do so, can bring added value to the risk management decision-making process by providing information to suggest and support a broader range of management options. However, moving from an organism-level to a population-level perspective will represent a major philosophical and technical shift in approaches to ERA. Any advantages notwithstanding, experience has shown that such shifts toward new, more informative, but also more complex, methods in support of risk-based decision-making inevitably present risk managers with challenges. The introduction of probabilistic methods into ERAs, which can give managers considerably more information than point estimate methods but are more demanding, are a case in point (Roberts 1999). This section discusses a several practical issues that are likely to challenge risk managers as they pursue the objective of greater use of population-level ERA.

Need to Define Appropriateness of Use

Although the use of population-level endpoints and population-level ERA can convey several advantages, risk managers should be aware that it is neither the only assessment tool available to them and or the tool of choice in all circumstances. Criteria for determining when population-level ERA is an appropriate tool are outlined in Chapter 10.

Defining the Assessment Population

How to define a population (or Assessment Population, as discussed in Chapter 3) was a source of much debate and discussion in all stages of the development of this book. This book does *not* dictate how and when a population should be defined as the biological population, or some subset of the population, because there is no single set of criteria or rules that can be applied across regulatory programs and risk management applications. A point of agreement among workshop participants and authors of this book is that, before implementing a population-level ERA, risk managers and risk assessors must agree on the assessment population that will be the focus of the management decision. Another point of agreement in the development of this book is that there are occasions when the exposed population is such a small subset of the biological population that it is obvious that impacts to the biological population cannot occur regardless of impacts to the exposed population. How much exposure or impact a population can sustain remains a challenge in the application of population-level ERA no matter how the population is defined.

Before implementing any of the ERA approaches described, risk managers and risk assessors must define the "population" that will be the focus of the management decision.

With that said, defining a population is based on a variety of factors that influence the risk management context (i.e., societal values, stakeholder concerns, governing policy, future land management goals, and biological considerations). Therefore, the definition of a "population" can vary greatly from one risk management application to another. Examples of the assumptions that risk managers consider when defining a population by risk managers and risk assessors include (Suter et al. 2007) the following:

1) The assumption that the exposed population represents the entire biological population (i.e., all individuals within a population are assumed to be exposed to contaminants).
2) The assumption that a small number of individuals out of the biological population are exposed (e.g., areas of suitable, contaminated habitat can multiplied by the species-specific density values to estimate the number of exposed or potentially impacted individuals. These values can be compared to the total estimates of species-specific population density within a geographic area).
3) The assumption of population effects foraging can be used to estimate the number of individuals within a local population that could be adversely impacted (i.e., an approach based on Freshman and Menzie 1996).

Need to Define Acceptable Population-Level Risk

For risk managers to use the results of a population-level ERA for decision-making, there must be some a priori determination of what constitutes an acceptable risk to a population. If we accept that the population will persist despite the loss of a few individual organisms, then how many individuals is it acceptable to lose? Is this a fixed number or dependent in some way on the life history of the receptor or its population characteristics? Is it acceptable to lose a few individuals if the population abundance is expected to recover within a reasonable period? How can we achieve flexibility when determining population-level risk while achieving the goal of making the choice of acceptable risk transparent?

At present, there are very few enforceable environmental standards or criteria based explicitly on population-level endpoints (an exception is ORDEQ 2001). From a purely legal perspective, there is also limited case law with which to draw precedents on the use and interpretation of population-level endpoints. Both regulators and lawyers have thus not been able to gain experience or familiarity with the use or interpretation of these endpoints. For many agencies and programs, the process of formulating policy on acceptable risk levels is a difficult one, in some cases requiring action by outside political institutions, such as state or provincial legislatures. Because the informed dialogue necessary to develop these policies will take some time, use of such endpoints may initially force regulatory risk managers to be pioneers in a new paradigm that is beyond any comfort zone they may have developed with other risk-based approaches. In addition, absent a common definition of acceptable population risk, there is the potential that independent policy decisions by multiple agencies or programs will lead to different agency- and program-specific definitions of acceptability. Such differences are likely to be confusing to stakeholders and undermine confidence that risk-based decisions are rational and fair.

Need for Additional Resources

The use of population-level ERA can be expected to entail, at least initially, additional resources (e.g., time, money, data) not typically associated with traditional organism-level based approaches. There can include those associated with the following:

- Obtaining ecological data necessary to support a population-level analysis, such as knowledge of the types, distribution, and quality of habitat; life-history characteristics of the selected assessment entities; or exposure data from multiple, spatially disjunctive areas. For many possible receptors and habitats, ecological data adequate to support even the simplest population-level ERA are lacking and the cost (in terms of both time and money) of obtaining them adds to the overall cost of the assessment.
- Conducting the population-level ERA, including having available practitioners with expertise in developing, calibrating, and validating any population models, undertaking empirical field investigations, and interpreting the results of these more involved analyses. There may also be additional costs associated with the extra time required to perform population-level analyses.
- Reviewing a population-level ERA, if it was prepared in support of a regulatory risk management decision, such as registration of a new pesticide. The primary cost here would be in obtaining and maintaining the expertise needed to review a population-level ERA, whether this expertise was held within the agency or within external contractors. In

either case, such costs would have to be borne directly by the regulatory agency, which may find it difficult, amid a number of competing budget priorities, to do so. Limited reviewer resources within an agency typically translate into additional time to complete a review, which appears as an additional cost to the regulated party.

Significant reductions in some of these costs could be expected if performing a population-level ERA was to eventually transition from a customized to a commodity service. This transition would be encouraged if methods were used with sufficient frequency to allow development of a pool of practitioners and dissemination of commonly understood and accepted techniques that are practical, cost-effective, and relatively transparent. In the interim, however, risk managers will likely have to martial additional resources to support use population-level results for decision-making.

Need for the Application of Lessons Learned

As the use of population-level ERA increases (or those previously conducted are analyzed), it should be expected that lessons learned in some applications may be applicable in other applications. This can be particularly true for the use of population-level ERAs at historic hazardous waste sites (Appendix 1). Identifying the common themes and elements that are exportable from one situation to the next will allow focus of management resources toward analyses that are truly needed. Chapters 8 and 9 provide detailed information and example of the application of empirical and modeling approaches to population-level ERA. In addition, Appendix 3 of this book provides a compilation of population-level ERA information that exists in the scientific literature but was not cited directly in this book. It will be a challenge to identify how and when these examples and others will be applicable to similar situations.

Need for Training and Guidance

There is no doubt that population-level ERA poses a higher level of complexity than organism-level risk assessments, and that widespread use of population-level ERA will necessitate additional education and training of both risk assessors and risk managers alike, as well as guidance for the use of population-level ERA and interpretation of its results.

A key factor in the success, in terms of its usefulness in support of risk management decision-making, of any ERA, population-level or otherwise, is the involvement of qualified risk assessment practitioners. Although there is no commonly accepted definition of what constitutes a qualified ecological risk assessor, it is generally understood to include some combination of formal education in relevant disciplines such as chemistry, toxicology, biology, ecology, and statistics, and practical experience in preparing, interpreting and presenting the results of risk assessments to risk managers. It may also include experience as a risk manager. Because of the traditional focus on risk assessments at the organism level, risk managers may be challenged to find practitioners with education and experience in population-level methods.

Training would be an affirmative response to any actual or perceived shortage of practitioners qualified to perform a population-level ERA. However, formal training opportunities in ERA, in general, are not widely available, even within academic institutions. Practical, hands-on training opportunities are more often on-the-job events or, less usefully, academic field or laboratory exercises. Within this narrow universe of training opportunities, there are few that explicitly address population-level endpoints. Risk managers may therefore have difficulty in overcoming any shortage of population-level ERA practitioners solely by means of training.

With any type of analysis in support of decision-making, a risk manager should consider whether there are tools available with which to perform the analysis, and guidance (e.g., technical, regulatory, legal) to direct both use of such tools and (more importantly) interpretation of their results within the context of the management decision. The availability of population-level analysis tools should not be a concern to risk managers. There are a variety of such tools, ranging from simple extrapolation methods (Suter et al. 2000, 2007) to sophisticated population models. Although many are basic research or academic tools, a number have been used in support of risk management decisions in situations ranging from hazardous waste sites to game management. A number of such tools are reviewed and demonstrated in Chapters 8 and 9.

A more significant concern is the apparent lack of either general or context-specific guidance for the selection of an appropriate population-level ERA tool or tools, and for interpretation of population-level ERA results. Particularly in a regulatory context, guidance is critical for articulating some common understanding between regulators and the regulated over matters of interpretation and decision-making. It is a way for a regulated party to have some assurance that the time and effort expended on an analysis will produce a product acceptable and useful to a regulator. In a business setting where population-level analyses might be used for strategic, nonregulatory, decision-making, guidance is useful for capturing and conveying lessons learned to other areas of the organization or between changing administrative structures. While this book is a partial step (see Chapter 11) toward guidance for population-level ERA, risk managers should be aware that a lack of guidance can place them, and their decision-making, in uncharted waters.

In addition, risk managers will need to develop, or have made available to them, criteria for assessing the scientific and technical quality of a population-level ERA with respect to the context in which it will be used (as has been suggested for probabilistic risk assessments by Burmaster and Anderson 1994). This will be of special concern for regulatory risk assessors and managers who must review and approve risk assessments created by others. The preparation of any risk assessment requires making a number of important decisions and assumptions. How is someone reviewing the population-level ERA to know if judgment exercised by the risk assessor was sound and defensible? What information should be included in a population-level ERA to justify critical decisions?

Need for Improved Risk Communication

There is a continuing appreciation of the need to make the risk assessment process and output understandable not only to risk managers but also to stakeholders

(Roberts 1999). This is a challenge in that, even in its simplest implementations (e.g., Suter et al. 2000), population-level ERA is a more complex and technically demanding approach than traditional organism-level–based approaches. Although the models for population assessment many be more complicated, it does not simply follow that the outputs of population-level analysis are too complicated for managers and lay persons to understand. For that matter, it is quite possible that communicating potential impacts to populations may actually be easier for nonscientists to accept as relevant.

The caution here is to achieve effective communications with the right level of technical detail. Extensive details about the intricacies of modeling efforts and assumptions may overwhelm the audience and be perceived by those to whom a risk manager is ultimately responsible (stakeholders, the affected community, the public, and higher political authority) as decreasing the clarity and transparency of the decision-making process. Complex models with numerous variables can be perceived of as less certain or reliable and shown statistically to embody higher uncertainty at the extremes (Regan et al. 2002a). Use of mathematically and procedurally complicated methods (such as risk assessment in general and population-level ERA in particular) may be perceived by stakeholders as an attempt to hide the truth behind a screen of numbers and convoluted processes. These perceptions, whether justified or not, can engender a lack of trust in risk managers on the part of stakeholders that can significantly challenge the success of the decision-making process. Risk managers are likely to face challenges when communicating the need for, and results of, more complex population-level ERA analyses. Although there is no guidance available on how to communicate population-level ERA methods and results effectively to lay audiences, carefully designed communication process can achieve a level of understanding that would support risk management.

Another communication challenge will be explaining, to both risk managers and stakeholders, why loss of one or more individual organisms is acceptable. Humans are deeply conditioned by cultural and personal experience to value the individual and to see the loss of an individual as a significant adverse event. For many people, placing such a loss into a larger population context, as in "you will die, but the species will go on," can be untenable, because common sense suggests that the presence of any dead or injured individuals is unacceptable in its own right or as evidence of a significant, perhaps hidden, problem that demands action. A population-level ERA, guided by its fundamental premise that a sustainable population does not require survival of all individuals, could easily reach a conclusion at odds with this deeply held perception. In addition, population-level ERA is an emerging discipline, and it is reasonable to expect that the science behind its models and methods will evolve and change over time. Although such change is normal to (and welcomed by) the scientific community, it can be unsettling to regulators, who often have a mandate for consistency and continuity, and stakeholders, for whom such changes may appear as an arbitrary and capricious changing of the rules (to their disadvantage).

It is reasonable to expect that these issues of complexity, acceptability, and change will lead, in many cases, to more vigorous discussion between risk managers, risk assessors, and stakeholders regarding meaning of risk assessments and reasonable levels of protection. This is not unhealthy, but requires a well thought-out risk communication strategy if such discussions are to be productive.

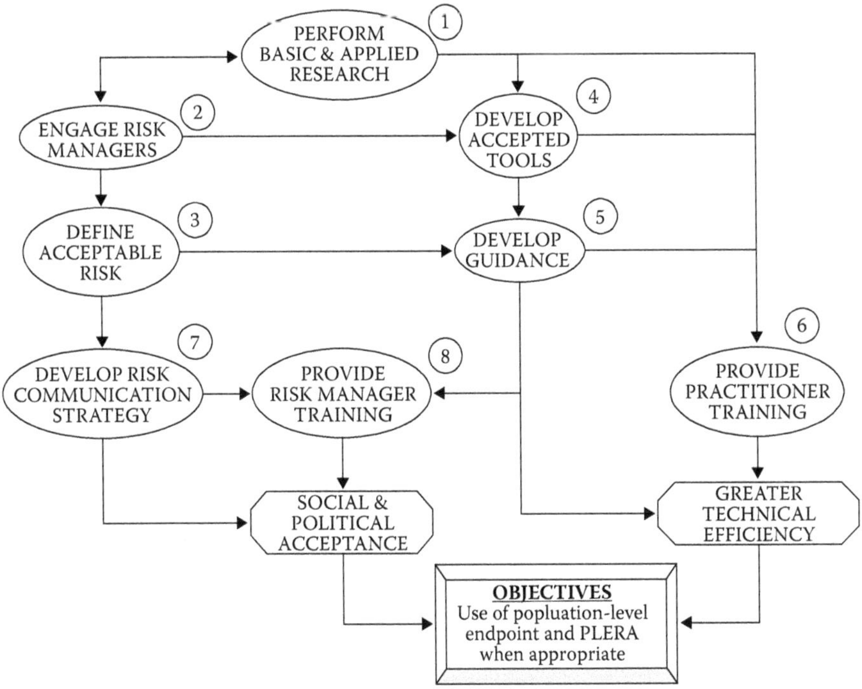

FIGURE 2.1 Attributes influencing the objective of achieving increased use of population-level ERA.

ACHIEVING RISK MANAGEMENT WITH POPULATION-LEVEL ERA

Added value can be brought to the risk management decision-making process by the use of population-level endpoints, when it is appropriate to do so. Therefore, this section outlines an approach for achieving greater acceptance of population-level endpoints, supported by use of population-level ERA methods. However, as was outlined in the previous section, there are a number of challenges facing attainment of these objectives. This section discusses how these challenges can be recast as deliverables that influence attainment of our objectives (Figure 2.1). By doing so, risk managers can be given specific recommendations on how to produce these deliverables, on the premise that by doing so they will achieve the stated objectives. Also discussed are some of the positive, and negative, consequences that could be associated with attainment of these objectives.

① Perform Basic and Applied Research

(Numbers in Figure 2.1 refer to specific objects). If risk assessment is defined as the scientific analysis and characterization of adverse effects of environmental hazards, then it must stand on a foundation of the best available science, which is typically provided by basic research activities. Basic research is necessary to increase our understanding of population biology and ecology in general, as well as the population biology of specific receptors. In addition, because risk assessment is primarily intended as a practical tool for the risk manager, there is also a strong role for applied science, whose purpose (in part) is to translate basic research findings into workable tools and methods. Information generated by basic and applied research activities directly supports tool development, guidance development, and practitioner training. These activities provide the factual basis for informing various management actions or risk communication strategies. Although defining acceptable risk at the population-level could be an entirely policy-driven exercise, a more defensible definition will be one that is informed by the best available science. Refer to Chapter 12 for specific recommendations regarding basic and applied research activities needed to support population-level ERA.

② Engage Risk Managers

If population-level ERA is to have an influence on decision-making, risk managers in various sectors (e.g., political, public policy, business) must engage in discussions regarding its development and use. This means an active dialog of issues affecting decision-making at the population level. In one direction, risk managers need to be kept apprised, by the science sector, of the results and implications of basic and applied research activities in population biology and population-level ERA. In the other, the science sector needs to understand how population-level information can be best conveyed to risk managers to support decision-making effectively. All sectors need to engage in a joint discussion of what constitutes acceptable population risk, which population-level ERA tools are most useful under certain circumstances, and how best to communicate risk in a population context. We recommend formation of a joint population-level ERA working group, composed of representatives from all sectors, to initiate and facilitate such discussions. Part of this engagement process would be a public information campaign to inform both the scientific community and the public as to the uses and benefits of population-level ERA.

③ Define Acceptable Risk

For risk managers to use the results of a population-level ERA for decision-making, there must be some a priori determination of what constitutes an acceptable risk to a population (and inherently understand how the population is defined). How a population is defined and what defines acceptable risk will be heavily conditioned by site- and context-specific legal and policy issues and somewhat less so by scientific possibilities and limitations. Nonetheless, there may be some fundamental, common aspects to such a definition that cut across site- and context-specific issues. We recommend attempting to identify and capture any such commonalities in a basic set of narrative criteria that operationally (i.e., in a way amenable to actual measurement) define acceptable risk. These could be used either directly as generic acceptable risk criteria or to guide development of context-specific criteria. Such criteria could bring a common basis to what otherwise might be independent policy decisions by multiple agencies or programs and forestall the proliferation of different agency- and program-specific definitions of acceptability. We also recommend considering development of regulatory structures based on these common criteria so that population-level ERA can be used within a commonly understood legal and policy context.

④ Develop Accepted Tools

Basic and applied research activities have generated a variety of population-level analysis tools, ranging from simple extrapolation methods (Suter et al. 2000) to sophisticated population models (see Chapter 9). However, the issue for risk managers is not one of availability but rather stakeholder acceptance of tools that support risk management decisions. Sophisticated tools that find favor with the scientific community may appear cumbersome, uncertain, or impractical by the regulated community. Conversely, tools seen as practical and expedient by the regulated community may be viewed by the scientific community as factually deficient. We recommend identifying, ideally as a joint exercise amongst various stakeholders, a basic set of tools acceptable for performing a population-level ERA from among those available. See Chapters 8 and 9 for specific recommendations on tools to support empirical and modeling approaches, respectively. To provide greater flexibility in tool selection as new tools are developed, we recommend extending this joint exercise to include defining a set of criteria to guide tool selection.

⑤ Develop Guidance

Guidance is critical in a regulatory context to provide a common understanding between regulators and the regulated over matters of interpretation and decision-making. In a strategic business, nonregulatory context, guidance is useful for retaining and disseminating information within the organization. Developing guidance for population-level ERA requires, at a minimum, knowledge of which tools are acceptable and some basic, but operational, definition of acceptable risk. Given these precursors, we recommend undertaking development of guidance for population-level ERA that explains, at varying levels of detail 1) what tools are available; 2) how they should be deployed; 3) how their results should be reported; and 4) how results from one or more tools should be interpreted, within various legal and policy contexts, to estimate risk.

⑥ Provide Practitioner Training

The historic and current focus of risk assessments on organism-level endpoints has limited development of a pool of practitioners with the education and experience in both risk assessment and population-level methods. Any actual or perceived shortage of practitioners qualified to perform a population-level ERA can be lessened by the availability of suitable training opportunities. We recommend increasing formal training opportunities in population-level ERA within academic institutions to include practical, hands-on training opportunities either through internships or field or laboratory exercises. We also recommend increasing the availability of short-course and workshop training opportunities for risk assessment practitioners desiring to perform population-level ERAs.

⑦ Develop a Risk Communication Strategy

The use of population-level ERA and articulation of what constitutes risk to a population will, in many cases, spark vigorous discussions between risk managers, risk assessors, and stakeholders regarding the suitability of risk assessment for decision-making and what constitutes a reasonable level of protection. Such discussions are not to be discouraged but must be guided by a well thought-out risk communication strategy if they are to be productive and not divisive. If it is to play an effective role in support of decision-making, the risk assessment process must be made more understandable to risk managers and to those to whom risk managers are ultimately responsible. Explanatory outreach is critical here. Use of risk assessment must also be shown as an opportunity to increase, not decrease, the clarity and transparency of the decision-making process. The perception that any use of risk-assessment methods, population-level ERA or otherwise, is simply an attempt to cloak some awful truth behind a convoluted and hard-to-follow process must be challenged. It is also important to address, for both risk managers and stakeholders, why and under what circumstances, loss of one or more individuals is acceptable. This may require stepping back from the risk aspects to address more fundamental issues such as how a population persists through time. We recommend developing a risk communication strategy, involving public and professional outreach that addresses, at a minimum, the issues of transparency and acceptability.

⑧ Provide Risk Manager Training

Although we recommend engaging risk managers early in the population-level ERA process (see 1), not all risk managers who do or might use population-level methods to inform their decisions will have access to these formative discussions, even if such discussions involve a fairly broadly constituted population-level ERA working group. Reaching a larger audience of risk managers will involve both increased communication (see 7) and training opportunities. Such training should provide risk managers with an overview of the technical aspects of a population-level ERA but focus more on how to interact effectively with risk assessors to obtain risk assessment results that they can use to make informed decisions. Risk managers need training as to the types of strategic questions they should ask, and receive answers to, before the risk assessment starts (e.g., Is the risk assessment problem understood in the context of the decision it will support? Is there already an answer?), while it is under way (e.g., Are available data necessary and sufficient?), and after it is completed (e.g., What is the answer? Has a sensitivity analysis been performed?). They also require training on how best to communicate the risk assessment process, and its results, to stakeholders. In short, the usefulness of a risk assessment is enhanced (or diminished) by presence (or absence) of a knowledgeable risk manager. We recommend increasing the availability of short-course and workshop training opportunities in risk assessment and population-level ERA for current and future risk managers. We also recommend extending risk communication outreach activities to specifically target risk managers, particularly those who are unfamiliar with population-level ERA.

Achieving the Objective

Achieving the objective of using population-level endpoints and population-level ERA when appropriate will require the availability of scientifically defensible methods that can be performed in a cost-effective manner by trained practitioners and social and political acceptance of these methods and whatever definition is offered of "protective." Coming, a priori, to some agreed-on definition of acceptable risk, coupled with an effective risk communication strategy to help explain acceptability and the process for discerning it, is necessary to achieve social and political acceptance of population-level ERA-based decisions. Such acceptance is critical to lessen or dispel any real and perceived concerns of stakeholders and to provide risk managers with a familiar comfort zone for the use of population-level ERA and other risk-based approaches. The availability of accepted population-level ERA tools, guidance for their use and interpretation, and practitioners familiar with both such tools and guidance is necessary if there is to be greater technical efficiency, and significant cost reductions, in the preparation and review of population-level ERAs.

3 Population Protection Goals

Charles Menzie, Nancy Bettinger, Alyce Fritz, Larry Kapustka, Helen Regan, Vibeke Moller, and Helen Noel

INTRODUCTION

This chapter offers insight into what it means to protect populations. To this end, the chapter explores the following questions:

- What is a population and how do attributes of a population differ from those of organisms?
- What does it mean to protect populations?

As we explored these questions, it became clear that while they can be approached from a technical (e.g., population ecology) standpoint, they are also contextual and are strongly influenced by management decisions and language differences among population biologists, risk assessors, managers, and the public. In short, the word *population* can mean different things to different people. For this reason, we considered these questions broadly to capture the population biology as well as the contextual aspects. In our view, treating protection of populations from a strict population biology standpoint can lead to outcomes that miss expectations of other interested parties. To avoid this undesirable outcome, it is essential that the problem formulation phase of a risk assessment be clear on what is being assessed and how that assessment will proceed.

The chapter begins with definitions and describes why populations are different than organisms. Here the term assessment population is introduced and attributes of populations and organisms described. We then discuss what it means to protect populations within the context of management goals. This is followed by examples of assessment and measurement endpoints appropriate for population-level risk assessments. This is followed by recommendations.

WHAT IS A POPULATION?

Population has a particular meaning to most ecologists, evolutionary biologists, and ecological geneticists. An ecological definition is:

> A population is a group of plants, animals, or other organisms, all of the same species, that live together and reproduce. Individual organisms must be sufficiently close geographically to reproduce. Sub-populations are parts of a population among which gene flow is restricted, but within which all individuals have same chance of mating with any other individual.
>
> **—Gotelli (1995)**

But the word "population" has several meanings to others and this can create a situation that can lead to confusion in risk assessments. This variety can be seen in Webster's dictionary.

> 1. a) All the people in a country, region, etc., b) the number of these, c) a specified part of the people in a given area. 2. populating or being populated. 3 *Biol.* All the organisms being in a given area. 4. *Statistics* the total set of items, persons, etc. from which a sample is taken.

Among biologists and regulators there exists a variety of ways to express or define populations:

- The total number of individuals of a single species in one place (Williamson 1972)
- Local population or deme: the community of potentially interbreeding individuals at a given locality (Mayr 1970)
- Local population: a group of individuals so situated that any two of them have equal probability of mating with each other and producing offspring (Mayr 1970)
- The individuals of a particular species in a particular group or in a definable place (Soule 1987)
- Any group of organisms coexisting at the same time and in the same place and capable of interbreeding with one another (Purves and Orians 1987)
- Group of interbreeding organisms of the same kind occupying a particular space. It is characterized by density, the number of organisms occupying a definite unit of space. It has an age structure, the ratio of one age to another. It acquires new members through birth and immigration and loses members through death and emigration (Smith 1990)
- A group of individuals of one species in an area, though the size and nature of the area is defined, often arbitrarily, for the purposes of study being undertaken (Begon et al. 1990)
- A species population includes all existing individuals throughout the species' range (Mitton 1997). This can be important to consider in a regional assessment (e.g., amphibian population declines perhaps influenced by widespread introduction of endocrine modifiers within the range of the species)
- Effective population size: $N_e = 4N_mN_f/N_m+N_f$ where N_m and N_f are numbers of breeding males and breeding females in the population (Primack 1998)

- Effective size of a population, N_e= the number of individuals in an ideal population that would give the same rate of random genetic drift as in the actual population (Lande 1988)
- For large mammals: suggested effective population size (the number of individuals contributing genes to the next generation) of at least 50 (Franklin 1980)
- For large mammals: suggested long-term minimum effective population size of 500 (Franklin 1980)

Experience shows that population biologists, managers, and the public may think of populations differently in the context of a specific management goal. For example, a management goal to protect wildlife populations might relate to populations as defined by population biologists, or alternatively to protect animals in general. Risk assessors should be aware, in particular, of the differences in meaning between popular and technical uses of the term. In popular usage, a population might refer to a group of organisms in a place at one time. Six robins in a backyard on a Tuesday describes such a population, and on Wednesday, the population might be made up of different individual robins. In contrast, a more technical (or ecological) definition denotes that the group is a separate, self-regulating, local unit of interacting, conspecific organisms that remains coherent in time. This ecological use was well developed by 1949, when Allee et al. (1949) explained that a population not only has attributes that come from the organisms as individuals, but also has attributes that are unique to the group. However, modern metapopulation biologists have added more refinement to the concept and these refinements are very pertinent to predicting effects to populations. Most modern ecologists would carry a metapopulation context for most (not all) populations and see the historical framework as a restricted context.

The term "*assessment population*" is advocated for use in risk assessments that evaluate risks to populations.

The ecological definition connotes separateness; that the organisms breed or otherwise interact more with individuals within the population than with those outside the population. The boundaries of a population are likely to be influenced by geography and other barriers, but are defined biologically by the distribution and interactions among individuals. The boundaries of a population defined in the popular sense may be drawn anywhere, such as around a Superfund site. Mayr (1970) recognized that actual populations deviate from the conceptual ideal and that they intergrade with other populations. Similarly, Wilson (1999) conceded, "Few such objectively defined populations exist in nature." Local populations may be grouped together into larger, regional populations that, when combined, encompass the species. Some biologists find the concept of a metapopulation (a group of loosely connected local populations of the same species that may continually become reestablished or be extirpated) helpful. The spatial dynamics of populations (isolated, patchy, continuous, and metapopulations) is discussed further in Chapter 6.

Given the potential for miscommunication about populations in the context of risk assessment and risk management, we adopt the term assessment population (US Environmental Protection Agency [USEPA] 2003) for use in population-level ecological risk assessments, reflecting the contextual nature of such assessments. The assessment population may be the same as the biological population as defined by ecologists but it might also be something different, reflecting regulatory or societal bounds on the assessment. Munns and Mitro (2004) have recently used this term in the context of risk assessments for Superfund-type sites in the United States.

The Assessment Population

During problem formulation, the assessment population must be defined explicitly. The management and problem context will often determine how the assessment population is defined. And, this will be further informed through discussions among risk assessors, managers, and other interested parties. As a result, the assessment populations defined for purposes of hazardous site management, pesticide registration, or regional and watershed assessments could be different. The identification of the assessment populations might also be influenced by the nature of the stressors. Obviously, this up-front thinking and communication will be important to the successful conduct of the assessment and to how it is used to inform management decisions.

The assessment population may be

- the biological population,
- a component of the biological population (e.g., the exposed population), and
- a component of the relevant metapopulation (e.g., a subpopulation).

The operational definition for an assessment population may be those members of the species residing, foraging, or otherwise using the specific area of interest in the assessment. Generally, with very large assessment areas or at regional scales, the differences between the definitions of assessment population and biological population should converge. Even so, for neotropical migrants, many waterfowl, or anadromous fish, defining the population is challenging. It may be helpful to use additional descriptors to clarify the assessment population for regulatory purposes. For example, terms such as local subpopulation, exposed subpopulation and study area subpopulation could be defined specifically for each assessment.

The operational definition of the term "assessment population" as used here avoids the problem of redefining a term such as "population," which may have a particular meaning, and it places the emphasis on the assessment process that leads to a case-specific definition. As with most biological studies, ERAs should develop clear, explicit, operational definitions of the population of interest. For assessments that involve 2 or more species, there may be 2 or more definitions of assessment populations so that the relevant ecological relationships and management objectives can be satisfied. For some situations, there may be no difference between the

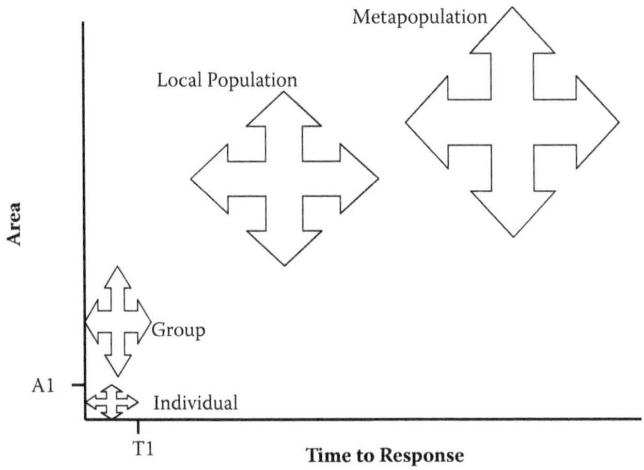

FIGURE 3.1 Space–time relationships for considering environmental effects along the continuum from organisms to metapopulations. This is a conceptual view to which there can be exceptions. [T1 = the life span of an individual; A1 = foraging range for an individual] Within an environmental setting, complexity generally increases along the space–time continuum.

definition of the biological population and the assessment population (e.g., a large site, where geographic or habitat boundaries are coincident with the likely biological boundaries for a particular species). In general, as the spatial area of the assessment increases, the distinctions between assessment populations and biological populations are less obvious.

Members of a species may be considered in various ways. A particular individual may be described in terms of its life-stage (e.g., larvae vs. juvenile vs. adult), size (e.g., body weight, length), age, genetics (e.g., expressed observations such as color, or less obvious characteristics, such as nucleotide sequence for a specified gene), or behavioral characteristics (e.g., home range, dietary preferences, and foraging strategies), and mortality (i.e., living or dead). Similarly, a collection of individuals may be described without distinguishing particular properties that pertain to or derive from the multiple interactions among individuals (e.g., average body weights and average clutch/litter sizes). As attention shifts from the consideration of organism-level characteristics toward group dynamics, different scales of space and time (Figure 3.1) take on greater importance in terms of both understanding the biological and ecological characteristics and interpreting responses to environmental stressors (biological, chemical, or physical factors). In general, at the organism and group level, the temporal response times are defined by the lifespan of the individuals for the area of interest (noting that life spans are different for different latitudes or ecoregions and that each individual will experience its own lifespan). As one switches from group considerations to natural or local populations and finally to metapopulations, the time to response and the area of interest increases. Response time generally refers to sublethal stressor conditions. For example, if the entire metapopulation complex is exposed to lethal conditions (e.g., a killing frost, a major chemical insult), the response time will be short across spatial scales affected by the stressors.

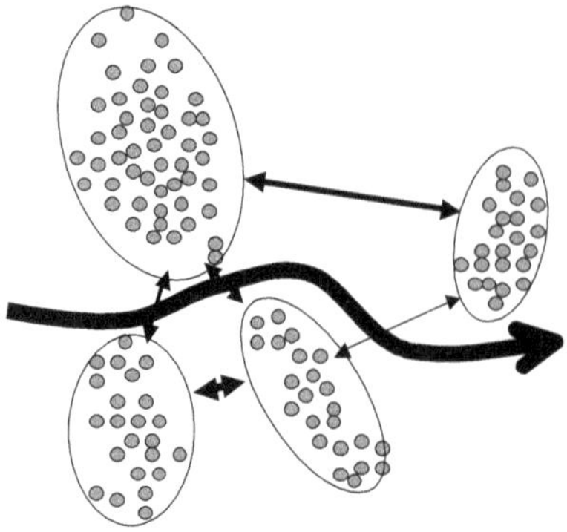

FIGURE 3.2 Representation of four satellite populations composing a metapopulation complex. The curved arrow traversing the middle of the diagram represents a medium-size stream, ravine, fence, road, or other physical structure that somewhat impedes movement of individuals between satellite populations on the opposite side of the impediment. The width of arrows depicts relative movements between satellite populations.

The existence of a metapopulation complex (Figure 3.2) implies that the landscape is heterogeneous. Shifts in vegetation cover or composition, impediments to movement (e.g., streams, ravines), or disturbances (e.g., human settlements, clusters of resident predators) limit free movement across the landscape, and therefore define the boundaries of two or more satellite populations that constitute the metapopulation complex. The degree to which these groups interact may vary. Exchanges between satellite populations may have a directional bias. Because interactions take place among the satellite populations, effects from exposures to stressors, ripple through the metapopulation complex. Spomberg et al. (1998) have demonstrated that in some circumstances the population effect may be observed only in unexposed satellite populations. This underscores the importance of framing the assessment questions properly as well as carrying out the proper study design and analysis in the risk assessment process. Further discussion of population spatial dynamics (isolated, patchy, continuous, and metapopulations) is provided in Chapter 6.

Attributes of Organisms and Populations

The assessment population can range in composition from individuals that are a part of the actual local biological population, to the biological population, to a metapopulation. It follows that risks to the assessment population might be expressed in several ways including risks to some number (or fraction) of individuals in the population (when the assessment population is not the biological population) to risks to the local biological population or metapopulation. It also follows that the methods used to quantify these risks will vary depending on the definition of the assessment population.

Attributes of Organisms

Attributes of organisms can roughly be grouped as demographic attributes, physiological attributes, habitat use attributes, and genetic attributes (Table 3.1).

Demographic Characteristics

Demography refers to the statistical study of individuals within populations with respect to reproduction, mortality, migratory movements, age, stage, and sex. Demographic attributes include the individual's age or stage class (as well as age or stage class at time of death), sex, and measures of reproduction and development. Measures of reproduction include fecundity (e.g., number of eggs produced per female per unit time, potential seeds, ovules per parent) and number of viable offspring per female (e.g., births per female, female offspring per female). Measures of development include time to reproductive maturity, time to first free feeding, weaning, ripening, and gestation time.

Body Size

The size and weight of individuals in the population are also important demographic characteristics. Individual size and weight can influence other characteristics including metabolic rates, growth rates, and locomotion (Peters 1983). Size and weight of individuals also influences population attributes such as abundance and density. In some applications (e.g., in fisheries) the relationship between the length and weight of individuals in the population is used to judge the health or condition of the animals.

Physiological Characteristics

Physiological data include vital rates such as respiration (liters of air per minute; mg O_2wet weighttime), food intake rate (mg food/mg body mass), and metabolism including the rate at which chemicals are eliminated from the body. Organism attributes also include size and somatic growth. Also related is dietary information including types, sources, and proportions of food items eaten by the focal species.

Ecology, Behavior, and Environmental Exposure

Other attributes of organisms relate to how they experience and/or use the environment. General ecology texts often refer to this as the species' niche. For example, individuals of some species, such as barnacles, would be described as fixed in space as adults but carried over great distances in ocean currents as planktonic larvae. Other individuals may be fixed in place throughout their life history, as is the case with many plants. Still other individuals may migrate at different temporal scales (e.g., seasonally as with many bird species or daily as with many zooplankton species). All of these behaviors and characteristics influence the types of exposures that the individuals receive to a spectrum of stressors. Some measures that might be used to describe the individuals within the population include range of animal per unit time (such as km/day of a coyote), or range of dispersal of gametes or offspring per unit time (such as seed movement from parent per season). Another common organism-level metric is home range. Home range is another measure of landscape use and represents the overall area of the landscape required by a given animal to fulfill feeding and reproduction requirements (Efroymson et al. 2004). Movement, dispersal, and home range are determined through various field observations including tracking the specific location of an animal at a specific time. This information may also provide insight as to whether the population is part of a larger metapopulation (Hanski and Gyllenburg 1993). Such location specific data are also useful in parameterizing individual-based models for species. An example is the Spatially Explicit

Exposure Model that tracks individuals exposed to stressors in a landscape and simulates their exposures in accordance with variations in organism characteristics (Wickwire et al. 2004).

Genetic Characteristics

Individuals each have a genetic code that may vary in some respects from other individuals in the population. In some cases, the genetic composition of an individual may enhance or reduce the ability of the individual to survive, reproduce, or cope with a stressor. Clearly, for humans, there is considerable interest in such genetic predisposition for each individual. For other species, the interest appears to be mainly at the population level. Still, there may be cases where knowledge of individual characteristics is important for environmental decision-making (e.g., in conservation biology).

Life History Strategies

Life history strategies relate to how populations reproduce themselves. Some aspects relate to reproductive strategies including the varied means for achieving fertilization. Animals and plants have evolved diverse means of accomplishing this. Other aspects relate to factors that result in dispersal of individuals; these can include transport of fertilized eggs and larvae as well as migrations of juveniles or adults.

Attributes of Populations

Populations can be characterized using attributes that can be evaluated either empirically through population studies or by using population models. Although there are many types of population attributes, certain ones give insight into the sustainability of the population and these are likely to be most useful for informing environmental decisions (Table 3.1). We describe these selected attributes and briefly discuss each attribute's utility for population-level risk assessment. Other attributes, which can be evaluated using population models, are identified on a model-specific basis in Chapter 9. Maltby et al. (2001) and Munns and Mitro (2004) provide more detailed treatments of population attributes.

Abundance

Abundance is the size of a population, in terms of the number of individuals, biomass, or expressed as some other quantity (e.g., carbon). Abundance can be used to describe the size of an entire population, the sizes of subunits of the population (subpopulations), or the sizes of various classes within the population (see Population Structure). Population abundance is determined either through direct observations (e.g., field studies) or from rates of birth, death, immigration, and emigration.

Population density, a related attribute, is simply the size of a population expressed on a per-unit basis (area or volume). Population density is a convenient way to describe abundance in that it facilitates comparisons among populations or sites.

The equilibrium abundance, or steady state abundance, of a population is that population size at which inputs to the population (births, immigration) are exactly balanced by losses (deaths, emigration). A related concept is carrying capacity, the number of individuals that environmental resources can support without increases or decreases. Equilibrium abundance can differ from carrying capacity if environmental processes depress numbers below that which the environment can support, and most

TABLE 3.1
Examples of attributes of organisms and populations

Attributes of organisms	Attributes of populations
Demographics of individuals • Mortality (e.g., living or dead) • Reproductive state and output (e.g., fecundity, births per female, potential seeds) • Development rate (e.g., time for larval development, time to maturity, weaning, ripening)	**Abundance** • Population size (number or biomass) • Population density • Equilibrium (steady-state) abundance • Carrying capacity **Extinction and recovery** • Probability of extinction • Time to extinction • Quasi-extinction • Minimum viable population **Recovery time (from disturbance)**
Physiologic characteristics • Individual growth rate • Respiration rate • Ingestion rate • Metabolism and excretion	**Population growth rate** • Intrinsic rate of natural increase • Finite rate of population increase • Birth, death, immigration, and emigration rates
Demographics of individuals (continued) • Age • Size • Sex • Locations and "home range" or dispersal of an individual	**Population structure** • Age class distribution • Size class distribution • Sex ratios • Spatial distribution of the population
Genetic characteristics • Individual genotypes • Presence of particular alleles • Heterozygosity	**Genetic structure and variation** • Genotypic frequencies • Heterozygosity • Genetic diversity
Organism "health" or condition • Condition factors (weight and length relationships) • Morbidity • Deformities • Tumors and other histopathologic anomalies	**Incidence (frequency, percent, or fraction) of the population or distribution thereof with respect to** • Specified conditions • Morbidity • Effects (e.g., percent killed) and exposures to stressors
Ecology, behavior, and exposure • Life history for an individual • Habitat and food "preference" or location in space • Locomotion, dispersal, migration (e.g., range), and spatial extent of activity (e.g., home ranges) for an individual • Individual environmental exposure	**Spatial distribution and habitat** • Spatial distribution across available habitat (may involve distributions of age and/or size classes as well as influences on genetic composition) • Critical patch size • Habitat requirements (quantity, quality, fragmentation)

populations fluctuate around some average abundance as a result of environmental variability and demographic stochasticity. Maltby et al. (2001) suggest that changes in equilibrium abundance have greater significance than do changes in population size, since the former implies some long-term effect, whereas the latter can imply a temporary effect.

Abundance is of central interest in population-level risk assessment because 1) the abundance of a population affects its functioning with ecosystems; 2) low abundance can increase the likelihood of population extirpation (as a result of environmental and demographic stochasticity, inbreeding depression and increased genetic load [see Chapter 5], and the Allee effect [1931]); and 3) society places value on population abundance. Thus adverse effects on assessment populations often are couched in terms of decreased population abundance.

Population Growth Rate

Population growth rate describes the temporal change in population abundance. It is generally denoted as r, the intrinsic rate of natural increase (also referred to as the Malthusian parameter), or as λ, the finite rate of population increase (also called the population multiplication rate). These two rates are related mathematically ($r = \ln \lambda$), and their use in the ecotoxicological literature to express population growth rate generally is a matter of preference (although the nuances of how they are calculated and how they are used in population models should dictate which expression to use that is, r is an instantaneous measure, whereas λ is finite; see Michod and Anderson 1980).

Generally, r ranges from $-\infty$ to $+\infty$, and is symmetric around the value 0, which represents zero population growth. Values less than 0 indicate a declining population abundance, whereas positive values of r indicate a growing population. λ, on the other hand, ranges from 0 to $+\infty$, with a value of unity representing zero population growth. Thus, at $r = 0$ and $\lambda = 1$, births and immigration into the population are exactly balanced by deaths and emigration out of the population.

Population growth rate can be thought of as an attribute related to population fitness (Leslie 1945; Demetrius 1975; reviewed in Caswell 2001). That is, populations with genotypes more suited to a particular environment should have greater rates of maximum population growth than do those less well suited, and as a result, the better suited (i.e., more fit) genotype will out compete all others (all else being equal). Conversely, the degree to which population growth rate is adversely affected by environmental conditions (e.g., contaminant concentrations) is a measure of risk to the population. In extreme cases, population growth rate may be reduced to such an extent that the population goes extinct. Less radical reductions might increase the risk of extinction when such impacts occur in conjunction with other environmental insults or with stochastic fluctuations in controlling processes.

If the population oscillates, there are other factors that influence the risk of local extinction, not just r. In such a case, the change in r might not capture all of the important information to express the risk. The probability of extinction is likely a more effective measure of risk under toxicant exposure.

Population growth rate, and the extent to which it is impacted by stressors, is therefore of central importance in ecological and evolutionary theory. It is generally acknowledged to be the key attribute linking organism-level effects to populations (Calow et al. 1997; Sibly 1999), is a fundamental underpinning to population regulation (Sinclair 1996) and underlies most fisheries' and wildlife management concepts and approaches (Sutherland and Reynolds 1998). Before the mid-1990s, use of population growth rate in ecotoxicology was limited (Sibly 1996), despite its usefulness as a measure of stress having been demonstrated some four decades ago (Marshall 1962). However, evaluation of the effects of stressors on population growth rate is becoming more commonplace (see Appendix 3: Supplemental Reading).

Population Structure

Variation in the attributes of individual organisms within a population creates internal structure. This structure can be defined in terms of the distributions of individuals (or biomass) among ages, individual sizes, developmental stages, sex, reproductive status, genotypes, and so on. Groupings or classes of like individuals can be delimited on a fairly arbitrary basis (such as years in calendar time), or can be tightly linked to the biology of the species (e.g., eggs, larvae, pupae, adults). Structure can also be characterized as a continuous variable, such as a sex ratio or distribution of phenotypic tolerances (Maltby et al. 2001). Additionally, structure can be defined in terms of the spatial distribution of individuals, as with subpopulations arrayed within a heterogeneous landscape (see Chapter 6).

Structure can both influence and be an indicator of the dynamics of the population, because the vital rates of survival and fecundity often vary across ages, sizes, genotypes, and so on. For example, the distribution of individuals typically is skewed toward younger age classes in populations that are growing rapidly. The opposite may be true for populations experiencing reductions in overall size (although see Chapters 4 and 6). To illustrate this, human populations in developing countries that are experiencing relatively rapid population growth generally have lots of young and few older adults (on a relative basis), whereas the distribution among ages in industrialized countries with more stable population sizes tends to be more even (Thomlinson 1965). Confounding this is that stressors can affect the structure of a population by modifying the processes of birth and death. Depending on how these effects manifest, the age, size, and stage structure (for example) can shift from preexposed conditions in ways that are inconsistent with the generalization previously discussed. If, for example, the susceptibility of organisms to a particular chemical increases with age, the resulting distribution might be skewed toward younger age classes, even though a potentially adverse population-level effect is occurring. Thus changes in structural attributes can convey diagnostic information about the causes and mechanisms of population-level effect, as when population declines are caused by substantially skewed sex ratios (e.g., Kalmus and Smith 1960; Hamilton 1967) from exposure to endocrine disruptors (e.g., Gibbs and Byran 1986; Moore and Stevenson 1991).

Genetic Structure and Variation

Although not typically considered in population-level ecological risk assessments (nor in population models), genetic structure can have profound effects on the responses of populations to stressor exposure (Chapter 5). Genetic structural attributes include genotypic frequencies and heterozygosity (diversity), and can be quantified using metrics like the genetic inbreeding coefficient, *F*, the degree of genetic differentiation among genetic markers which is informally interpreted as relatedness among individual organisms and groups. The adaptive value of genetic diversity of a population is related to its reflection of heterogeneity in alleles at loci which are important for fitness. Theoretically, ample genetic diversity should provide an ability to withstand a wider range of environmental conditions, including the presence of anthropogenic stressors.

Population genetic markers can be used to characterize genetic variation. Selection of specific approaches and markers depends on the goals of the study (Sunnucks 2000; van Tienderen et al. 2002) and the technologic constraints of measurement and availability of markers for the species of concern. As long as markers are neutral or near neutral, they can be used to assess population processes such as gene flow, genetic drift, population bottlenecks, and mutation rate. Population parameters typically estimated to describe population health include effective population size, allelic diversity, and mean population heterozygosity. However, the extent to which the diversity of genetic markers reflects ecologically important variation is the subject of continuing controversy (e.g., Reed and Frankham 2001; Edmands and Harrison 2003). In this regard, the relative amount of genetic diversity among populations can be a predictor of risk for persistence in the face of future stress. It also can be an indicator of current and past exposure to stressors, because such stress can reduce genetic heterogeneity when population size decreases rapidly in response to disturbance (the so-called bottleneck effect; Wiens 1977). For example, some nonmigratory populations at sites that have experienced pollution or other forms of disturbance have reduced genetic diversity (Lavie and Nevo 1982; Lavie et al. 1984; Nevo et al. 1986; Benton and Guttman 1990).

Populations experiencing small sizes or low genetic diversity for extended periods may demonstrate detrimental changes in birth and death rates, i.e., inbreeding depression (e.g., Hedrick and Kalinski 2000). Recognition of these factors suggests that among populations of similar size, populations with reduced genetic diversity are at additional risk from exposure to other anthropogenic stressors, disease, and environmental stochasticity. The importance of genetic diversity in the persistence of wild and captive populations has been recognized by conservation biologists and is incorporated into management strategies for the preservation of threatened and endangered species (see Chapter 7).

Extinction and Recovery — Probabilities, Rates, and Time

Several population attributes pertain to the persistence of the population and its ability to withstand and/or recover from stressor-related disturbance. Probability of extinction is the likelihood that population abundance will go to zero within some predefined time interval given the demographic, genetic, and environmental conditions during that interval. Time to extinction can be derived directly from the probability of extinction (Foley 1994; Gillman and Hails 1997). Examples of the use of these attributes are described by Harrsion et al. (1991) and Foley (1994) for butterflies and Gillman and Silvertown (1997) for plants. Snell and Serra (2000) recently suggested that probabilities of extinction could be used to interpret the ecological significance of toxicity test results. Because they are difficult to quantify in natural populations, and are only slightly less so in laboratory experiments (because of the replication required), these attributes may best be estimated using modeling and simulation techniques.

Related to probability of extinction is quasiextinction, defined as the probability that the population will fall below some critical abundance (Ginzburg et al. 1982). Critical abundance can be thought of in terms of specific population densities below which adverse effects are known or suspected to occur. For example, critical densities of puma may exist below which individuals are no longer able to find mates for reproduction. Maltby et al. (2001) describe a method by which the quasiextinction probability curves could be used to evaluate the ecological significance of estimated risks.

Minimum viable population, an attribute used frequently in conservation biology, is defined as the smallest population abundance that will persist for some specified length of time with a given probability. At least at low abundances, the expectation of a negative relationship exists between absolute population abundance and the likelihood of extinction (from demographic and environmental stochasticity). Examples of minimum viable population (MVP) as an attribute are given by Samson et al. (1985), Shaffer and Samson (1985), and Goldingay and Possingham (1995).

When a population is disturbed from its equilibrium or preexposed abundance, the time it takes to return to that abundance is called its recovery time. Recovery time as an attribute is readily appreciated, and has been advocated in the consideration of adversity of effect (e.g., USEPA 1998). However, several conceptual and methodology issues can confound its use in risk assessments. Because it may be unrealistic to expect a population to return exactly to its predisturbed abundance, criteria are necessary to define when recovery has occurred. As summarized in Maltby et al. (2001), these might include aspects of absolute population abundance (e.g., 90% of original population size; see Sherratt et al. 1999), natural variability in that abundance (e.g., within 2 standard deviations of long-term mean abundance; see Wiens 1996), its abundance relative to a reference or control population (e.g., 90% of reference population density; see Thacker and Jepson 1993), or perhaps its population growth rate relative to that of a reference or control population (e.g., 90% of the growth rate of an unaffected population; see Kareiva et al. 1996). Additional research and deliberation is needed to clarify the value of any of these in supporting risk management decisions.

Spatial Distribution and Habitat

Populations may also be viewed and assessed in terms of their spatial distribution. Related to this is the spatial distribution and quality of habitat. The spatial distribution of a population provides information on migration rates and possible metapopulation dynamics. By explicitly considering spatial distribution, a risk analyst can bring together related information on the distribution of stressors and landscape features that influence habitat availability, fragmentation, and quality.

Critical patch size is the quantity of area required to maintain a population. Although this concept seems straightforward, it is quite difficult to parameterize. Most data on critical patch size come from studies on different patches of habitat resulting from fragmentation. In these studies, critical patch size has been interpreted as 1) the minimum patch size below which the species is never found, 2) the minimum patch size below which the species is not present in 100% of the patches, 3) the minimum patch size that can sustain a viable population, or 4) the percentage of habitat that must be remaining for the species to be found in the landscape (Efroymson et al. 2004).

Incidence of Occurrence

Incidence refers to the frequency, percent, or fraction of a condition, effect, and/or exposure within the population. It may be expressed as a discrete variable such as the numbers of individuals or fractions of the assessed population that either do or do not have the characteristic. It might also be expressed as a continuous distribution for the variable. During and after the workshop, there was considerable discussion about whether such properties relate to organisms or populations. While an individual animal or plant may have a specific characteristic (i.e., it is an attribute of the individual organism being examined), "incidence" is a population attribute as it reflects the degree to which that characteristic is present within the population. Examples of incidence include the distribution of fish condition factors, presence of tumors, and numbers of animals (or fraction of the population) exposed to different levels of a stressor.

Incidence of a condition, effects, or exposure is considered a population attribute. This attribute might be articulated in an assessment endpoint or might serve as a measurement endpoint.

Scales for the Assessment Population

As discussed previously, Figure 3.1 through Figure 3.3 provide insight into scale considerations for assessment populations that might be defined in different ways. This is illustrated best in Figure 3.3, that identifies 4 types of assessment populations ranging from the organism through the metapopulation. It also illustrates the relationships among these.

In many cases, the scales of the assessment population will be determined during the problem formulation phase of the ERA. In some cases, there is guidance on how to define space and time scales for assessment populations. With regard to temporal and spatial scales, the US Fish and Wildlife Service developed guidelines for classifying species as either endangered or threatened, as defined by the Endangered Species Act of 1973 (USFWS 1983). The Endangered Species Act of 1973 states that the following factors determine whether or not a species should be listed as endangered or threatened (Nicholopoulos 1999): the present or threatened destruction, modification, or curtailment of the species' habitat or range; overutilization for commercial, recreational, scientific, or educational purposes; disease or predation; the inadequacy of existing regulatory mechanisms; and other natural or man-made factors affecting the species' continued existence. This classification system is primarily subjective.

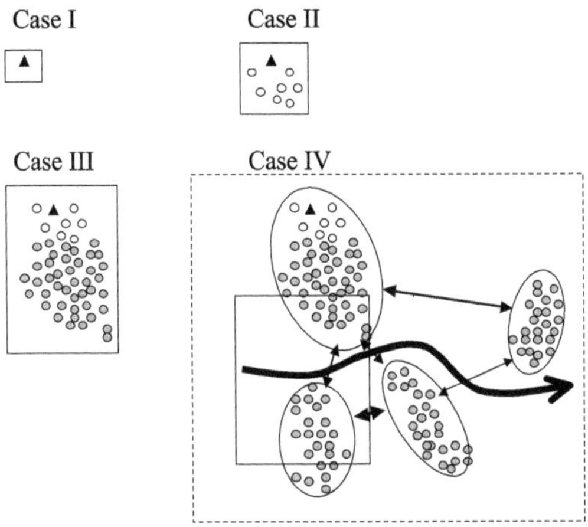

FIGURE 3.3 Representation of different assessment populations overlain on a metapopulation complex. The traditional boundary for a site assessment for chemicals would be defined by property boundaries or extent of contamination (solid-lined rectangles). Note that the individual in Case I (solid triangles) is included in Cases II to IV), the group of individuals (open circles) is included in Cases III and IV. If the metapopulation were used as the assessment population, the assessment boundary would be extended (dashed-line rectangle).

This system is intended to be quick and straightforward to expedite determination of a species' status. These protocols take into account current and future threats and management activities that may cause population decline. It is important to note that there are no prescriptions for consideration of temporal or spatial scale.

The scientific working group assigned to evaluate methods for conducting population viability assessments in the United States under the National Forest Management Act addressed the issue of temporal and spatial scale of risk assessments (Andelman et al. 2001). Population viability is defined in terms of self-sustaining population well distributed throughout their range (Federal Register, November, 2000, pp. 67580–67581). Their recommendations are worth noting here. Two factors that enter into the choice of species for viability assessments include 1) the spatial and temporal scale of proposed activities and 2) the potential for cumulative effects as a result of combinations of activities through time and across the landscape, including actions on lands adjacent to National Forest land. How do the species distributions match up with the spatial scales of proposed actions? It is rarely appropriate to conduct viability analyses at the spatial scale typical of project-level decisions; assessments at broader scales will usually be both more meaningful biologically and more cost effective. For species whose distributions are much larger in scale than the proposed actions, regional scale assessments may be needed. For species with distributions of about the same scale as the proposed actions, assessments that match that scale are most helpful. For species whose distributions are much smaller than the scale of the proposed activities, as may be the case for narrow endemics or species specializing on small, patchily distributed habitats, site-specific assessments might be considered for management activities that will affect areas where the species occurs.

PROTECTION GOALS FOR THE ASSESSMENT POPULATION

START WITH MANAGEMENT GOALS

Ecological risk assessments are used to inform policy and management decisions by numerous agencies, ministries, and programs operating under diverse regulations and mandates (see Chapter 2). Each operates with a view toward its own management and policy goals and risk assessments must be planned accordingly. In planning an assessment, questions about whether a risk assessment should focus on population rather than organism attributes and what is meant by protection must be answered. For the assessment to be relevant, those answers must be drawn from the management goals of the agency or program being supported.

The management goals influence the selection and definition of the assessment population. For example, the management goals for some environmental programs clearly call for a focus on effects or risks with respect to the local biological populations. For example, the management objectives for the European Water Framework Directive are stated as achieving good ecological quality for aquatic systems (see Appendix 1, Scenario 1 for elaboration). ERAs with population metrics may be used as a gauge for judging good ecological quality. Biological populations (and metapopulations) are usually the focus of many watershed and regional assessments.

At the other end of the spectrum are programs that clearly focus on protection of individuals or groups of individuals (note these may be how the assessment populations are defined). For example, with respect to protection of populations of species under special protection at a global, national or local level, the management goals could involve the protection of organism-level attributes as a basis for protecting populations.

In other cases, the goal may involve maintaining healthy thriving populations of several species within the aquatic system, and maintaining a functioning community and supporting habitat. Thus protection of the populations in the latter case involves management of the chemical, biological, and physical aspects of the environment that are intricately associated with sustaining the populations. It follows that evaluating risk to the population may include evaluating risks to components of the ecosystem that are necessary to support assessment populations. The development of water quality or sediment quality criteria may be examples of regulatory efforts that are directed at protecting communities and the populations that comprise those communities. Scenario 2 in Appendix 1 provides an example of such an approach.

Evaluating risk to the population may include evaluating risks to components of the environment that are necessary to the support of the populations.

As noted, definitions of protection and population depend on the context of the problems being considered (i.e., the species, the type of activity restricted, habitat characteristics required to sustain or enhance the population of the species under protection, the relevant time frame of management, feasibility of management action, and the spatial scale of the threat, among other considerations). Protection may reflect a variety of situations:

- Cultural mores serve to place some species into special categories (e.g., cattle in India, owls in some Native American tribes) in which each individual is revered (i.e., all individuals of the species have protected status).
- Management goals for fish and game species that are harvested commercially or for sport are aimed at sustaining sufficient populations within designated units to support specific harvest levels. Such goals are adjusted to meet the particular values of the most influential stakeholder groups (e.g., Trout Unlimited, promoting catch-and-release or slot fisheries, put-and-take operations). (See, for example, Scenario 4 in Appendix 1.)
- Habitat management may be targeted at improving reproductive success (e.g., old-growth forests for northern spotted owl, spawning redds for salmonids, prairie pot holes for several dabbling ducks), or to support behavioral patterns and predator avoidance (e.g., open sandbars for greater sandhill cranes, large open meadows for elk calving).
- In some regulatory contexts, protection of species means no unauthorized taking is permitted. When this is applied to risks estimated from exposures to hazardous waste, protection means zero mortality.
- For watershed and regional assessments, population protection may be guided by broad goals to protect and/or enhance the ecological services provided by the systems.
- For hazardous waste site management, the goal is often focused on establishing chemical concentrations that will be protective of organisms. These may be individual organisms, groups of individuals, or even the biological populations depending on the scale of the site. Appendix 1 describes a typical scenario that could be encountered at a large hazardous waste site. The management objective stated for this scenario is: protect wildlife populations exposed to hazardous chemical releases, or maintain sustainable wildlife populations. These objectives are consistent with USEPA's principle number 1 of Superfund to reduce ecological risks to levels that will result in the recovery and maintenance of healthy local populations and communities of biota (USEPA 1999).

Clearly State Protection Goals

Protection goals will vary depending on how the assessment population is defined. When the assessment population is a part of the biological population, protection goals may be expressed in terms of individuals or in terms of a fraction of the exposed individuals. Often the protection goals for such an assessment population are considered as follows:

> The individuals comprising the assessment population are able to carry out biological functions that influence their ability to maintain themselves within the area of evaluation

> and enable them to contribute to the larger biological population. These biological functions typically include but are not limited to survival, growth, and reproduction. There are also cases where certain biological functions (e.g., reproduction in stocked fish) would not be a goal.

Various forms of the protection goal stated previously are commonly used (or implied) in ERAs for evaluations on individuals or groups of individuals. Although these expressions of protection goals often require the specification of spatial and temporal scales, they differ somewhat from protection goals expressed in terms of the biologic populations. We explore some of these differences in the following discussion. Recognizing these distinctions are important as they can lead to confusion over what is being assessed and the methods to be used.

When the assessment population is the biological population, protection generally means providing or ensuring conditions that permit the existence of that species or population at a self-sustaining level with maintenance of sufficient genetic diversity to minimize risk from periodic natural stressors. This implies an active engagement in the management of agents that otherwise would harm the population or species. The active engagement may be focused directly (as in controls on takings or mitigation strategies to limit exposure to hazardous substances) or indirectly (such as habitat management or predator control) on the species or population.

Clearly, the concept of protection can be vague, with degrees of protection or borderline cases between protected and not protected (Regan et al. 2002a/b). Some species' numbers are so low that they no longer have a viable population (e.g., California condor). For such species, protection is dominated by policy, with minimum need for scientific input. For such species, the need for mandated protection is clear. As one moves from rare and endangered species to threatened species to species that have not yet been specifically identified for special protection, protection policies may shift from a focus on organisms — as the target of protection — to the population as a whole. When the focus is on entire populations, it is acknowledged that a stressor may affect the survival, growth and/or reproduction of some members of the population but that the acceptability of the stress is judged in terms of how it affects the population as a whole. With this shift in focus, population biology becomes especially important as it provide information on likely fates of populations across relevant time and spatial scales.

As suggested previously, protection of the assessment population may require explicit consideration of the habitat as well as protection of populations of other organisms that are ancillary to the assessment population. For example, to protect species of special concern (e.g., trust[1] or endangered species such as migratory fish, certain terrestrial mammals and birds, and shellfish, various marine mammals, and sea turtles) the focus may be on their supporting habitat. Habitat would include the critical portions of the biological community and the physical structures that are required or preferred by the species.

Just as populations and habitats are inseparable when considering protection, they are often closely linked in risk assessment. The quality, quantity, and degree of

[1] A term used in the United States by resource management agencies and departments.

fragmentation of habitat can all be important for populations. Chemicals and other stressors that influence these aspects of habitats can pose or influence risks to populations that use the habitats. Although these habitat-related aspects of population risk are not considered as commonly as are demographic effects, they are clearly important when considering the spatial and temporal aspects of population dynamics. A habitat-based approach to addressing risks to populations has been described by Menzie and Wickwire (2001). Influences and risks of stressors on habitat quality have been receiving increased attention. For example, habitat suitability indices were used to establish qualitative relationships between the levels of metals and wildlife use patterns in wetlands (as cited in Kapustka et al. 2004). In the same project, in testing the hypothesis that metals were impeding the development of complex vegetation, plant community composition, and plant succession were evaluated.

Spatial scale and the nature of the problem appear to influence how the assessment population is defined and also the development of protection goals. On one end of the spectrum may be small, medium, or large hazardous waste sites where the issue is how best to clean up the site and what is appropriate for site management for current and foreseeable future conditions. The stressor and the site scale drive these types of problems. Other situations such as watershed management have a regional management context that ideally attempts to balance the uses in the watershed to achieve a desired outcome for the resources. Often the resources (e.g., species' populations) drive these types of problems and the spatial scales considered are chosen to be relevant to the scales and biological characteristics of these resources.

Smaller Scale Assessments

Where the spatial scale of exposure to a stressor is relatively small, evaluating risk to populations of assessment species, per se, may not be as valuable to risk managers as characterizing the availability of high-quality habitat for that species or habitat use patterns across a landscape. Populations are influenced by many physical, biological, and chemical factors, and it can be difficult to control for each important factor extraneous to the stressor of interest. Although the stated goal may be to evaluate dynamic properties of the biological populations, the reality of risk characterization at smaller scales restricts analysis to an evaluation of individuals or groups of individuals and an evaluation of available habitat on site. The available habitat on site may be reduced by risk to ecological receptors (other than the assessment population) because of exposure to the stressor. However, the site may be a small component of the larger landscape.

Hazardous waste sites typically (but not always) have exposures over relatively small spatial scales. Although management goals vary to an extent, and some programs such as that in the State of Oregon may encourage population ERA, others focus on effects commensurate with the scale of the contaminated area. In Massachusetts, for example, waste site risk assessments generally identify the assessment populations as the groups of organisms actually exposed at the site, which for many wildlife species may be a small fraction of the biological population. The focus of the regulatory action is remediation of the contamination and the risk assessment is designed to help guide clean-up actions. If a contaminated area is not functioning as a normal component of the habitat because of direct toxicity or indirect food web effects, the assessment may reach a conclusion that there is a significant risk of harm to the environment.

Nevertheless, smaller scale assessments may benefit from at least a cursory characterization of available habitat of the adjacent landscape that extends to encompass the generalized dispersal distances of individuals within the assessment population. Potential habitat loss at the site may place the biological population at increased risk if the critical area is diminished. There may be certain large sites that contain sufficient habitat to actually support an entire biological population, but in most cases, hazardous waste sites are but one part of the larger mosaic of habitat available to a population.

Finally, even for small sites, population-based approaches can provide insight into the proper balance of factors when considering remediation. Considering the problem in a broad way may minimize unforeseen consequences of remedial options. In some cases, where the net effect of habitat removal can have a greater impact on the population than the predicted effects of the hazardous substance, a broad perspective that considers populations and their habitats can guide the decisions in a way that leads to net environmental benefit.

Larger Scale Assessments

Protection goals converge with protection of biological populations as the scales of the problem increase. These larger scale problems can be either stressor or resource (i.e., population) driven. Examples of the former include evaluations of pesticide applications such as that described in Appendix 1, Scenario 4. This case study explores how populations of a bird species would respond to applications of pesticides. For this problem the population-level effects of pesticide application on extinction potential or population size are approached through modeling population responses. These responses are not the same as organism-level responses to applications and would be missed if the assessment focused solely on risks to individual birds.

Examples of resource-driven assessments are the evaluations of stressors influencing the sustainability of marine species in the Mediterranean Sea, Chesapeake Bay, and Pacific Northwest. In these resource-driven cases, the assessment must take into account the factors that influence the biological populations and metapopulations to formulate management strategies for the water bodies of interest. Such strategies might include management of nutrient inputs, toxics control, and preservations of habitats. Such approaches are commonly applied at the scales of watersheds, lakes, estuaries, and coastal areas. The framing of this particular problem gives insight on how a resource-driven assessment might proceed.

In this example, a prospective population ERA will be used to protect the commercial harvest of the spawning run of pacific herring, Clupea harengus pallasi, at a specific coastal region. This process will be used to project over 20 years the potential impacts of 2 specific stressors associated with a municipal and industrial complex adjacent to fishery spawning grounds. Results will be used to inform management decisions that may be required to maintain this fishery at current levels, 1000 tons per annum (±20% per 2-year rolling average)

Assessments of the effects of land management decisions are also commonly accompanied by assessments of the vulnerability of wildlife populations at regional scales. Such decisions could relate to physical alternations that directly affect habitat as well as changes in geochemical cycles and the introduction of toxic chemicals. Because the endpoint in such cases is protection of populations and possibly habitats, these larger scale assessments are more likely to focus on protection goals that are directed at protection of assessment populations that are the same as the biological population.

Use Conceptual Models

Communication will be critically important for proceeding with population-level ecological risk assessments. All of the scenarios provided in Appendix 1 relied on conceptual models to help develop the analysts understanding of the problems and to communicate to other members of the Pellston Workshop. The experience gained in ERA over the years and during the course of the Workshop underscores the importance of developing a good conceptual model. Chapter 11 provides further discussion on the development of conceptual models for population-level risk assessment.

ASSESSMENT AND MEASUREMENT ENDPOINTS FOR THE ASSESSMENT POPULATION

Selecting Assessment Endpoints

Management and protection goals for the assessment populations are made operational in the ERA through the selection of assessment endpoints and the measurement endpoints used to evaluate these. Assessment endpoints are chosen to reflect environmental values that are protected by law, that provide critical resources, or that provide an ecological function that would be significantly impaired (or that society would perceive as having been impaired) if the resource were altered (USEPA 1992c, 1997c, 1998bc). From the set of ecological receptors identified for the assessment, specific receptors (habitats, local populations, individual threatened and endangered (T&E) organisms, keystone species, soil ecosystem processes/functions) are selected as the entities of assessment endpoints. A variety of criteria may be used to select these entities including, but not limited to

- species vital to the structure and function of the food web (e.g., principal prey species or species that are major food items for principal prey species);
- rare, endangered, or threatened species or those protected under various legal statutes;
- species that exhibit a marked toxicological sensitivity to the stressor;
- economically important or societally valued species;
- species with unique life histories (those that fill unique ecological niches) and/or feeding habits or representatives of a particular guild; or
- species common on or near the site.

The assessment endpoints should capture what is important to protect within the context of the problem at hand. For a particular assessment, this is often the most critical ERA decision considered by risk assessors and managers. Without clear assessment endpoints, an ERA may fail to inform decisions properly. As population-level ERAs become increasingly used to inform decisions, care must be taken to develop and articulate the assessment endpoints to reach a shared understanding of the problem with other interested parties.

Ecological risk assessment may have one or many assessment endpoints depending on the needs of that assessment. Some may be framed in terms of populations.

Ecological risk assessment may have one or many assessment endpoints depending on the needs of that assessment. Some of these assessment endpoints may be framed in terms of populations and others might be framed in terms of other levels of biological organization.

Examples of assessment endpoints framed to guide assessments of risks to populations include

- sustainability of the warm-water fish populations in a particular water body,
- abundance of mink in a particular watershed,
- amount of suitable habitat available for sustaining a population of a rare salamander, and
- genetic diversity within a metapopulation of a particular mammal species.

This list indicates the broad range that assessment endpoints may take. Common attributes for assessment endpoints for populations are given in Table 3.1.

Selecting Measures of Exposure and Effect

The model estimates and measurements that are used to evaluate the assessment endpoints are commonly referred to as measures of exposure and/or effects (USEPA 1998). For population-level ERAs, these can be quantified using empirical methods such as those described in Chapter 8 and modeling approaches such as those presented in Chapter 9. The scope of any ERA flows from the statements of assessment endpoints. The type of information needed to evaluate risks for the assessment endpoints will govern scope elements and will typically include various empirical measures and model applications. In some cases, the assessment will examine an existing problem where the stressor is already present (retrospective ERA). In other cases, the assessment will be used to predict or compare the relative risks of a future action or of various alternatives (prospective or predictive ERA).

The ability to reach and support a decision is also influenced by the uncertainty associated with the analyses carried out for each assessment endpoint. Some decisions can be made in the face of greater uncertainty if the consequences are small or if the uncertainty can be managed through bounding analyses.

Uncertainties, where they arise in an assessment, reflect lack of knowledge about aspects of the analyses (Hoffman and Hammond 1984; Morgan and Henrion 1990; Regan et al. 2002). As with other ERAs, it is anticipated that those that include a population focus will also be subject to uncertainties. However, because the issues that arise in risk assessment that includes population-level endpoints may be more complex and perhaps more difficult to communicate than assessments on organism-level endpoints, the treatment and management of uncertainties is likely to be an especially important aspect of the assessment. Adequate treatment and management of uncertainties will be a critical aspect of achieving population protection goals that are understood and accepted by managers and other interested parties.

Uncertainties in ERA are often best reduced and/or managed by developing better information on a particular line of evaluation (e.g., making site-specific measurements rather than relying on generic information) and also by developing more than one line of evaluation to provide additional perspective on the problem. When more than one line of evidence is used for an assessment endpoint, the term "lines of evidence" or weight of evidence is used to describe the evaluation process (Menzie et al. 1996; Suter 2007; USEPA 1998). Experience has indicated that such approaches can reduce uncertainty and increase the level of confidence in the assessment.

With respect to population-level ecological risk assessments, the lines or weight of evidence will likely include one or more of the following types of analyses:

- empirical studies conducted in the laboratory,
- empirical studies conducted in the field,
- modeling estimates of population processes, and
- published literature on the specific species and related species life histories and autecology.

These are not independent activities and laboratory and field data are often used to guide model development. However, these 3 categories capture most of the available lines of evidence described in Chapters 8 and 9. In addition, a weight of evidence approach can be considered for integrating lines of evidence (Table 3.2).

In some cases, evaluations may proceed through a tiered format that can begin with a model and proceed to laboratory and/or field data. The case study for pesticides (Chapter 10) provides an example of a tiered application involving models. In other cases, decisions may be made to use the approaches simultaneously.

Clarifying Definitions, Endpoints, and Approaches

We have adopted the term "assessment" population to ensure that what is meant by population is defined within the context of the assessment. However, selection of this term requires further effort to sort out appropriate assessment endpoints as well as metrics. The challenge comes from possible differences between assessment populations and biological populations. The assessment population is an inclusive term, whereas the biological population is more specific. In general, population-level types,

TABLE 3.2
Examples of weight-of-evidence approaches

Scenario	Brief descriptions of analyses
Scenario for Genetic Modified Organisms (GMO)	It may initially be sensible to develop a model on the spread of pollen and seed from the GMO in a single field and test this model by collection of seedlings and pollen. The results of this exercise in combination with the distribution of the target species should help in predicting risks and might dictate plantings of GMOs.
Herring stock scenario	Empiric information would be derived on the herring stock (recruitment and yield); relationships between egg production and spawning stock can be developed; the role of eelgrass and the effects of pier structures would be ascertained; and, the effects of discharges on herring and eelgrass beds measured.
	Modeling approaches could include simple nonspatially explicit fisheries-type models, species-specific demographic models, and/or detailed spatially and temporally explicit hydrologically accurate models.
Population risk assessment of pesticide (Appendix 2)	This was a modeling effort. However, the manager's perspective is instructive with respect to relying on weight of evidence.
	In principle, the type of modeling risk assessment performed here would be useful as a tool for higher tier risk assessment of pesticides. In reality, however, the assumptions and uncertainties mentioned would need to be explored further before this type of assessment could be used in the actual decision-making.

effects, and risks are considered with respect to biological populations including metapopulations. Where the assessment population is defined as the same as the biological population, risks to the population has a clear and consistent meaning. However, where the assessment population is not the same as the biological population, what is being assessed must be made clear along with the methods.

The balance of this book largely focuses on assessing risks to the assessment population defined as the biological population. However, before proceeding to subsequent chapters we share workshop efforts to distinguish among approaches for evaluating risks to assessment populations defined in various ways.

Table 3.3 illustrates how various measures might be used for the various types of assessment populations defined earlier. It begins with evaluating risks to organisms and proceeds to evaluations for populations and metapopulations. It is illustrative and there may be many examples of approaches. However, it should be clear that assessment endpoints and the approaches taken in ERA will be linked to how the assessment population is conceptualized. The table implies distinctions between effects on organism and population attributes. The characteristics measured in toxicity tests (survival, mortality, fecundity) are arrived at by observations on individuals; with respect to the individuals there exists a specific effect (e.g., the organism is either alive or dead or grew at a specific rate). When the results for the individuals in the test (or in the field) are compiled across all individuals, the incidence (e.g., percent mortality) is obtained. As described previously, incidence is a population

TABLE 3.3
Examples of assessment approaches dependent on how the assessment population is defined

Types of assessment populations	Assessment focus	Examples of applications	Assessment properties	Examples of assessment approaches
1. Individual organism	Effects on probability of survival, or probability of effects on growth or reproduction for the individual	Protected species (in some jurisdictions)	Entities are particular species or habitats; attributes are organism-level biological characteristics	Extrapolate from toxicity or field test to probability of effects on survival, growth or reproduction of organisms
2.a. Group of individuals (fraction of larger population) exposed area of concern	Effects on survival, growth, and reproduction rates for exposed group	Area of regulatory concern (site) encompasses only a fraction of population	Same as above	Extrapolate from toxicity or field test to survival, growth, repro changes for the group of individuals
2.b. Population of which only a subgroup is exposed to stressor	Change in measures of size, persistence and/or structure of the population	Only a fraction of population is exposed to the stressor; site-specific assessments	May be the same as above, but may extend to population-level assessments	Extrapolate from toxicity or field test to effects on exposed group, then model effects on population size
3.a. Local subpopulation (unit of a metapopulation)	Effects on size, persistence or structure of local subpopulation (that could lead to effects on metapopulation)	Generic or site-specific assessments	Entities are usually the local population of the species; attributes include those of populations	Empirical method or population model to evaluate effects on subpopulation
3.b. Metapopulation	Effects on size, persistence and/or structure of metapopulation	One or more subpopulations exposed to stressor; Generic or site-specific assessment	Entities are usually the population of the species; attributes include those of populations	Empirical methods or model to evaluate effects on subpopulations, model to evaluate metapopulation effects

attribute. In some cases, it will be an attribute of a laboratory population while in other cases it may reflect conditions in the field (e.g., a distribution of condition factors or frequency of tumors).

If the assessment endpoint for the assessment population has been framed in terms of incidence of a particular type of effect or level of exposure, the measures of incidence derived from the assessment provide a means for judging risk with respect to that assessment endpoint. This approach is commonly taken in risk assessment where an incidence level (e.g., 20% mortality) is specified as the basis for judging the significance of an effect or where a joint probability analysis is used to integrate a distribution of exposure with a distribution of effects (dose response curve). Examples of these approaches can be found in Suter (1993).

However, if the assessment endpoint is framed in terms of population attributes more commonly associated with the biological population (e.g., abundance, growth rate, quasiextinction), then the information on incidence of a particular condition, effect, or exposure does not provide a direct measure for the assessment endpoint. It may, however, influence the assessment endpoint (e.g., as might occur from changes in the incidence of mortality). Typically, additional work is needed at the population biology and/or population modeling level to translate measures on organisms (and expressed as incidence or distributions) into population outcomes (e.g., changes in abundance or probability of extinction).

To consider a concrete example, elevated amino-levulinic acid dehydrase (ALAD) incidence in a bird population would be a legitimate measure of effects (risks) on that population only if the assessment endpoint were defined in terms of elevated ALAD. If, however, the assessment endpoint were defined in terms of the abundance or productivity of the population, then elevated ALAD incidence would be only an indirect indicator of potential effects.

The terms "organism-level" and "population-level effect" are often used to characterize different types of studies or assessments but their meanings are not clear. This was evident at the workshop and in numerous discussions that followed. Given the ambiguity in these terms, it is suggested they not be employed as a basis for classifying effects or risks. Instead, the management goals, definition of the assessment population, and selection of assessment endpoints should be the tools employed for framing the risk assessment and the language used to present it and communicate about it. Use of terms such as "organism-level" and "population-level" can be useful within the context of the specific assessment but probably are less useful for differentiating among broad categories of effects or broad categories of risk assessments.

Although we acknowledge the ambiguity around the term population-level effect, we recognize that when it is used it most commonly refers to population attributes such as size, growth rate, and structure of the population with respect to age, size, sex, or genetic composition (Barnthouse 1993). We also recognize that organism-level effect is commonly understood to mean effects that are observed on organism attributes such as death or reduced fecundity. Such organism-level effects as they are commonly understood may or may not translate into population-level effects as those are commonly understood.

Confusion can arise when organism-level effects (e.g., the death of individuals) are translated into incidence and where incidence has been selected as the basis for judging effects of risk to the assessment population. In such cases, incidence of a particular effect may or may not result in a significant change in such population attributes as abundance, population growth rate, or potential for extinction. Nevertheless, incidence may be judged in terms of an effect on or risk to the population. Obviously, clarifying the meaning of generic terms such as population level is partially a problem of semantics. This potential problem is resolved by clearly defining the assessment population and endpoints and by consistently relying on definitions to analyze effects and risks and for communicating these to managers and other interested parties.

Protection Goals and the Biological Population

We have described the various ways an assessment population may be defined and how this influences an assessment. However, it is useful to consider why an assessment of risks to the biological population provides insight regarding protection goals that may be missed when the assessment is limited to organisms or assessment populations defined in terms of groups of organisms.

Biological populations have unique characteristics that are not captured or taken into account when the ERA focuses only on characteristics of organisms or groups of organisms that do not comprise a definable biological population. The importance of density dependence on the demographics of biological populations is one example and is discussed in Chapter 4. Assessments that focus only on organisms can easily miss important spatial and temporal factors important for the decision. For example, when alternatives are being compared, a focus on endpoints constructed for organisms or groups of organisms (e.g., incidence) rather than population attributes (e.g., abundance or population growth rate) can actually mislead the assessors and decision-makers concerning the outcomes of particular actions.

Recently, it has been recognized that processes that influence attributes of the biological population may affect the attributes of organisms. An example is population compensation, in which the reproductive potential of surviving individuals in a population increases because of an increased availability of resources (Stark et al. 1997); when individuals are removed from the population after toxicant exposure, survivors have more resources available and may reproduce at a greater rate. Alternatively, characteristics of the biological population (such as density) can influence the effects of stressors in ways that are not apparent when considering effects only on individuals (Sibly 1999).

Finally, populations exist in relation to other populations of species as well as the habitat and sources of nutrition. This fabric — comprising relationships within and among species — responds to perturbations in ways that cannot be easily foreseen from evaluations on individual organisms. In the scenario on Hazardous Waste Sites (Appendix 1), an argument is advanced that remediation focused on risks to organisms can actually result in significant risk to populations. Although this is a simplification, it does underscore the importance of considering biological responses that are at organizational levels greater than the organism or group of organisms. It may be the case that such refinements are most useful when decisions are being made in a gray area where risks to organisms may occur but where risks to populations are uncertain. In such cases where the management decision brings its own potential population-level risks the assessment should properly proceed to a level of organization above the organism so that risks of management alternatives can be properly compared.

Accounting for interactions among populations is also important for considering the potential for cascading effects. These are most prominent when keystone species is placed at risk. Such species might be a critical food source of habitat former. Risks to populations of these species would likely translate to risks to populations of other species that depend on them for food or habitat.

RECOMMENDATIONS

We offer the following recommendations:

1) Recognize that an ERA may involve a number of assessment endpoints, only some of which may be framed in terms of populations. Therefore, there is not a distinction between population and other types of ERA. Rather, ERAs may or may not include population-level assessment endpoints.
2) Define the assessment population clearly in terms of space and time as well as its relationship to the stressor.
3) Recognize that protection of the assessment population may involve protection of environmental components required to support the population. Examples include habitat and supporting (e.g., prey) species. These may be translated into complementary assessment endpoints.
4) Assessment endpoints for the assessment population are most commonly framed in terms of demographics but may also be framed in terms of habitat needed to support the populations or in terms of population genetics. The latter type of endpoint is discussed in Chapters 5, 8 and 9.
5) The choice of measures of exposure and effects should flow from the definitions of the assessment population and the associated assessment endpoints. Choices of measures would lead to various combinations of empirical and modeling approaches.
6) A weight-of-evidence or lines-of-evidence approach will likely reduce the uncertainties associated with population-level ERAs and would increase the confidence that decision makers have in the underlying analyses.

4 Density Dependence in Ecological Risk Assessment

S. Jannicke Moe

INTRODUCTION

The purpose of this chapter is to summarize the existing state of knowledge concerning the quantification of density dependence for ecological risk assessment. Density dependence is a fundamental concept in population biology, and it seems clear that density dependence in some form is exhibited by most animal species. The process is fundamental for sustainable exploitation of populations, and also plays an important role in conservation biology and pest management. Theoretical and empirical studies have demonstrated that toxicant effects on the population level may be modified by density-dependent processes. This implies that information about organism-level responses cannot always be extrapolated directly to the population level without considering density dependence. Density dependence should therefore be a more integral part of ecotoxicology and ecological risk assessment. One reason why density dependence is often ignored may be that there are many practical problems related to applying this concept in practice; it has even been termed "an unwelcome complication for ecotoxicologists" (Walker et al. 2001). This chapter will focus on interactions between density stress and toxicant stress and how such interaction can give indirect toxicant effects. Different methods for quantifying density dependence will be presented toward the end of this chapter, and related problems discussed. Finally, some recommendations will be given for the incorporation of density dependence in risk assessment.

HISTORY OF THE DENSITY-DEPENDENCE CONCEPT

The inclusion of density-dependent effects in population models began in a biologically ad-hoc manner that largely persists today. Malthus (1798) recognized that a population growing at a constant positive rate would ultimately outstrip its resource base. Forty years later, Verhulst (1838) introduced density dependence in the mathematically simplest way: he assumed that the instantaneous population growth rate was a linearly declining function of population density, limited by the carrying capacity of the environment and scaled by the intrinsic population growth rate. Verhulst's "logistic equation" was later rediscovered by Pearl and Reed (1920), and is still widely used in population models (Getz 1996). See Chapter 9 for models that describe density-dependent population growth rate.

The importance of density dependence as a driving force of population dynamics was advocated by Nicholson (1950), who demonstrated that competition for food could lead to sustained oscillations in densities of laboratory insects. Nicholson concluded that "there is a single dominant controlling mechanism, namely, density-dependent reactions" (Nicholson 1950), whereas Andrewartha and Birch (1954) argued that density-independent factors such as climate may play a more important role in limiting natural populations. There have been intense debates regarding density dependence ever since. The disagreement in part concerns the interpretation of the density dependence concept itself. For instance, it can be argued that the true limiting factor is not the number of individuals itself but the underlying biological processes limiting population growth (e.g., the population's food supply) (Murray 1999; White 2001). Population density can therefore be regarded as a unit representing the population size in relation to food quantity or other limiting variables.

There has also been much disagreement about methods for detecting density dependence in natural populations (Holyoak 1994; Wolda et al. 1994; Williams and Liebhold 1995; Berryman and Turchin 1997; Hunter and Price 1998; Turchin and Berryman 2000; Solow 2001). The discussions concern both the validity of the different tests and the interpretation of the results. According to Hanski et al. (1993), "animal-population ecologists have somehow managed to muddle the basis of the very existence of their study object, to the extent that the field must look baffling to newcomers and outsiders." It is nevertheless a commonly accepted view that most animal populations are regulated by density-dependent feedback to some extent and at some scale (Cappuccino and Price 1995; Turchin 1995). For applied purposes such as risk assessment the debates regarding significance may be less relevant, and a more important objective may be to obtain reliable parameters for description of density dependence and for population-level prediction (Langton et al. 2002).

DENSITY-DEPENDENT PROCESSES AND POPULATION-LEVEL PATTERNS

Density dependence can be defined as the tendency for death or emigration rate in a population to increase, or the birth, growth, or immigration rate to decrease as the density of the population increases, and vice versa as the density decreases (after Begon et al. 1990). In other words, density dependence results from negative feedback mechanisms that affect the demographic rates, and thereby tend to regulate the population density within bounds. A variety of biological processes may produce density dependence operating both within a species and among species.

Intraspecific interactions often imply competition for limited resources. The term "competition" is used both about direct encounters between individuals (interference competition) and about indirect effects from depletion of shared resources (exploitation competition). Hence the concept refers more to the population-level effects than to the actual biological process. Competition typically results in reduced individual growth, higher mortality rates, or higher emigration rates. Competition among different age groups or size classes will often be asymmetric in favor of the largest individuals (De Roos et al. 2003). Size-structured interactions also include cannibalism in some species. Negative density-dependent feedback in general has a stabilizing effect on population densities. However, there may be a delay between a change in density and the effect on demographic rates (e.g., related to the development time from birth to maturity). Strong density dependence combined with a delayed response may rather have a destabilizing effect and give rise to cyclic or fluctuating densities. This is the mechanism underlying the cycles in the classical "Nicholson's blowflies" experiment (Nicholson 1954): when only adults compete for resources, there is a time lag corresponding to the development time before the adult population shows a response to high density. Other examples of laboratory systems that have been thoroughly studied by both experiments and theoretical modeling are the flour beetle *Tribolium* (Park 1937; Mertz 1972; Cushing et al. 1996), the cladoceran *Daphnia* (McCauley et al. 1996), and the moth *Plodia* (Bjornstad et al. 1998). Such systems with only one or a few species are obviously unrealistic, but can help us understand how density-dependent processes interact with external factors.

Interspecific interactions encompass a range of trophic interactions, such as predation, herbivory, parasitism, and infectious diseases, as well as competition. Interspecific competition is generally stabilizing, in a similar manner as intraspecific competition. Trophic interactions, on the other hand, may produce many different patterns, depending on, for example, the strategy of the predator. Generalist predators that choose the most abundant prey species may have a stabilizing effect on the prey populations, whereas predators that specialize on one prey species may generate cyclic dynamics. A classic example of predator-prey cycles is the Canadian lynx and the snowshoe hare (Stenseth et al. 1998). Another well-studied system is the lemming cycles; in this case, however, there is disagreement about whether the cycles are run by predator-prey interactions (weasels and lemmings) or herbivore-plant interactions (lemmings and vegetation) (Turchin et al. 2000).

Density dependence normally refers to negative effects, but positive effects of density (termed positive or inverse density dependence) are possible at low densities. This implies that individuals in some way benefit from the presence of conspecifics, and accordingly that individuals will suffer from reduced densities. This phenomenon is known as the "Allee effect" (after Allee 1938), or as "depensation" in fisheries literature. The most cited mechanism is shortage of mates at low densities, but the factors involved in generating the Allee effect are numerous and have been described for most major animal taxa (Fowler and Baker 1991; Stephens and Sutherland 1999). The factors can be classified into three major categories — genetic inbreeding, which reduces genetic variation and general fitness; demographic stochasticity (random variations in vital rates); and facilitation by group behavior (including both active cooperation and passive mutual benefits) (Courchamp et al. 1999a). Species with naturally low fecundity or with naturally low population sizes may be particularly vulnerable to demographic stochasticity. An example is the threatened giant parrot Kakapo (*Strigops habroptilus*), which also suffers from low proportion of females from random variations (see Courchamp et al. 1999a). Other species characteristics linked to the Allee effect are various forms of group behavior, such as cooperative breeding (e.g., mongoose; Courchamp et al. 1999b), gregarious feeding (e.g., larvae of many insect species), and cooperative predator defense (in fish schools and ungulate herds). Species that benefit from mutualistic interactions with other species may also suffer from reduced densities, because the mutualistic relationship may be weakened. Many plants species depend on pollinators for reproduction; reduced plant density may make the population less attractive to pollinators and thereby reduce the plants' reproductive success (Knight 2003).

Because most populations in nature interact with several other species, it is usually difficult to disentangle effects of the many possible density-dependent factors from each other and from external factors such as climate. One case in which this has been possible is the study of Soay sheep populations at the St. Kilda archipelago, which constitute a simple herbivore-plant system without predators and competitors. Long-term observations from this system have been used to quantify the relative importance of density dependence versus climatic influence (Coulson et al. 2000).

Population dynamics also has a spatial aspect, and it can be important to consider the spatial distribution of the species to estimate the density dependence. For example, in territorial ptarmigan populations, high density will result in higher emigration of young individuals to new territories outside their natal area (Pedersen et al. 2003). If population growth is measured in the natal area only, there will be a density-dependent loss of individuals, but measured on a larger scale, the population growth may not be so strongly affected by density. The spatial structure of populations and implications for risk assessment are considered in more detail in Chapter 6 and in Chapter 9.

Stochastic variations in demographic rates may also modify the density-dependent feedback and influence the population dynamics. Purely deterministic density-dependent models therefore usually give an oversimplified picture of the real population. Analyses of Atlantic cod have demonstrated that the dynamics are controlled by stochastic variation in reproduction and by age-structured competition (Bjornstad et al. 1999b; Stenseth et al. 1999). Hence, the relationship between population-level processes and patterns may be quite complex. A certain type of biological process may give rise to various population-level patterns, depending on the strength of the interactions, the external forcing, and degree of stochastic variations (Figure 4.1). Density dependence will in general be stabilizing, but strong density dependence in combination with high reproductive rate may give rise to cyclic population dynamics or even chaotic fluctuations. On the other hand, different types of density-dependent biological processes (e.g., competition, predation) may give rise to similar patterns of population dynamics (e.g., regular cycles). For a review on density dependence in theory and practice, see Murdoch (1994). a) A single density-dependent mechanism may lead to qualitatively different types of population dynamics, depending on the demographic rates and the strength of the density dependence. Population-level predictions from mechanistic models may thus be sensitive to parameter values. b) Several types of density-dependent mechanisms may produce a similar population-level pattern (e.g., sustained cycles). It is therefore not straightforward to identify the mechanisms that generate an observed pattern.

DENSITY DEPENDENCE IN NATURAL RESOURCE MANAGEMENT

Density dependence plays important roles in conservation biology, pest management, and harvesting. Conservation biology often focuses on risks related to low densities:

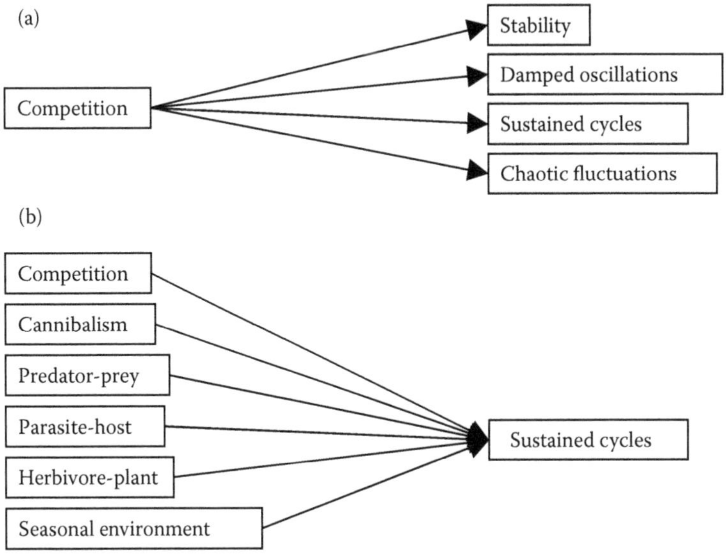

FIGURE 4.1 Examples of density-dependent processes and population-level patterns, based on theoretical population models.

low-density populations are particularly prone to extinction risks for several reasons, including the Allee effect (Bruno et al. 2003). Low-density population are also subject to higher degree of demographic stochasticity (random variation), which increases the possibility for extinction by chance. The relevance of conservation biology and resource management for ERA is discussed in Chapter 7. Pest managers also need to consider the consequences of density dependence in actual and potential pest species. Biological control strategies that use natural enemies as control agents need knowledge about the responses of the enemies to the pest density (Mills 2001). Many unwanted effects of pesticide use are related to density-dependent interactions, such as target pest resurgence (Van 1994) and secondary pest outbreaks (Godfray and Chan 1990). The numerous examples of pesticide-density interactions from pest management should, in principle, be relevant for chemical risk assessment. However, the species we consider as pests usually have particular life-history strategies, such as high reproductive rate; therefore, the relevance for species with different properties may be limited.

Sustainable exploitation of populations relies completely on the capacity for density-dependent compensation. That many species have persisted in the presence of sustained exploitation indicates the necessity of density dependence. The concept of compensatory mortality entered the wildlife literature as the Compensatory Mortality Hypothesis, which stated that loss from hunting would be compensated by higher winter survival of the remaining individuals (Errington 1946). The Doomed Surplus Hypothesis stated that individuals that were not able to hold territories were more vulnerable to diseases and predation, so that victims of hunting would otherwise have become victims of another mortality factor (Errington 1946). The relationship between natural and hunting mortality has been well studied in the management of

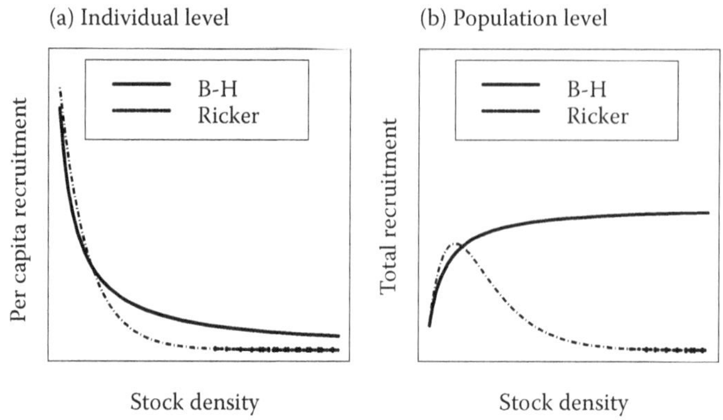

FIGURE 4.2 Models for density-dependent recruitment: the Beverton–Holt (B–H) model and the Ricker model.

waterfowl populations. Based on banding and recovery data in North America of 410 000 Mallard ducks between 1950 and 1980, it was concluded that annual hunting mortalities are largely compensated for by a decrease in other forms of mortality (Burnham et al. 1984).

Ricker (1954a,b) and Beverton and Holt (1957) introduced density-dependent population models, termed "stock-recruitment" models, which have been widely used in fisheries management (Figure 4.2). In these models, the number of young fish produced annually (recruitment) is expressed as a nonlinear function of the number of spawning adults (stock). The Beverton–Holt model expresses partially compensatory density dependence: an increase in stock size reduces the per capita recruitment, but still increases the total recruitment. The Ricker model expresses overcompensatory density dependence: an increase in stock size reduces not only the per capita recruitment, but also the total recruitment. Figure 4.2 illustrates that the structure of density-dependent functions may have large impact on population-level predictions: apparently small differences at the individual level (Figure 4.2a) are translated into large changes at the population level (Figure 4.2b). The original models proposed by Ricker and by Beverton and Holt are too simple to capture the dynamics of most populations; however, the density-dependent functions used in these models are widely used in more complex population models (Zheng 1996).

In both models, the per capita recruitment (i.e., the number of recruits per spawner; a), decreases with increasing stock density. However, the total number of recruits b) increases with stock density the B–H model (partial compensation), whereas it decreases with stock density in the Ricker model (overcompensation). Although the per capita recruitment functions look quite similar for the two functions a), the differences in degree of density dependence translate into large differences for total recruitment b) (after Caswell 2001).

The concept of maximum sustainable yield in fisheries management is based on an assumption that up to some maximum harvesting rate, increases in the survival or reproduction of the remaining fish compensates for mortality from fishing, so that populations can sustain themselves. However, simple density-dependent concepts such as the maximum sustainable yield concept have not worked so well in practice: models may overestimate the population's capacity for compensation if they assume a wrong form of density dependence or ignore stochasticity. Indeed, harvest strategies based on concepts such as the maximum sustainable yield has led to some famous ecological disasters, such as the decline of arctic whale populations in the 1950s and 1960s (Begon et al. 1990).

Can population models for chemical risk assessment include density dependence in a similar way as harvesting models? The applicability of harvest models to ecological risk assessment depends on the extent to which effects of chemicals on organisms are similar to effects of harvesting. As long as a chemical mainly affects survival or fecundity, the demographic effects will be similar to harvesting. However, if there are sublethal effects such as reduced growth or behavioral modifications, which are not directly translated into effects on survival and reproduction, the analogy may not be valid. This problem will be considered in more detail in the next section.

POTENTIAL IMPORTANCE OF DENSITY DEPENDENCE IN ECOLOGICAL RISK ASSESSMENT

An important problem for population-level risk assessment is how toxicant effects may be modified by population-intrinsic processes. Different types of population models can be used to predict toxicant effects on populations from information about toxicant effects on individual organisms or life stages (see Chapter 9). These models have considered different population-level endpoints such as population growth rate (e.g., Levin et al. 1996; Munns et al. 1997; Walthall and Stark 1997; Kuhn et al. 2000; Jensen et al. 2001; Salice and Miller 2003), reproductive potential (Stark et al. 1997), delay in population growth rate (Wennergren and Stark 2000), carrying capacity (Sibly et al. 2000b), extinction time (Tanaka and Nakanishi 2000), equilibrium density (Laskowski 2000), or population dynamics (Rose et al. 2003). However, only a few models have included density dependence (Barnthouse et al. 1990; Laskowski 2000). In some of the studies based on laboratory experiments, density-dependent responses are actually reported, but not included in the models (Hansen et al. 1999a; Kuhn et al. 2001). Chemicals can exert various toxic effects on organisms which propagate in complex ways throughout populations, and including functional processes such as competition for food resources can help to make population models more realistic (Bartell et al. 2003). There is a risk that the exclusion of density dependence weakens the predictive power of such models, at least for long-term predictions (Barnthouse 1993).

Two important questions regarding density dependence and risk assessment are 1) Does a toxicant affect the population independently of the population density (additive effect), or does the effect of a toxicant vary with density (interactive effects)? and 2) When there is an interaction, does higher density strengthen the negative toxicant effects (synergistic interaction) or weaken the effects (antagonistic interaction, indicating density-dependent compensation)? The different types of responses are illustrated in Figure 4.3 (see also Underwood 1989; Forbes and Calow 1999). Whether there will be a synergistic or antagonistic interaction depends on several aspects of the density dependence and the toxicant effects. External mortality

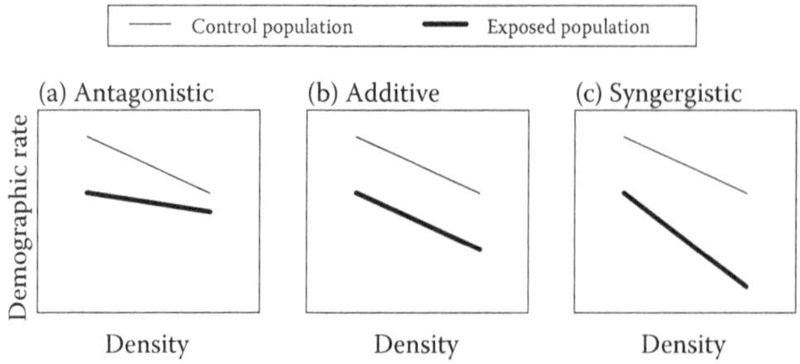

FIGURE 4.3 Main types of interaction between density and a toxicant, in which both density and the toxicant separately have negative effects on population growth rate.

factors may show complex interactions with density dependence: mortality caused by harvesting generally tends to be compensatory at high densities, but additive when the population is at a low level (Sutherland 2001). Because toxicants are likely to have more subtle effects on organisms than harvesting does, they may give even more complex interactions with density. The following sections discuss how various aspects of these 2 stress factors may determine the outcome of an interaction. a) Negative interaction (antagonistic effects): effects of the toxicant are weaker at higher densities, which indicates that the population has density-dependent compensatory responses. b) No interaction (additive effects): toxicant effects are not affected by density. c) Positive interaction (synergistic effects): effects of the toxicant are stronger at higher densities, which indicates a lack of density-dependent capacity and high sensitivity to multiple stress factors.

Aspects of Density Dependence

The strength of a population's response to a change in density may be particularly relevant to risk assessment. One way to quantify this strength is by the degree compensation (i.e., to which degree a decrease in density is compensated by a decrease in density-dependent mortality or by changes in other demographic rates). Theoretical population models show that the degree of compensation may have great importance for the population dynamics: a high degree of compensation may, under certain circumstances, contribute to cyclic and chaotic behavior. More important for risk assessment, the degree of compensation may also represent a population's capacity for compensatory responses to toxicant-induced mortality (i.e., for an antagonistic density-toxicant interaction). This compensatory capacity may be an important mechanism for ameliorating toxicant impacts at the population level.

The "degrees of compensation" to changes in density may be categorized as partial compensation, exact compensation and overcompensation (see Figure 4.4). The Beverton–Holt model and the Ricker model (Figure 4.2) are examples of partially compensating and overcompensating density dependence, respectively. Partial or exact compensation to toxicant-induced mortality has been reported from laboratory studies with various taxa and toxicants — such as trichopterans exposed to 4,5,6-trichloroguaiacol (Petersen and Petersen 1988), copepods exposed to dieldrin (Grant 1998), polychaetes exposed to fluoranthene (Linke-Gamenick et al.

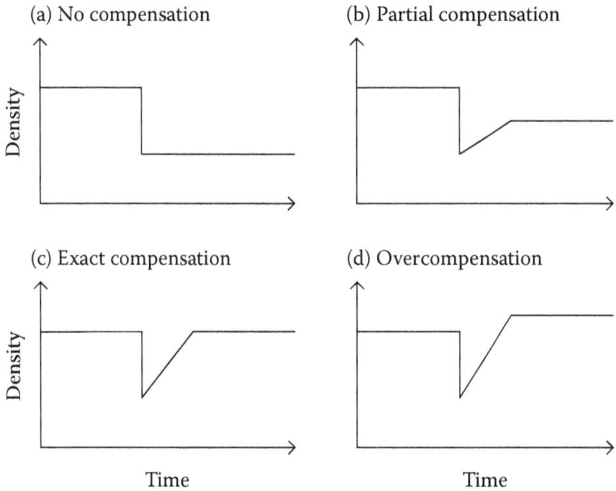

FIGURE 4.4 Different degrees of density-dependent compensatory responses (antagonistic interactions, see Figure 4.3).

1999), mysids exposed to para-nonylphenol (Kuhn et al. 2001), daphnids exposed to cadmium (Barata et al. 2002), and trichopterans exposed to fenvalerate (Liess 2002). Documentation of compensatory responses from field studies is more rare. A detailed study of the Hudson River striped bass population showed no population-level effect of polychlorinated biphenyl exposure, although it is obvious from laboratory experiments that this chemical has toxic effects on individual organisms (Barnthouse et al. 2003). The explanation may be that density dependence in larval survival, which is known to compensate for fishing mortality, also compensates for polychlorinated biphenyl-related mortality. The examples of density-dependent compensation imply that the realized population-level risk may be lower than expected from the organism level.

For simplicity, Figure 4.4 illustrates a single pulse disturbance, such as an acute toxicant exposure, resulting in a sudden drop in density. A compensatory response can result from increased survival, growth or fecundity of the surviving individuals, or increased immigration from surrounding populations. The graphs describe theoretical responses to a perturbation by populations with a) no density-dependent compensation, b) partial compensation, c) exact compensation, and d) overcompensation.

If a population experiences "overcompensating density dependence," then a toxicant-induced density reduction may in theory give an indirect benefit at the population level, at least during some stage. A long-term study of blowflies exposed to cadmium showed that the toxicant group had higher biomass production than the control group (Daniels 1994; Smith et al. 2000). A cohort experiment with the same study system demonstrated that the toxicant reduced reproduction and survival of young larvae, and thereby increased survival and growth of the remaining larvae to a level exceeding the control group (Moe et al. 2002b). Similar indirect benefits of toxicants are also reported for chironomids (Postma et al. 1994) and snails (Salice and Miller 2003) exposed to cadmium at high population densities. The phenomenon is also known from pest management: an initial density reduction from pesticide application may be followed by a "pest resurgence" reaching higher levels than before the application. However, because overcompensatory responses to toxicants require a very strong density dependence and high reproductive potential, this mechanism is not likely to be common in natural populations.

The compensatory mechanism described above assumes a direct, lethal effect of density, so that the number of competing individuals is reduced. However, density stress may also have "sublethal effects" that affect growth or other aspects of organism condition (Altwegg 2003). Under these circumstances, a population may be even more susceptible to a toxicant. Hence, the combined effect of toxicant and density may be additive or even synergistic, and therefore stronger than expected from density-independent assessment. Synergistic effects of food limitation and toxicants have been demonstrated by several laboratory studies (e.g., with cladocerans exposed to cadmium [Chandini 1989] and polychaetes exposed to fluoranthene [Linke-Gamenick et al. 1999]). An elegant field experiment has demonstrated a synergistic interaction between density stress and physical stress (sedimentation) for bivalves: individuals in high-density groups suffer more from sedimentation than those in low-density groups (Peterson and Black 1988).

The bivalve experiment also demonstrated that the density-toxicant interaction was caused by the "delayed effects" of past density experience, whereas present density had no effect. The authors therefore warn that the density history of individuals may affect the results of toxicant testing (Peterson and Black 1988). Sublethal density stress may also induce delayed effects on the density of the next generation (e.g., reduced fecundity will result in lower offspring density). Such delayed density effects may therefore produce even more complex interactions with toxicants: a toxicant-induced density reduction may not result in immediate compensatory responses in terms of survival, but a compensation may be seen in the next generation (e.g., by an increase in offspring production or offspring survival [Marshall 1978]). Because laboratory experiments often measure only the short-term effects on survival and fecundity, and not the subsequent effects of reduced densities, there may not be many examples of these mechanisms. In natural populations, delayed responses to density changes may be even more common, because interactions with other species tend to give delayed density effects (e.g., Hansen et al. 1999b). Sublethal and delayed numerical responses to density will obviously make it more difficult to predict the outcome of toxicant-density interactions.

Density dependence may be confined to certain "stages or age groups" in a species' life history. For instance, juvenile individuals may suffer more from food limitation than adults because of poorer energy reserves. If density dependence is limited to a certain stage, then the compensatory capacity for toxicant effects will also be linked to this stage. Juvenile stages are often more sensitive to toxicants than adult stages, but this sensitivity can be outweighed by greater capacity for compensation, as demonstrated in laboratory studies (Kammenga et al. 1996; Stark and Banken 1999; Moe et al. 2001). Hence, the population-level response does not necessarily correspond to the response of the most sensitive stage. This situation is referred to as "the weakest link incongruity" (Newman 1998). An example from a natural system is the 1988 toxic algal bloom in Skagerrak, which killed 60% of the youngest age class of cod. Nevertheless, this disturbance had no detectable long-term effect on the population (Chan et al. 2003). One explanation is that if the density of this group had not been reduced by the toxic algae, it would instead have suffered density-dependent losses from cannibalism and competition by older age classes. Such stage-specific impacts (both lethal and sublethal) will also involve a delay before the changes are manifested at subsequent stages (e.g., from juvenile mortality to effects on adult size).

Another relevant aspect of density dependence is "individual heterogeneity," which may result in uneven resource partitioning. Degree of resource partitioning may affect both stability properties of the population dynamics and the capacity for density-dependent compensation. The two extreme versions of resource partitioning are referred to as "contest competition" and "scramble competition" (Nicholson 1954). In pure contest competition, all resources are allocated to a limited number of survivors. An increase in density will not affect the total number of survivors, because any increase will be compensated by an equivalent increase in mortality. In scramble competition, the resources are more evenly distributed among individuals, and therefore more of the resources will be wasted on the "losers" that do not obtain enough resources for survival. This type of competition results in overcompensation at high densities. At the population level, contest-like competition may be more beneficial than scramble-like competition, because a limited resource may support more individuals this way (Uchmanski 2000). Toxicant exposure may result in enhanced variability in individual fitness, both because already weak individuals will suffer more from the additional stress, and because individual sensitivity to the toxicant may be variable. By increasing the variation among individuals, the toxicant exposure may shift the competition from scramble-like to more contest-like. Such a shift implies that the limited resources can support a higher number of individuals (i.e., an increase in "carrying capacity"). The net effect of the toxicant, however, will depend on the balance between this indirect benefit and the direct cost. This mechanism was suggested to explain why toxicant-exposed blowfly larvae had higher survival than control larvae for any given density (Moe et al. 2005). Grant (1998) performed a density-dependent sensitivity analysis on data from a population toxicity experiment, specifying both contest-type competition (juvenile survival) and scramble-type competition (reproduction). The result of the analysis differed between the 2 types of competition. In his model, there was also an interaction between competition type and toxicant level: the population with contest-type competition was more strongly affected by low toxicant concentration, whereas the population with scramble-like competition was more strongly affected by high concentrations. Hence, the degree of resource partitioning may at least in theory influence density-toxicant interactions.

Populations that experience "positive density dependence" (Allee effect) may be particularly sensitive to density reductions by toxicants. This type of density dependence may therefore produce a particular type of toxicant-density interaction: below some density threshold, further reductions in density may lead to extinction without any capacity for compensatory responses (Figure 4.5). In addition, small populations are intrinsically more susceptible to extinction because of higher demographic stochasticity. This means that random variations in birth and death rates play

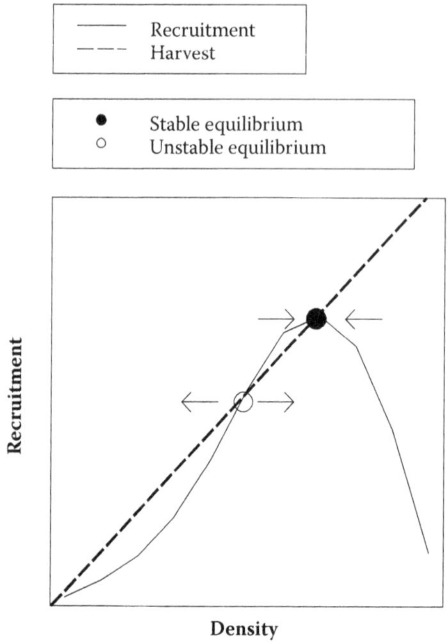

FIGURE 4.5 Positive density dependence in low densities (the Allee effect).

a relatively larger role than in larger populations and increase the chance of extinction (e.g., Wang et al. 1999). Models for prediction of toxicant risks to small populations should therefore consider also demographic stochasticity.

A simple harvest model shows that a population with Allee effect in the per-capita recruitment rate may in theory display multiple equilibria, when it is subject to density-dependent toxicant impacts (harvest). As long as the density is above a threshold (the unstable equilibrium), density-dependent feedback will push the population density toward the stable equilibrium (illustrated by arrows). However, if density declines below this threshold, the population will continue to decline toward extinction. Hence, a population with these properties may be more susceptible to external mortality factors than expected from models that do not consider the Allee effect (after Begon et al. 1990).

There are few examples of density-toxicity studies that consider "interactions between species." One example is given by gammarids exposed to both competitive stress from asellids and toxic stress from lindane (Blockwell et al. 1998). Increasing the toxicant dose has a directly negative effect on the gammarids but also an indirect benefit (reducing survival of asellids). This resulted in a nonlinear dose-response function: intermediate toxicant doses gave the best feeding response of the gammarids. Another example considers amphibian tadpoles exposed to carbaryl: certain concentrations had an indirect benefit on the tadpoles both by reducing the number of competing zooplankton and by reducing the efficiency of the predatory newts (Boone and Semlitsch 2001). Other examples of density-dependent interactions between species are given by Fleeger et al. (2003) in a review on indirect effects of contaminants in aquatic ecosystems, and by Chapman et al. (2003) in a review on risk assessments of inorganic metals and metalloids. In principle, results from single-species tests may be extrapolated to multispecies settings, but the number of possible density-toxicant interactions will probably grow very rapidly with the number of species involved and make predictions highly uncertain.

Aspects of Toxicant Effects

The density-toxicant interactions discussed previously assume that the toxicant has direct and lethal effects. However, toxicant stress may also affect organisms in various ways, as described for density stress above sublethal, delayed, or age- and stage-specific effects, and individual variations in susceptibility. Unlike many other stress factors, chemicals may also accumulate in organisms and produce adverse effects after the exposure has terminated. These aspects of the toxicant effects may also affect density-toxicant interactions.

The organism-level effects of a toxicant are likely to increase with the "concentration" and duration of exposure. The population-level effects, however, do not necessarily follow the same pattern. The potential for density-dependent compensations is highest when the toxicant dose is strong enough to reduce density, but not strong enough to affect all individuals. An example is given by trichopterans exposed to the pesticide fenvalerate (Liess 2002): high-density populations initially had lower survival than low-density populations, and low toxicant concentrations further reduced survival slightly. At modest toxicant concentrations, however, the high-density populations showed increased survival, demonstrating an overcompensating response to the toxicant. At even higher toxicant levels, no individuals survived. A similar response is reported for polychaete populations exposed to fluoranthene at different densities and toxicant concentrations: interactions between the toxicant and density were antagonistic at low toxicant concentrations, but became synergistic at high concentrations (Linke-Gamenick et al. 1999). These experiments demonstrate that density-dependent interactions may contribute to a nonlinear and nonmonotonous relationship between toxicant concentration and population-level response.

The "duration" of the exposure will also affect the capacity for compensatory responses. An acute pulse disturbance is more likely to be compensated by density-dependent responses than a chronic disturbance. The 1988 toxic algal bloom (Chan et al. 2003) is an example of a pulse disturbance that was rapidly compensated. A chronic disturbance with a constant effect on demographic rates is more likely to interact with density dependence until a new, lower equilibrium size is reached (harvest models).

Toxicants with only "sublethal" effects that do not cause an immediate reduction of density, will not readily be compensated by density-dependent responses. Still, if there are "delayed effects" on density, compensatory mechanisms may operate in later generations (delayed effects of density discussed previously). For example, a toxicant-induced reduction in fecundity may be compensated by higher survival of offspring. An example from management of free-ranging rabbit populations may illustrate this mechanism (Twigg et al. 2000). Different proportions of the female rabbits were sterilized surgically to investigate the effects of reduced reproduction on population sizes. For all sterility levels up to 60%, the reductions in offspring production were fully compensated by increased survival of both offspring and adults. In this example, however, the outcome of the "exposure" was either complete sterility or no effect. Toxicants are likely to have more intermediate effects on fecundity, and may also reduce the offspring fitness by reduced size or other maternal effects. Moreover, toxicants may accumulate in the organism during the juvenile stage and later reduce fecundity, even if the reduced fitness is not reflected in reduced individual growth (Moe et al. 2001). Hence, compensatory responses to sublethal or delayed effects of toxicants may be more difficult to detect.

"Individual heterogeneity" in toxicant susceptibility or exposure is also likely to affect density-toxicant interactions in a similar way as described for susceptibility to density effects. A higher degree of heterogeneity may increase the possibility for compensatory responses in 2 ways: the most susceptible individuals will have higher mortality and thereby reduce the density, and the least susceptible individuals will have better chances for increased survival and reproduction.

STATISTICAL METHODS FOR QUANTIFYING DENSITY DEPENDENCE

There are several approaches to quantifying density dependence, and the choice of method should depend both on the type of questions to be answered and on the type of data available or obtainable. The relevant data for a population will often consist of repeated measurements of abundance or density, and density dependence can be estimated using a statistical model that includes some density-dependent function. The density-dependent function will typically represent the relationship between density and population growth rate or vital rates (survival, growth, or reproduction). The appropriate statistical model depends on, for example, how detailed the data are, how often the population is measured, and whether data for other species or other environmental factors are available. (Note that the term "model" here is used in a broader sense than it is in Chapter 9. In this chapter, "model" also includes statistical models for parameter estimation.)

Density dependence can also be predicted from mechanistic models that are based on organism-level parameters, such as individual energy budgets, physiological rates, and behavior (Kooijman and Metz 1984; DeAngelis and Gross 1992; Gutierrez 1996; McCauley et al. 1996). Density dependence may then be predicted as an emergent property of the model rather than being specified by certain functions. However, this approach requires very detailed knowledge about the system, which is normally available only for laboratory populations, and the models will be quite species specific. Using statistical models to estimate density-dependent parameters from abundance data is a more versatile and efficient approach for quantifying density dependence, and I will therefore focus on this approach here. See Kendall et al. (1999) for a discussion of mechanistic vs. statistical models.

Population data can be broadly categorized into 2 types, which are referred to here as cohort data and time-series data. "Cohort data" are obtained from experiments that follow even-aged groups of individuals through a part or the whole of their lifecycle, and consist of repeated counts or estimates of population size. Reproduction may also be measured, but offspring are removed. These types of experiments are called life-table response experiments. This design is suitable for studying effects of density and other stress factors on exponentially growing populations, but not for studying equilibrium dynamics. Cohort data may also be obtained from field populations, but this requires that the species have synchronized reproduction and that the different cohorts can be distinguished from each other. "Time-series data" consist of repeated counts or estimates of a whole population over a long period. The population may reproduce continuously and the offspring remain in the population. These data are normally obtained from observations of field populations. Time-series data may also be obtained from laboratories, but this may be very time-consuming, and the data are often limited to small invertebrate species. The mesocosm design is appropriate for testing effects of density and other stress factors on populations at equilibrium and stable age distributions, or the effects of these factors on population dynamics. The 2 types of experiments may yield different results (van Leeuwen et al. 1986; Moe et al. 2002a), so it is important to consider the design most appropriate for the actual problem.

Cohort Data

Life-table response experimentation is a traditional way to measure demographic rates. Because there is no recruitment to the cohort, all changes in density can be measured as age-specific mortality. This experimental design is the basis for "key-factor analysis" (Varley and Gradwell 1960; Varley et al. 1973), which is a traditional and popular way for estimation of density dependence. For each age-specific mortality factor, a *k*-value ("killing power") is calculated from the proportion that has died from this factor. The factor with the highest k-value is then the "key-factor." To test whether a mortality factor is density-dependent, each *k*-value is analyzed as a function of density. If the *k*-value increases with density, the factor is (negatively) density-dependent and does thereby contribute to regulation of the population. This method has the benefit that it is easy to calculate, but it has limited application (being developed for populations with non-overlapping generations) and is also considered too conservative (i.e., it sometimes fails to detect density dependence). Another test that has been widely used was developed by Morris (1959); this is considered too liberal. Both tests have been applied to analyze marine survival in Pacific salmonids from 12 different systems: density-dependent survival was found in 7 of the systems with Morris' test, but in only 1 case with Varley and Gradwell's test (Peterman 1978). Other tests for density-dependence tests are proposed by Maelzer (1970), Dennis and Taper (1994), and Pollard et al. (1987); see also review by Murdoch (1994).

Key-factor analysis was initially proposed as a method to pick out 1 single key factor, but the method has often been extended to investigate several *k*-values other than the most important one. This has caused some difficulties when different mortality factors are correlated (Smith 1973). A key-factor analysis of Tawny owl populations concluded that the most important mortality factor was loss through failure to breed, and that the second most important factor was the failure to achieve maximum clutch size (Southern 1970). However, Smith (1973) argued that the second factor did not contribute much to mortality, but was simply correlated with the key factor. A modified version of the analysis that accounts for correlations between factors indicated that overwinter mortality is the true second most important factor (Smith 1973). Another modification of key-factor analysis considers effects on population growth rate rather than just on mortality (λ-contribution analysis); this method had been applied to analyze red deer populations (Sibly and Smith 1998). The λ-contribution analysis gives more weight to factors acting early in the life history than the traditional key-factor analysis does and thereby gives a more correct assessment of the contributions from different mortality factors to the population growth rate.

Organism responses such as growth, development times, and fecundity as well as survival are often analyzed by ordinary linear methods, such as linear regression (for continuous covariates) or analysis of variance (for categorical covariates). Linear methods make certain assumptions about the quality of the data (e.g., normal distribution of the response variable, which are not always met by ecotoxicological data; Kennedy et al. 1999). One way to solve this problem is to use a nonlinear transformation of the data, but transformations should be used with caution. An alternative may be to use a "generalized linear model" (McCullagh and Nelder 1989). A generalized linear model can use different types of "link functions" to connect the response variable to the predictor variables, and thereby allows for more types of data distributions. The model is therefore more applicable than an ordinary linear model. As with linear models, it also allows for interactions between predictor variables. An example of a generalized linear model is logistic regression, which can be used to analyze a bivariate response (e.g., effect/no effect) as function of density in interaction with other factors, such as toxicants. Tanhuanpaa et al. (2001) used logistic regression to estimate effects of density and predation on the proportion of surviving moth larvae in a field experiment with predator exclusion. They found a negative effect of density on survival, but no effect from predators.

Effects of density and/or other factors on survival can also be estimated by a class of methods called "survival analysis;" for instance, the proportional hazard regression (Cox and Oakes 1984). Survival analysis considers not only the number of deaths, but also the individual longevities, and can give predictions about life expectancies and relative risks (proportional hazards) for different groups. It may also estimate interactions between variables, such as density and toxicant effects. In the literature, survival is often analyzed as the number of survivors by the end of the experiment, but survival analysis makes more efficient use of the individual information. However, because the analysis require data on individual longevities, this method is most suitable for studies with even-aged individuals or with marked individuals. Tanhuanpaa et al. (2001) also used this method to estimate effects of density and predation on survival times of moth larvae. The analysis demonstrated that predator exclusion increased survival times of larvae in high-density treatments. However, the benefit from predator exclusion was later compensated by density-dependent mortality in later instars, so that the final proportion of survivors was the same as for predator-exposed larvae. Hence, the survival analysis revealed more details about the survival responses than could be obtained from analyzing of the number of survivors.

Time-Series Data

An introduction to analyses of time-series data is given by Chatfield (1996). Ecological time-series data typically consist of repeated counts or estimates of a whole population. Because the change between each count represents a combination of mortality and recruitment (and possibly migration), there is no direct way to measure the demographic rates. A population model is then needed to estimate the demographic rates. The simplest models consider only the total population, whereas more realistic models may include age or stage structure. The population may be modeled by a set of population transition equations or by matrix formulation (Leslie 1945; Caswell 2001), and the methods described below may also be applied to the stage- and age-specific time-series.

There are several simple models for testing the presence of density-dependence based on whole population counts. For example, Pollard et al. (1987) proposed a model with the density effect proportional to the log-transformed density in the previous time-step. This model tests the presence of direct density dependence (i.e., an effect of the density 1 time-step ago only). Slightly different versions of tests for direct density dependence are given by Maelzer (1970) and Bulmer (1975).

However, it is important also to consider effects of previous densities. If one can identify and distinguish between contributions from direct and delayed feedback, then this may indicate what type of intra- and interspecific interactions affect the population, such as predation or competition (see Bjornstad et al. 2001). To include delayed density-dependent effects, it is conventional to use an "autoregressive model," in which the time series is regressed against previous densities with different time lags (e.g., years) (Box and Jenkins 1970). A critical problem is to determine the number of lags relevant for the model. A commonly used diagnostic in conventional time-series analysis is the "autocorrelation function," which reveals periodic patterns. To decide the number of relevant lags, one can estimate the partial autocorrelation function, which measures how much each term contributes to the explained variation when added subsequently.

An autoregressive model can also be extended to include density-independent effects of external factors, even if these factors are not measured. A perturbation with effects that lasts through several time-steps can be incorporated as a so-called "moving-average process" (MA). This may be a suitable method for estimating effects of acute exposure incidents for field populations or for perturbation experiments. The density-dependent parameters and the density-independent parameters can then be estimated simultaneously by a so-called autoregressive-moving average process (ARMA) model. Although this type of time-series analysis is a good framework for exploring density dependence, it requires regularly sampled observations, and it is constrained by the underlying assumptions of linear relationships. There are many alternatives for specifying nonlinear density-dependent functions, but for exploratory analysis it is preferable to make as few assumptions as possible in advance. One way of allowing for nonlinearity in an autoregressive model is to include a threshold density, which defines 2 regimes (high and low density) with separate density-dependent parameters. The "threshold autoregressive" model will

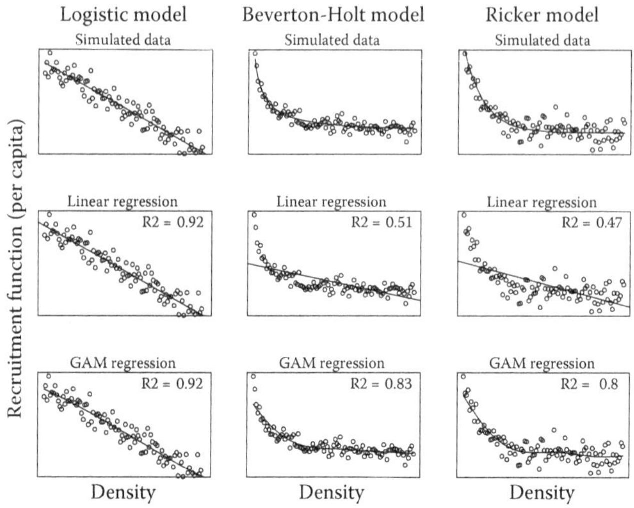

FIGURE 4.6 Comparison of GAM (generalized additive models) regression with linear regression for fitting different recruitment functions. The graphs in the upper panel show data simulated by the logistic model (a), the Beverton–Holt model (b), and the Ricker model (c); the curves show the model predictions, whereas the points show stochastic simulations of the respective models (with normally distributed noise). The middle panel and the lower panel show the 3 simulated data sets fitted by linear regression and GAM regression, respectively, with the R^2 values as a measure of the goodness of fit. For the logistic model, where density dependence is linear, GAM and linear regression give similar goodness of fit. However for the two models with non-linear density dependence (B–H and Ricker), the more flexible GAM regression gives considerably better fit than linear regression.

estimate both the threshold density and the parameters separately on either side of the threshold density. Threshold autoregressive models have been used for analyzing density dependence in populations of, for example, lynx and hare (Stenseth et al. 1998), voles and lemmings (Stenseth 1999), and Soay sheep on islands in the St. Kilda archipelago (Grenfell et al. 1998). The Soay sheep analysis revealed different mechanisms on either side of the threshold density; below the threshold, there is a constant population growth, whereas at higher densities the population may increase, remain constant, or fall, depending on environmental conditions (Grenfell et al. 1998). In risk assessment, this type of modeling might be useful for cases in which there are density thresholds that modify toxicant-density interactions.

A more flexible way of exploring nonlinear structures in density dependence is by “generalized additive modeling” (GAM; Hastie and Tibshirani 1990). This method can be used for so-called nonparametric population modeling, in which the density-dependent relationships are represented by unspecified functions. The unspecified function can be estimated as a curve, which is actually a combination of many short linear pieces, and which can have any shape (see Figure 4.6). A great benefit of GAM is that it may estimate nonlinear relationships (e.g., between density and survival) without the need for specifying a function in advance. Another strength of this method is that several functions can be added, for instance delayed-density effects, stage-specific density effects, or effects of other variables. A GAM-based test for density dependence is developed, in which density dependence is demonstrated by a significant convexity of the estimated survival curve (Bjornstad et al. 1999a). This model was applied to describe density dependence in juvenile survival of coastal cod populations, where suggested mechanisms are cannibalism, competition for habitat, and food limitation. Nonparametric or semiparametric approaches have also proved useful for field data from various laboratory systems (e.g., flies [Lingjærde et al. 2001], moths [Bjornstad et al. 2001], copepods and daphnids [Wood 2001], as well as measles epidemics [Ellner et al. 1998]). GAM is also considered a suitable tool for analyzing fisheries data (Daskalov 1999). A weakness with GAM, however, is that it cannot estimate interactions between different factors.

A quite different approach is "Bayesian analysis," which can make more efficient use of available knowledge from different sources (described in Chapter 7). This method is being used increasingly for risk analyses in conservation and management (Johnson 1999). Bayesian analyses of capercaillie populations (Kangas and Kurki 2000) and Atlantic salmon populations (Rivot et al. 2001) show that these model predictions are also sensitive to the specifications of density dependence.

Important parts of the estimation procedure are "model selection and model evaluation," which will be mentioned only briefly here. A simple measure of how well a model fits the data is the squared correlation coefficient (R^2), which measures the proportion of variance in the data that are explained by the model. A weakness with this measure is that it will always increase with the number of parameters in the model; the model may then be overfitted to a specific data set and give a poorer fit to new data. A commonly used alternative is Akaike's information criterion, which penalizes for increasing numbers of parameters in the model (see Chapter 7). Cross-validation may be used to test the model's ability to predict data that is not used for estimation (Wood 2001).

The range of statistical models available provides very powerful tools for data analysis, but they do not necessarily give an explanation for the density-dependent mechanisms. For instance, several types of interspecific interactions may give rise to similar cyclic patterns (Figure 4.1). A way of dealing with this problem is to formulate a different mechanistic version of the model, and judge how well the simulated data correspond to the observed data. Although this is a common way of evaluating mechanistic models, it has been criticized for being too subjective (Wood and Thomas 1999). Because both mechanistic and statistical modeling approaches have their inherent strengths and weaknesses, a combination of the 2 approaches may be the best strategy (Cushing et al. 1996; Kendall et al. 1999).

PROBLEMS WITH APPLYING THE DENSITY-DEPENDENCE CONCEPT

Detecting and Estimating Density Dependence

It has been argued that shortage of relevant of data is part of the problem with the density-dependence debate (Murray 1999). Vast amounts of data have been analyzed for density dependence, but the data are not necessarily collected in a way that is suitable for answering the actual questions. There are different inherent weaknesses with laboratory and field data. Laboratory populations are subject to various unrealistic conditions and lab artifacts — for instance, lack of competing species and natural enemies. Populations in long-term experiments may also be affected by changed selection pressure (Stokes et al. 1988). The initial age or stage structure of a population may influence the effects of both density and toxicants on the populations (Stark and Banken 1999), and it is therefore likely to influence the interaction between density dependence and toxicity. These interactions may also be influenced by factors such as the physiological condition of the individuals and whether the population is growing or declining (Forbes and Calow 1999). Field observations are more realistic, but suffer from lower precision than laboratory data. Population densities are often estimated from samples rather than measured by a complete census. Furthermore, field data often consist of observation of certain age groups (e.g., only the adults or the most sedentary stage). This introduces "noise" to the data in terms of sampling error and other types of observation error.

The simplest analysis of density dependence is the detection of presence or absence, but even this may be problematic. For estimation from time-series data, there are several factors that may affect the result. An inherent problem with these tests is that the observations are temporally correlated, and therefore do not conform to the assumption of independent observations. Moreover, each observation is used both in the response variable and as a predictor variable. As a consequence, both temporal trends in abundances and observation errors may lead to higher possibility of falsely detecting density dependence when it is not present (type I error) (Solow 2001). Methods exist that aim at separating the various sources of error (state-space methods; e.g., de Valpine and Hastings 2002), but these often require large amounts of data. The presence of delayed density dependence may also increase the probability of spuriously detecting direct density dependence (Holyoak 1994). Other conditions may reduce the probability of detecting density dependence when it is actually present (type II error), such as spatial population structures (Ray and Hastings 1996). It may also be harder to detect density dependence that occurs only infrequently, only at some spatial scales, and only in some stages of the population. Finally, the chances of detecting density dependence seem to increase with the length of the time series (Hassell et al. 1989). As for quantification of the strength of density dependence, the conditions that lead to type I and type II error will contribute to overestimation or underestimation, respectively.

Statistical models can be very flexible regarding shapes of density-dependent functions, but the popular saying that they "let the data speak for themselves" is probably overly optimistic. If a model is used to estimate stage-specific density dependences, then the model will still need certain assumptions about, for example, development times, which may have considerable effects on the estimates. Another limitation for many of these flexible statistical methods (e.g., GAM, autoregressive-moving average process model (ARMA), threshold autoregressive (TAR)) is that they cannot estimate interactions between variables. Thus it is not straightforward to quantify interactive effects among several species or other environmental factors.

Predicting Effects of Density Dependence

Mechanistic population models are powerful and popular tools for describing biological mechanisms and predicting population dynamics, but it is important to be aware how much the modeler's assumptions constrain the model output (Wood and Thomas 1999). Assumptions regarding functional form of density dependence may have large effects on the population-level predictions, as illustrated by the Ricker and Beverton–Holt models (Figure 4.2). Moreover, for a given functional form, the parameter values may have a strong impact on predictions regarding growth rates and stability. Models with strong nonlinear relationships (such as overcompensating density dependence) are also inherently more sensitive to initial population conditions than linear models.

Population models that incorporate density dependence will normally predict some degree of compensation to loss of individuals. However, natural populations that appear to have density-dependent growth do not necessarily show compensation to losses. In the case of ptarmigan populations, the density-dependent emigration of juveniles gives a negative relationship between density and population growth. However, when these populations were harvested experimentally, only 33% of the losses were compensated for by reduced emigration (Pedersen et al. 2003). For some species, it can therefore be necessary to find the appropriate spatial scale for modeling the density-dependent processes.

Stochasticity can be included in mechanistic models to make the simulations more realistic (Bjornstad and Grenfell 2001). However, the choice of where and how to include stochasticity may be quite subjective. Stochasticity may also make model predictions more sensitive to density dependences and thereby more diverging—the mean output of several stochastic model runs does not necessarily correspond to the output of the deterministic version (Nachman 2001; Lande 2002).

RECOMMENDATIONS FOR TREATMENT OF DENSITY DEPENDENCE IN ECOLOGICAL RISK ASSESSMENTS

This chapter has pointed out different theoretical possibilities for interactions between density dependence and toxicants, which may have unexpected population-level consequences. However, it is not obvious to what degree these predictions are relevant to risk assessment. Model simulations indicate that incorporating density dependence does not necessarily improve long-term predictions, because natural variation may mask the effects of density dependence (Barnthouse 1993). It has been argued that density dependence will usually give a population higher compensatory capacity; therefore, ignoring density dependence will give more protective estimates (Forbes et al. 2001a,b). However, life-table response experiments are often carried out at low densities and may not reveal synergistic effects of toxicant and food limitation that may occur in the field (Petersen and Petersen 1988; Linke-Gamenick et al. 1999). Moreover, for special cases such as the Allee effects, ignoring density dependence may have fatal consequences. Because such cases may turn out to be more common than previously recognized (Bruno et al. 2003), precautionary advice would be to pay special attention to the density-dependent structures at low densities and under multiple stress factors.

There is general agreement about the need for more experimental testing of density-toxicant interactions. However, the experimental design may constrain the types of interactions that can be detected. Detailed recommendations for experimental design are given by Sibly (1999) and by Forbes et al. (2001a). They recommend life-table response experiments for obtaining information on the mechanistic bases of density-toxicant interactions, and mesocosm experiments for examining density-toxicant interactions at the population level (corresponding to cohort data and time-series data, respectively, in Statistical Methods for Quantifying Density Dependence).

Density-toxicant interactions may be quite species-specific, so one should be careful to extrapolate results to species with different life histories. Species that are commonly used for toxicological testing (such as cladocerans and rotifers) are suitable for this purpose in part because of their low juvenile mortality. However, low juvenile mortality also implies a low capacity for compensatory responses. Hence, standards based on test organisms such as *Daphnia* may not be representative for populations that normally experience higher juvenile mortality, and therefore may have higher compensatory capacity (Petersen and Petersen 1988).

It is tempting to recommend that density-toxicant experiments be performed in the field under more natural conditions, but this will often not be feasible. An alternative approach is to perform perturbation studies (i.e., to manipulate densities in order to simulate the effects of toxicants). Perturbation experiments have often been recommended by ecologists for studying density dependence under natural conditions, yet surprisingly few such experiments have been performed (Harrison and Cappuchino 1995). Simulation of toxicant effects by density manipulations can be performed as, for example, a single density reduction to imitate effects of an acute contamination incident or repeated "harvesting" to imitate effects of chronic exposure (Pedersen et al. 2003). Nicholson (1954) demonstrated compensatory responses to "harvesting" by regularly removing fixed proportions of the fly populations. This type of experiment should also be feasible for field populations, by using a fixed "harvest" effort to obtain a constant proportion of the population. However, this procedure simulates only the lethal effects of a toxicant, and not the possible effects on growth and fecundity, which will be more difficult to imitate. This approach is therefore most relevant to risk assessment for toxicants with mainly lethal effects. Several questions should be considered, for example: Can results be generalized beyond the particular experimental system and chemical in question? Can these types of experiments be performed to support risk assessment of pesticides?

For species in which density manipulations are not possible, the best approach may be to obtain reliable estimates of density dependence from observation studies and try to combine this with information on toxicant effects from laboratory studies. Because time-series analysis and other statistical modeling methods require long time-series (30 points generally; Chatfield 1996), it may be difficult to obtain enough data for reliable estimates. Power analyses may also be used to estimate the amount of data needed to detect significant effects. More detailed demographic information, such as age or stage structure, may nevertheless compensate for shortness of data (Smith et al. 2000). To obtain as much information as possible from the data, the appropriate statistical models for analysis should be considered before the data collection, so that the sampling scheme can be designed accordingly. For instance, time-series analyses will make most efficient use of the data if the sampling interval is regular (which is often not the case with existing data sets). The sampling intervals should ideally not be too long relative to the generation time of the species. Furthermore, it is useful to measure or otherwise obtain information on as many covariates as possible (regarding, for example, individual condition, habitat, climate, or other disturbances), which may help explain the observed variation. Consequences of naturally occurring perturbations can also be analyzed to obtain a better understanding of interactions between density dependence and external factors, such as the effects of toxic algae bloom on cod population dynamics (Chan et al. 2003).

There is much information on density dependence and interactions between density and abiotic factors available from ecological literature, but the selection of taxonomic groups is often biased. For instance, theoretical studies of density-dependent mechanisms have focused on species with cyclic or outbreak dynamics, such as insects and rodents. The estimated parameters and mechanistic models for these species may be too specific to be used directly for other species in risk assessment. Still, the wide range of tools developed for analysis of population data and population-level predictions in these fields could be exploited more for risk assessment. For example, conservation biology often deals with limited data sets on rare species, which has stimulated development of methods for risk prediction based on sparse and noisy data (Dennis and Taper 1994; Holmes 2001).

Population models used for risk assessment have mostly been density independent, and constructed for predicting effects on population growth rate or extinction time. Density-independent predictions of toxicant effects may be misleading for populations that have density-dependent regulation (Grant 1998). Population-level risk assessment may therefore be improved by including density dependence. In density-dependent models, neither the population growth rate nor the extinction time are very useful measures of toxicant effects: if the compensatory mechanisms are strong enough, the population will not go extinct and population growth will approach zero in both control and exposed populations (Laskowski 2000). A more relevant measure for population-level effects of a toxicant is the equilibrium density, which is likely to be lower in toxicant-stressed populations. This endpoint may also be easier to measure in field population. An objection against density-dependent models is that they require additional assumptions about the form of the density dependence, which is difficult to measure. However, a more general (phenomenological) model may be able to give reasonable predictions, even without specifying the biological mechanisms in detail (see Kooijman and Bedaux 2000). At any rate, different density-dependent scenarios can be explored to see how much their predictions deviate from the density-independent predictions, and to analyze what consequences this will have for the risk assessment.

CONCLUSIONS

Density dependence is a controversial issue in both theoretical and applied ecology, but density-dependent processes appear to regulate natural populations at least in some life-history stages and at some spatial and temporal scales. Effects of toxicants on the population level are likely to be affected by density dependence, but the short-term outcome of the interaction may depend on several factors. Density-dependent compensatory responses to a toxicant (antagonistic interactions) are most likely to occur if the population has high reproductive potential and strong (overcompensating) density dependence, if the toxicant is lethal to parts of the population, and if there is large individual variation in sensitivity to the toxicant. Synergistic interactions are most likely to occur if the toxicant and/or density have mainly sublethal effects, by chronic toxicant exposure, or if the toxicant is strong enough to outweigh any benefit from reduced density. But even in these cases, the population may have some benefits from relieved density stress in the long term. Allee effects (positive density dependence in low densities), however, may result in a particular type of synergistic density-toxicant interaction that cannot even be compensated by relief from density stress.

Information on both toxicant effects and density effects on different life-history traits may improve model predictions. Experiments for investigating density-toxicant interactions should ideally cover a wide range of densities and concentrations to reveal all possible types of interactions (both synergistic and antagonistic) and measure responses over more than one generation. In field experiments, toxicant stress can to some degree be simulated by density perturbations or by harvesting. Density-dependent structures and responses to perturbations in natural populations can be explored by various statistical methods. These methods can be powerful but may require large amounts of data.

When density dependence is built into a population model, then the theoretical population is more likely to sustain losses, because density dependence often serves to buffer population sizes. Uncritical use of strong density-dependent functions can therefore result in underestimation of the real risk. Conversely, models that ignore density dependence are more likely to overestimate population-level risks than to underestimate them. Hence, such ignorance may be seen as conservative and harmless to the environment. However, overprotective risk assessments may divert limited resources away from other populations that are at higher risks. Appropriate inclusion of density dependence may make population-level risk assessments more realistic and may thereby contribute to making better risk management decisions.

5 Genetic Variation in Population-Level Ecological Risk Assessment

Diane Nacci and Ary A. Hoffmann

INTRODUCTION: CHALLENGES AND OPPORTUNITIES FOR GENETICS IN POPULATION-LEVEL ECOLOGICAL RISK ASSESSMENT

Genetic variation among individuals serves as the foundation for processes by which species and populations adapt to local conditions and evolve (Hartl and Clark 1997). Ample genetic variation is required for adaptation to stressful and changing environments, including pesticide applications and chemical exposures. Genetic variation can be used to describe current and historical population attributes such as size and structure. But the inclusion of genetic methods and processes in population ecology has been slow (e.g., Jackson et al. 2002), and mechanisms for incorporating genetics into population dynamics are relatively unexplored (but see Chapters 8 and 9). Thus, despite its fundamental importance, genetic variation has rarely been considered quantitatively in predictions of population persistence (e.g., Beissinger and McCullough 2002), and is almost never considered in ecological risk assessments (ERAs). However, genetic information is becoming increasingly accessible and the relevance of genetic changes better understood as more gene functions become known (e.g., Collins et al. 2003; Jackson et al. 2002). More specific to the needs of ERAs, these advances are contributing to improved understanding of relationships among stress, changes in genetic variation, and population persistence.

Genetic variation has rarely been considered quantitatively in predictions of population persistence, and is almost never considered in ecological risk assessments.

In this chapter, we review and synthesize ecological applications of genetic information, and consider how genetic information can improve predictions of risks from environmental stressors to aquatic and wildlife populations. We first discuss how genetic variation has been considered relative to fitness (see Genetic Information: Neutral, Adaptive, and Detrimental), and then review how the analysis of genetic variation has been used to describe populations. Selectively neutral genetic variation is used to describe population attributes such as current and historical size and structure (see Neutral Genetic Variation and Population Condition). Adaptive genetic variation is explained as a mechanism by which populations respond to environmental conditions (see Adaptive Genetic Variation and Fitness). Examples of adaptation to chemical stressors are provided (see Genetic Variation and Risks from Chemical Exposures). We suggest how genetic information can contribute to empirical and modeling components of population risk assessments and recommend some directed research areas to enhance this inclusion (see Genetic Contributions to Population-Level ERA and Future Research Needs and Challenges). In conclusion, we emphasize the potential for genetics to contribute unique information about population condition, vulnerabilities, and persistence to ecological risk assessments.

GENETIC VARIATION: NEUTRAL, ADAPTIVE, AND DETRIMENTAL

Historically, genetic variation has been categorized as ranging from neutral to being fitness-related and therefore subject to selection (e.g., Hedrick 2001). In practice, the relationship between genetic markers and fitness is conditional, and is often unknown. However, this categorization provides a useful structure to discuss how genetic information may contribute to population-level ERAs. Working definitions and examples of genetic marker types for neutral and fitness-related forms of genetic variation follow.

Neutral genetic variation reflects the effects of temporal and spatial nonselective processes and can provide one way of defining groups biologically (Hartl and Clark 1997). In general, neutral variation in populations is reduced by random genetic drift and is increased by mutation and gene flow. Analysis of neutral genetic markers provides a means to define population structure, estimate effective population size and movement rates, and can infer historical and geographical relationships among groups. Genetic markers where the varying genotypes are considered to be selectively equivalent or neutral include allozymes, microsatellites, random amplified polymorphic DNA (RAPD), amplified fragment length polymorphisms (AFLP), and single nucleotide polymorphisms (SNPs) in segments of DNA that are not under strong selection. Of these, SNPs offer some advantages, including abundant variation (representing 90% of the genetic variation in the human genome; Brumfield et al. 2003), suitability for automated analysis systems, and mutation dynamics that are more easily modeled and less variable among loci (Manel et al. 2003). Hundreds of thousands of SNPs have been identified and characterized for model species, and development of these markers for nonmodel species is becoming more common as technology improves (Brumfield et al. 2003).

Genetic loci under selection can have patterns that deviate from those of neutral markers, reflecting forces other than population history and demography. Low variation at unique genetic loci in the context of high level genome-wide genetic variation can be diagnostic of directional selection on functionally important genes. Adaptive genetic variation is often assessed at the phenotypic level rather than the genetic level because the genes underlying adaptive shifts are often unknown. For example, genetic variation among and within populations can be assessed using morphological traits that are measured easily on individuals (Merilä and Crnokrak 2001). In addition, phenotypic variation is often more easily assessed because adaptive shifts often involve the combined effects of genetic changes at many loci, each of which may have only a small impact on the adaptive change. The extent and pattern of adaptive variation provides the potential for long-term survival of populations in variable environments.

Detrimental genetic variation refers to alleles with small or potentially large adverse fitness effects, including lethality. These alleles can lead to inbreeding depression, recognized to be of significant concern in small populations. Population viability analyses (Brook et al. 2002) suggest that inbreeding effects can increase the risk of extinction even in populations of 1000 individuals, or more. In addition, there is speculation and some evidence that the effects of detrimental variation may be more severe when combined with other stressors (reviewed in Sih et al. 2000), even in large populations (Frankham et al. 2002; Hedrick 2001).

Although genetic markers such as microsatellites are often considered neutral, patterns of differentiation within and between populations for these markers can also be influenced by selection. For example, selection acts directly on the markers for allozymes linked with temperature tolerance in *Fundulus heteroclitus* (reviewed in Mitton 1997). Allele frequencies at neutral loci can also be influenced by selection at linked loci, or hitchhike along with other processes. For instance, in *Drosophila*, microsatellites linked to inversion polymorphisms can exhibit geographic patterns reflecting selection because these markers cosegregate with other genes under selection but contained within the inversion (Weeks et al. 2002). Indeed, if the density of marker loci is high enough, some of them must be linked to genomic regions under selection. Similarly, the distribution of mitochondrial markers in insects can reflect selection by *Wolbachia* bacteria, which occur in the cytoplasm of insects and have strong phenotypic effects on their hosts (which in turn affect bacterial infection rate; Ballard et al. 2002). In addition, SNPs can also be identified in loci of known function, including (selection-) targeted loci, with potentially important functions (Tabor et al. 2002).

Recently, many researchers have advocated multiple marker and trait strategies to take advantage of complementary information provided by different types of genetic variation. For example, a comparison of neutral markers with selected markers and quantitative traits can be useful for inferring population demographic history or for estimating population parameters, whereas loci under selection are needed to understand adaptive shifts (Manel et al. 2003).

NEUTRAL GENETIC VARIATION AND POPULATION CONDITION

Neutral genetic variation has been used to describe current and historical population condition — for example, size (abundance), richness (diversity), and uniqueness (spatial distribution and differentiation). This application of genetic information is reviewed here because these attributes may be important in diagnosing and predicting population impacts of stressors in risk assessments.

Size and Richness: Effective Population Size and Genetic Diversity

Genetic information is necessary to estimate effective (breeding) population size (N_e). N_e defines population size as the number of individuals contributing uniquely to reproduction (Futuyma 1986), and differs from the census size because of variation in reproductive success among individuals. In the wild, N_e may equal only about 10% to 20% of total adult abundance, or less (e.g., Franklin and Frankham 1998; but see also Waples 2002). Genetic diversity is important for the long-term health of a population because it determines ability to respond to environmental variation, and threatened species have lower levels of genetic variation than related species that are not threatened (Spielman et al. 2004). However, a universal, quantitative relationship between the erosion of genetic diversity and reduced fitness is lacking. For many years, researchers have sought a direct connection between genetic diversity within individuals (i.e., the level of heterozygosity of individuals at multiple loci) and fitness. Although there are some studies that have shown an association between individual level heterozygosity and fitness, including associations with toxic responses (Kopp et al. 1992; Schlueter et al. 2000), it is weak or nonexistent in most cases (Booy et al. 2000). Genetic diversity can also have direct effects on fitness, such as in the case of self-incompatibility in plants. Many plants exhibit incompatibility systems in which pollen needs to carry a different incompatibility allele to that of the parental plant to achieve fertilization. In small populations, plants may all carry similar alleles at the incompatibility locus, resulting in failure of fertilization and seed set (Young et al. 2000).

Nevertheless, small population size can directly affect the health of populations by influencing the expression of deleterious genes. This may occur because inbreeding depression from matings between related individuals leads to increased homozygosity and the expression of rare recessive deleterious genes. Crnokrak and Roff (1999) found evidence of inbreeding effects in approximately 50% of (nonrandomly sampled) wild populations. Inbred populations may be more vulnerable to subsequent stressors because of low levels of variability, as reviewed in Keller and Waller (2002), who cite 7 studies where "increased competition, disease or harsher field conditions can all magnify inbreeding depression." Also, early life stage and fecundity traits seem more vulnerable to inbreeding effects than is survival (Keller and Waller 2002). Several field studies also provide evidence that inbreeding depression can affect the dynamics and extinction risk of small populations (Keller and Waller 2002).

Although there has been a great deal of theoretical and experimental work with model species, little is known about the effects of inbreeding on population dynamics of wild populations that are not very small in size (but see Brook et al. 2002 for mechanisms and results for predicting inbreeding effects). There is some evidence that inbreeding increases the risk of extinction in natural populations (e.g., Saccheri et al. 1998). In addition, it has been shown in experimental populations (Newman and Pilson 1997) that extinction probabilities are increased even under relatively low levels of inbreeding (Bijlsma et al. 2000). One finding to emerge from the recent empirical literature is that inbreeding depression is environment-dependent. Thus alleles that have deleterious effects when homozygous in stressed environments may not have these effects under favorable environments (e.g., Bijlsma et al. 1999). This can mean that the effects of inbreeding in populations have often been underestimated, because studies normally consider only 1 set of conditions.

Genetic variation may be more informative than size in predicting population health.

Apart from inbreeding, small population size can also lead to an increasing genetic load (the reduction in fitness of a population from a theoretical maximum) because of the chance accumulation of mildly deleterious genes that arise via mutation, which can lead to increased probability of extinction (Lande 1994; Lynch et al. 1995). Although evidence for this process is limited and accumulation effects can be difficult to separate from inbreeding, there is evidence in the natterjack toad (*Bufo calamita*) that genetic load in an isolated population is responsible for a decrease in fitness (Rowe and Beebee 2003). Lande (1994) has argued for the importance of this process even in populations with effective population sizes of several hundred individuals. Because of these effects, genetic variation may be more informative than size in predicting population health (Keller and Waller 2002). For instance, Schmidt and Jensen (2000) used molecular markers (AFLPs) to examine 13 populations of a short-lived plant. They showed an association between fitness traits such as fecundity and genetic variation, and demonstrated that this association can be stronger than between fitness and population size alone.

Uniqueness: Geographic Distribution and Differentiation

Populations can be defined geographically, but this may not be meaningful from an ecological and evolutionary perspective. Specifically, spatial boundaries may not reflect true connectivity among interacting subunits. In conservation genetics, important groups are often defined as "evolutionary units" (i.e., those with independent evolutionary dynamics) (Luck et al. 2003). Because there has been limited gene flow between these groups for some time, they are likely to have independent evolutionary trajectories when responding to selection pressures. Although gene flow can prevent the development of measurable genetic differences among groups, its effects can be countered by strong selection that can maintain differences at specific loci and for specific traits.

Populations can be defined geographically, but this may not be meaningful from an ecological and evolutionary perspective.

Metapopulation ecology describes population interactions and dynamics associated with local dispersal patterns, which can sometimes be defined genetically (Hanski and Gilpin 1997). Landscape genetics builds broadly on this concept, describing patterns of gene flow and local adaptation with special reference to the influence on population structure of landscape structures (i.e., environmental factors and landscape features). Recently developed geospatial statistical techniques provide powerful tools to resolve "genetic discontinuities" that signal interruptions in gene flow, and can sometimes be best understood in the context of spatially structured information about the environment (i.e., geographic information systems (GIS)). To do this, Manel et al. (2003) recommend using a multiple genetic marker strategy within a population genetic framework to infer interpopulation structure (e.g., distance-isolated and metapopulations). This strategy involves analysis of neutral markers to describe population genetic structure as well as adaptive markers to analyze factors that shape populations and their evolution. Examples of how this strategy has been applied on very large geographic scales include interpretations of the origins of cattle in Africa (Hanotte et al. 2002) and the evolution of pesticide resistance and population structure in mosquitoes (Chevillon et al. 1999).

Historical demography has been described as a means to estimate population parameters such as genetic diversity, divergence times, growth rates, and gene flow between populations through the use of unlinked loci (reviewed in Brumfield et al. 2003). By analogy to spatial analysis, estimation of population history using DNA sequence information provides an opportunity to infer demographic history and reveal temporal patterns of population size, growth rate, migration, and subdivision (reviewed in Hare 2001; Emerson et al. 2001; Brumfield et al. 2003). Coalescent theory provides a theoretical foundation for this approach: the probability of common lineage (as revealed through sequence diversity among individuals) is a function of population abundance (assuming neutral evolution, no recombination, random mating, and random sampling). Therefore, changes in direction and rate of population size can be estimated by analysis of genetic data. Although this approach provides great promise for understanding the impacts of anthropogenic stressors on populations, assumptions and analyses required to assign temporal patterns to historical patterns remain controversial (e.g., Lubick 2003 with respect to Roman and Palumbi 2003).

ADAPTIVE GENETIC VARIATION AND FITNESS

Understanding adaptive shifts in response to toxicants at the genetic level requires information about fitness, the genes underlying adaptive shifts, and the impact of population processes, such as gene flow, on the rate of adaptation. In addition to understanding adaptation, an important issue in the context of contaminated environments is how often (and why) adaptation does not occur. Adaptive variation is discussed here with relevance to the prediction of population vulnerabilities and persistence in ecological risk assessment.

ESTIMATING FITNESS

From a population genetics perspective, relative fitness is the viability and reproductive output of one genotype relative to other genotypes under specific conditions, associated with changes in the frequency of genotypes over generations. This fitness concept is different than the one used in risk assessment, where it is defined in terms of the growth and persistence of a population. Often, researchers investigate components of fitness and then use the results to compute overall fitness. For instance, Sayyed and Wright (2001) measured the fitness of genes encoding resistance to *Bacillus thuringiensis* toxins in a population of diamondback moth, *Plutella xylostella*. They measured growth rate of larvae, fecundity of paired males and females, and viability of the eggs laid. These measures were then converted to overall fitness measures by computing the net replacement rate (R_0) as the average number of female progeny per female, and the intrinsic rate of increase of the population (r_m) as natural log (R_0/development time).

From a population genetics perspective, relative fitness is the viability and reproductive output of 1 genotype relative to other genotypes under specific conditions, associated with changes in the frequency of genotypes over generations. This fitness concept is different than the one used in risk assessment, where it is defined in terms of the growth and persistence of a population.

In field populations, fitness is also measured by monitoring changes in the frequency of genes under varying conditions. A well-documented example is the evolution of resistance to organophosphate pesticides in *Culex pipiens* mosquitoes. Three loci control resistance, 2 coding for detoxification esterases, and 1 coding for an altered form of acetylcholinesterase, the target of organophosphate inhibition. In France, 4 resistant alleles at these loci have appeared sequentially since 1972. To characterize fitness of these mutant alleles, Chevillon et al. (1997) followed survival of adult females, and compared the composition of the population before and after overwintering. Whereas the first insecticide resistance allele to arise was deleterious in some conditions, and the third allele had a survival effect in all environments, the allele that conferred the most insecticide tolerance did not have any adverse survival effects. Therefore, sequential replacement of the alleles was inferred to be related to their costs in the absence of the pesticide.

To assess adaptive costs, fitness is normally measured in the presence of stress and in its absence. Fitness studies often indicate that individual fitness can depend subtly on environmental conditions. In the pesticide resistance literature, there are numerous examples of laboratory studies that fail to demonstrate fitness costs associated with resistance genes (Roush and McKenzie 1987). However, under field conditions associated with stressful periods, costs of resistance can appear. The mosquito example discussed previously, in which a sequence of different alleles for organophosphate resistance has been favored at least partly because of differential costs, also illustrates the effect of costs on genetic changes in populations. Costs associated with pesticide resistance genes may depend on the nature of the mechanism underlying resistance. Taylor and Feyereisen (1996) have suggested ways in which specific mechanisms of resistance are associated with different costs and provide supporting evidence. For instance, high costs are expected where structural mutations result in increased output from detoxification systems. Knowledge of molecular mechanisms can be used to hypothesize about costs associated with resistance to toxicants (Coustau et al. 2000).

Establishing Causal Connections between Fitness Effects and Genes

It can be difficult to associate fitness effects with a specific gene. For instance, one of the most extensive data sets on fitness costs concerns the responses of plants to heavy metals. Numerous plants have adapted to heavy metals that are found in high concentrations in mine tailings. Both the low incidence of metal resistance in unselected populations and the steep clines in tolerance at boundaries between tailing areas and uncontaminated areas argue for costs to tolerance (Shaw 1999). However, selection experiments in the plant *Mimulus guttatus* suggest that there are no physiological costs associated with resistance (Harper et al. 1997). Other factors may account for the poor performance of plants from tailings in other habitats. For instance, mine tailings are low in nutrients, which can lead to slow growth rates being favored independently of metal presence. There is also only limited evidence for tradeoffs in other cases of metal adaptation being specifically associated with the genes responsible for contaminant responses (van Straalen and Hoffmann 2001). Some of the strongest empirical examples of gene-fitness linkages have resulted from research based on an understanding of probable selective agents, and likely mechanisms of biological response. Classic examples include lactate dehydrogenase (LDH) variation and temperature response in the estuarine fish *Fundulus heteroclitus* (reviewed in Mitton 1997), and major histocompatibility complex (MHC) variation and parasitism in Soay sheep (Coltman et al. 1999, reviewed in O'Brien 2000).

When mechanistic relationships between genes, stressors, and responses are unknown, it can be difficult to make connections between a particular gene and an adaptive response. One problem is that genetic differences between the populations have to be strong enough to implicate selection. Differences between groups can arise because of the effects of other evolutionary processes such as genetic drift, or local adaptations to factors other than the stressor of concern. This problem can be tackled by considering variation at a number of neutral loci, such as those coding for allozymes (which are not always neutral) or for microsatellites. An example of this approach involving a quantitative trait is a study by Gockel et al. (2001) examining clinal variation in microsatellite variation in *Drosophila melanogaster*, and comparing this pattern with variation in body size. Because the clinal variation in size was much steeper than that in microsatellite variation, there was evidence for the cline in the quantitative trait being maintained by selection. An example involving a specific genetic polymorphism is selection on mannose-6-phosphate isomerase in the barnacle *Semibalanus balanoides* (Schmidt and Rand 1999). Genotype frequencies at this locus are associated with temperature and desiccation stress, whereas no such patterns exist for other loci, suggesting that selection acts on or near the mannose-6-phosphate locus. This approach can be valuable as long as a number of neutral loci are considered. Moreover, monitoring variation at a number of loci can also provide information about historical processes associated with contaminated sites, such as patterns of gene flow and historical patterns. Theodorakis (2003) has identified some suggested criteria to support the interpretation of causal relationships between genetic effects and environmental contamination.

A combination of laboratory and field studies can often indicate potential candidate loci. For instance, Heagler et al. (1993) showed that in mosquito fish the time to death under mercury exposure was related to genotypes at a glucose phosphate isomerase locus, and also related this finding to the distribution of genotypes in a contaminated environment. This suggests but does not prove that the locus (or another closely linked locus) is a candidate for tolerance of mercury. However, a series of studies, which included long-term mesocosm experiments on polymorphic populations, has also implicated the same locus in response to mercury and other toxicants (e.g., Mulvey et al. 1995; Tatara et al. 1999; Tartara et al. 2002). Ultimately, some type of genetic manipulation is needed to directly establish a causal relationship between a locus and a selection response. An example of this approach is provided by the work of Feder and Krebs (1997) on one of the heat shock proteins (hsp70) in *Drosophila melanogaster*. By using an engineered strain of flies with extra copies of the hsp70 gene, these authors were able to demonstrate that there was an increase in larval heat resistance associated with this gene as well as a cost in terms of an increase in development time and decrease in viability. These patterns matched natural variation in the activity of hsp70 (Krebs et al. 1998).

Predicting Adaptive Shifts

How often does adaptation occur, what limits its occurrence, and can it be predicted? In the case of pesticides, it has been possible to use mutagenesis under laboratory conditions to predict the likely evolution of resistance. The approach, outlined in McKenzie and Batterham (1998), is illustrated with respect to the evolution of pesticide resistance in the sheep blowfly, *Lucilia cuprina*. The genetic basis of field resistance to dieldrin and diazinon has been determined in detail in *L. cuprina*. Resistance to dieldrin involves an allelic substitution at the *Rdl* locus, whereas resistance to diazinon involves a substitution at the *Rop-1* locus, which codes for a carboxylesterase. In the laboratory, chemical mutagenesis was used to generate strains resistant to dieldrin and diazinon. In these strains, resistance mapped to the same locus as in the field and even produced the same amino acid change of the encoded protein. The evolution of resistance to these chemicals was therefore highly predictable from laboratory studies. The mutagenesis approach was also used to generate strains resistant to the insect growth regulator cyromazine, even though *L. cuprina* had not developed resistance to this chemical in the field. Strains with only a low level of resistance to this chemical were generated by mutagenesis and viability of these strains was low, helping to explain the absence of resistance evolution in the field (Yen et al. 1996).

It is also possible to predict evolutionary change by screening for resistance in the field before selection starts. Toxins produced by *Bacillus thuringiensis* (Bt) are being used in transgenic plants to generate tolerance to insect pests, but resistance to these toxins is starting to appear in the target insects. In the poplar beetle *Chrysomela tremulae* (Coleoptera), alleles for resistance to one *Bacillus thuringiensis* toxin (Cry3A) were shown to exist at a low frequency (0.4%) in natural populations, despite the absence of transgenic plants carrying the toxin (Ginissel et al. 2003). Resistance could therefore develop fairly rapidly in these populations and would be maintained (presumably) by mutation-selection balance. Mutation-selection balance probably also controls the incidence of resistance to environmental toxicants. Garcia-Villada et al. (2002) examined evolutionary changes in a microalga exposed to the contaminant 2,4,6-trinitrotoluene (TNT). They found TNT resistant variants segregating at a low frequency in the population before selection. The resistant variants had a low fitness in the absence of TNT, including a low photosynthetic rate and low competitive ability, suggesting that resistant mutations were being maintained in the base population at a very low frequency by mutation selection balance. The TNT stress acted to increase the frequency of the resistant variants, serving as an efficient selection mechanism.

This approach is likely to be successful when resistance is associated with single genes, but becomes more complicated when multiple genes are involved in responses to toxicants. When genes underlying a trait are unknown, the heritability of a trait can still be assessed, which provides a measure of the proportion of phenotypic variation in a trait that is genetic (not the genetic basis of a trait). Heritabilities are normally computed for traits that vary continuously. Posthuma et al. (1993) provide an example for metal resistance in the springtail *Orchesella cincta*. They investigated the heritability of cadmium excretion in individuals in 2 populations, one from a contaminated site and the other a reference site. The heritability for this trait based on parent-offspring comparisons was 33% (with a standard error of 0.10) in the reference site, but heritable variation was not evident in the contaminated site. This may be the result of selection for tolerance to cadmium decreasing genetic variation in the contaminated site. This type of approach is likely to be particularly applicable to cases in which variation in responses to pollutants involves changes in life history traits to allow organisms to evade stresses, as in the case of *Chironomus riparius*, which showed changes in development time and hatchability in response to cadmium stress (Postma et al. 1995). Heritabilities can also be computed when traits are treated as thresholds. This is often carried out for pesticide resistance and can be applied to studies on any toxicant (Firko and Hayes 1990).

Heritabilities depend on environmental conditions. Although estimates for some morphological traits are often similar across environments (Roff 1997), estimates for other traits can vary markedly between the field and laboratory environments and stressors can directly influence the expression of genetic and environmental variation. As reviewed in Hoffmann and Merila (1999), there are good cases in which genetic variance appears to be increased under stressful conditions, as in the expression of genetic variation in bristle number under temperature extremes in *Drosophila*. However, in other cases the opposite trend is apparent. For instance, in many bird species, there is a decrease in genetic variance and an increase in environmental variance for size related traits under nutrition stress. These inconsistent trends are unsurprising because they likely reflect the many ways that unfavorable conditions can interact with genetic variation. Hoffmann and Merila (1999) list a number of hypotheses about how environmental conditions can influence the heritability and other measures of the ability of traits to evolve. These include changes in levels of genetic variance due to effects on recombination and mutation, changes in levels of environmental variance, different histories of selection in environments, and effects of environmental conditions on phenotypic variance as conditions become limiting. As a result, heritable variation ideally needs to be investigated under conditions in which selection is imposed. As our understanding of candidate genes underlying variation in traits improves, it should be possible to examine the genetic markers that underlie quantitative traits rather than relying on changes in the traits. Variation in these markers can then be linked to selective effects of contaminants or other stressors.

GENETIC VARIATION AND RISKS FROM CHEMICAL EXPOSURES

Evolution, including recent or "contemporary" evolution (Stockwell et al. 2003), is influenced by interacting factors including genetic effective population size, gene flow, genetic variation, and the strength of selection. Chemical exposures can affect each and all of these factors. Chemical exposures result in increased morbidity and mortality, with varying genetic outcomes and population effects. As an evolutionary agent, chemical stressors can significantly reduce local population size, with or without selection on adaptive traits. In addition, chemical pollution may also affect gene flow, reinforcing the reproductive isolation of exposed populations. Although a review of published literature suggests that the effect of environmental pollutants on genetic diversity (e.g., genetic erosion) is variable (van Straalen and Timmermans 2002), a better understanding of the relationship between chemical stress, changes in genetic variation, and population persistence could be of great benefit to population-level ecological risk assessment.

Evolution, including recent or "contemporary" evolution (Stockwell et al. 2003), is influenced by interacting factors including genetic effective population size, gene flow, genetic variation, and the strength of selection. Chemical exposures can affect each and all of these factors.

In general, certain characteristics of both stressors and candidate species may predispose certain evolutionary consequences: a combination of stress severity, duration, and consistency dictates population response. A stressor is predicted to act as a directional selective evolutionary force when its action is long relative to generation time and large in geographic extent relative to population location. These conditions occur in environments that receive continuing toxic chemical input or are characterized by environmentally stable toxic chemical reservoirs (e.g., contaminated sediments). Chemicals with the potential to act in this way because of their long range of action include those identified as persistent, bioaccumulative, and toxic. However, chemical stressors with unknown mechanisms or those of special concern, including endocrine disruptors, may also act as selective agents. Genotoxic chemicals require special consideration, in that by altering mutation rate, they directly influence evolutionary rate. Rates of adaptive evolution could be increased if the mutations have favorable rather than deleterious effects on fitness (Shaw et al. 2002). An increase in mutation rate can also lead to an increase in the genetic load of populations. For instance, in the fern *Onoclea sensibilis*, aberrant gametophytes were substantially more common in environments that were more likely to be polluted (Schoen et al. 2002). This effect could decrease the rate of evolutionary change. Lynch et al. (1995) estimated that mutation rate (and genetic loading) can have a significant effect on population extinction for small populations.

Although a review of published literature suggests that the effect of environmental pollutants on genetic diversity (e.g., genetic erosion) is variable (van Straalen and Timmermans 2002), a better understanding of the relationship between chemical stress, changes in genetic variation, and population persistence could be of great benefit to population-level ecological risk assessment.

Even less predictable are population effects produced by more variable chemicals exposures (i.e., patchy environments and mixtures of toxicologically disparate chemicals co-occurring with other nonchemical stressors). When combinations of chemicals occur and different mechanisms are required for tolerance, adaptive shifts will occur more slowly because individuals with the right combination of alleles may be rare in populations. Kovatch et al. (2000) suggested that the complex of contaminants in sediments may have accounted for the absence of adaptation in populations of the copepod, *Microarthridion littorale*, although this species is present at large population sizes and individuals with resistance to multiple stressors might occur if there are positive genetic correlations between resistances to different stresses.

In addition to direct effects of chemical contamination, chemically contaminated landscapes may also affect rates of adaptation indirectly by altering immigration and gene flow. If levels of gene flow are too high, adaptive shifts can be prevented because favored genes are swamped by an influx of genes that do not increase fitness. Gene flow is sometimes invoked to explain the lack of adaptation to contaminants (Groenendijk et al. 2002), but direct estimates of gene flow are usually lacking. There are cases in the literature demonstrating the impact of gene flow in limiting local adaptation. For instance, in the spider *Agelenopsis aperta* there are distinct behavioral morphs associated with arid and riparian habitats that differ in levels of aggression, and the distribution of these genetically based morphs is dictated by patterns of gene flow between the 2 habitats (Riechert and Hall 2000).

Selection and Adaptation to Chemical Exposures

For chemicals intended (i.e., pesticides) and unintended (e.g., manufacturing byproducts including dioxins) to produce adverse biological effects, selection can be very strong relative to other environmental factors acting as evolutionary agents (reviewed in Kingsolver 2001). Rapid and intense selection can lead to the development of evolved resistance in a few generations, as long as alleles encoding resistance are segregating in populations at an appreciable frequency before selection (i.e., ample genetic variation at specific target loci). These adaptive responses are often based on changes in one or few genes (Macnair 1991).

For chemicals intended (i.e., pesticides) and unintended (e.g., manufacturing byproducts including dioxins) to produce adverse biological effects, selection can be very strong relative to other environmental factors acting as evolutionary agents (reviewed in Kingsolver 2001). Rapid and intense selection can lead to the development of evolved resistance in a few generations, as long as alleles encoding resistance are segregating in populations at an appreciable frequency before selection.

Target or candidate genes of chemical adaptation include those whose functions involve toxicological or defense mechanisms. Adaptive chemical variation often results in downregulation of toxicological pathways or upregulation of defense pathways (Taylor and Feyereisen 1996). While few of the biochemical or genetic mechanisms of inherited chemical tolerance are known, several mechanisms of organophosphate pesticide resistance in insects have been characterized because of intensive effort in this area (reflecting the human health implications associated with evolved pesticide resistance). Several biochemical mechanisms of adaptation to metals in animals and plants have been reported, often associated with upregulation of metal defense pathway via the sequestration protein, metallothionein (Shaw 1999). In vertebrate species, investigations have focused on genes of the aryl-hydrocarbon receptor pathway and their potential roles in tolerance to highly toxic dioxin-like compounds (Hahn 1998; Hahn et al. 2004). However, neither the specific biochemical nor genetic mechanisms are understood currently (although see the following section for more detailed discussion).

It is sometimes assumed that there is an association between the rapidity of a response and the number of loci controlling a trait. However, for quantitative traits controlled by many genes with small effects, responses to selection can also be rapid and dramatic. This is illustrated in the many artificial selection experiments that have been undertaken for quantitative traits and generated substantial and rapid responses (Falconer 1989). For stress resistance, these experiments can result in laboratory strains whose responses fall well outside the range of biological response in unaffected populations.

Evidence of Chemical Adaptation

Klerks and Weis (1987) reviewed the early literature on genetic changes in response to heavy metals in aquatic organisms. They cite numerous studies in a variety of organisms demonstrating differences between polluted and relatively clean sites. However, in many instances, including most studies with metazoa, they found that it was difficult to distinguish between genetic adaptation and individual acclimation not involving genetic changes. Unequivocal data for genetically based differences between sites were only obtained in a few cases. Even in these instances, comparisons were normally confined to 2 habitats, and differences might result from random drift rather than selection. Klerks and Weis (1987) also reviewed experiments where selection was undertaken to increase heavy metal resistance. These studies show increases in resistance in several experiments with algae and fungi, but the invertebrate and vertebrate data were limited. Even more recent examples, reviewed in Grant (2002) and Klerks (2002), suggest that evolutionary changes leading to resistance of toxicants might be limited to a few cases.

However, the observation of adaptation to pollutants in wild populations may be limited by available means for detection. Specifically, heritability studies require very large sample sizes if standard errors are to be small. For instance, when 1-parent 1-offspring comparisons are based on 400 families, the standard error of the heritability estimate will be around 0.1 (Falconer 1989). It is possible to reduce the standard errors by using other designs such as a half-sib analysis in which several females are each mated to a series of males, and by collecting data across several generations and applying models to estimate heritable variation. However, these types of designs are often difficult to implement unless the organism can be easily reared in the laboratory. Another option is to artificially select for traits under controlled conditions for a number of generations and compute heritabilities from the response to selection. This can lead to precise estimates based on values from several generations, but ideally replicate lines need to be selected and population sizes during selection need to be fairly large. For instance, Hoffmann et al. (2003) demonstrated a lack of response to selection in replicate populations of a fly exposed to 30 generations of selection with replicate lines maintained at 100 individuals during selection. Findings were further supported by parent offspring comparisons with around 100 families. Heritabilities for toxic responses may be low as recently suggested for sheepshead minnows responding to zinc and phenanthrene contaminants as well as complex mixtures (Klerks and Moreau 2001). However in this study, parent-offspring comparisons were based on only 20 to 32 families and the large standard errors of the heritability estimates meant that even moderate heritabilities would not have been detected.

Because evidence for adaptation is largely restricted to heavily contaminated areas where the biota are already impoverished (although see Nacci et al. 2002a describing occurrence at less contaminated sites), Grant (2002) has argued that evidence for adaptation has a limited role to play as a monitoring tool. Instead, he argues that community level monitoring (i.e., of species assemblages) is likely to be more productive. In fact, community monitoring will only work if there is limited adaptation; otherwise, changes in the distribution of specific species cannot be used to infer increased levels of contamination. However, tests of adaptation that have been undertaken so far are fairly crude, and involve tests of whether populations can persist at polluted sites, or whether populations from polluted sites have relatively higher fitnesses when they are compared with populations from uncontaminated sites. Specifically, tests of adaptation have mostly involved differential survival under contaminated conditions, whereas subtle life history changes may be relatively more common. Such changes may be difficult to detect and more difficult to ascribe to specific stressor conditions. The identification of target genes (i.e., involved in pollution responses and subject to selection) could provide a sensitive, diagnostic method of detecting adaptive shifts.

Two relatively well-documented examples provide evidence of adaptation and suggest genetic and biochemical mechanisms in an aquatic invertebrate and vertebrate species. Specifically, evolved resistance to heavy metals was demonstrated in populations of the oligochaete *Limnodrilus hoffmeisteri* resident to a site highly contaminated with metals (reviewed in Levinton et al. 1999). Interestingly, tolerant and sensitive populations of this species reside in patches, consistent with the spatial heterogeneity of contamination. The biochemical mechanism of tolerance results in lowered bioaccumulation of toxic metals, and is probably accomplished through a few genetic loci.

In the other example, *Fundulus heteroclitus* a nonmigratory fish resident to estuaries of varying quality along the East Coast of the United States, demonstrates evolved tolerance to local pollutants that act toxicologically through the aryl-hydrocarbon receptor (AHR). Chemical pollutants that act through this mechanism include highly toxic dioxin, as well as some polyaromatic hydrocarbons, and polychlorinated biphenyls (PCBs). One particularly well-studied population resides in New Bedford Harbor (NBH), Massachusetts, USA. These fish demonstrate inherited tolerance to the toxic effects of local contaminants (polychlorinated biphenyls) that is profoundly different from populations outside of NBH (Nacci et al. 1999). The occurrence and maintenance of these adaptive phenotypes suggests that strong divergence has occurred in response to recent and intense chemical contamination, despite potentially high gene flow. These results suggest that strong directional selection for chemical tolerance may have caused a genome-wide reduction in genetic diversity due to a population bottleneck, or localized reductions in genetic diversity at specific chromosomal regions (reviewed in Nacci et al. 2002b).

To test this idea, researchers used multiple, complementary techniques to characterize the genetic structure of *F. heteroclitus* populations residing in and around NBH: AFLPs, a multilocus DNA fingerprinting method, allozymes, and SNPs at target loci. On a genome-wide basis, genetic diversity in NBH *F. heteroclitus* was only slightly reduced at allozyme loci and was not significantly lower based on AFLP analysis (Roark et al. 2004; McMillan et al. 2006). SNPs of the major histocompatibility complex (MHC) antigen-binding loci were used to test whether intense selection for chemical tolerance has resulted in reduced genetic diversity at specific loci, which might have implications related to pathogen and/or parasite susceptibilities. Although genetic diversity at major histocompatibility complex loci was high, sequence patterns were unique in NBH versus *F. heteroclitus* populations indigenous to a noncontaminated site (Cohen 2002). Laboratory bacterial challenge studies were conducted to evaluate the functional significance of this change, but demonstrated minimal differences between populations (Cohen et al. 2006; Nacci et al., manuscript in preparation).

Candidate genes for adaptation in NBH fish also include those specifically associated with toxicological responses (i.e., those of the AHR signal transduction pathway) (reviewed in Hahn 1998; VanVeld and Nacci in press). For example, in this species, AHR1 is highly polymorphic in loci of the ligand-binding regions (Hahn et al. 2004). The adaptive implications of these genetic changes have been investigated using in vitro transcription and translation systems. However, simple and clear differences between tolerant and sensitive fish populations have not been revealed (Hahn et al. 2004). For example, although regions of AHR1 vary between NBH and reference fish, these variants do not differ in binding capacity, affinity for dioxin, or ability to support dioxin-dependent transactivation (Hahn et al. 2004). Ongoing studies are investigating genetic variation and their functional implications in other regions of the AHR1 (i.e., regulatory region; S. Cohen, personal communication), and other proteins of the AHR signaling pathway (M. Hahn, personal communication).

GENETIC CONTRIBUTIONS TO POPULATION-LEVEL ERA

In summary, a varied body of scientific literature suggests that genetics can inform population-level ERAs with respect to issues of population condition, vulnerabilities, and persistence, and has the potential to contribute uniquely and effectively to empirical and modeling approaches. In addition, biomonitoring efforts provide an opportunity to acquire molecular genetic information on a relatively large spatial scale, and might be useful for development, standardization, and verification steps that will be necessary as molecular genetic methods are applied more broadly.

Genetics can inform population-level ecological risk assessment with respect to issues of population condition, vulnerabilities, and persistence, and has the potential to contribute uniquely and effectively to empirical and modeling approaches.

Genetic Contributions to Empirical Assessments

Genetic information can contribute to empirical assessments by improving definition of the current condition of the population and detecting causes of impairment (see Chapter 8). Fundamental to the former task is the interpretation of neutral and adaptive genetic variation to characterize aspects of population condition. Factors contributing to strategies for marker selection include availability (i.e., species-specific development), ease of interpretation, and expectations of selection. Theoretical frameworks and statistical analytical methods are well-developed for the classic biochemical markers, but will require further development for anonymous molecular markers such as AFLPs, microsatellites, and SNPs. Similarly, specific theoretical and analytical frameworks are proposed or in use that employ genetic information to define various population attributes: population size and richness (conservation genetics), and spatial boundaries and intraspecific interactions (meta-population and landscape genetics), and to infer population histories (historical demography). Although a unified framework might encourage a broader implementation of genetics to field assessments, the goals of the assessment may be satisfied with more focused approaches (i.e., historical analysis may not always be relevant). Theoretically and experimentally derived predictions and examples relating genetic information and population condition and history are useful but limited (e.g., landscape genetics applications, Manel et al. 2003; contaminated sediment effects, Mulvey et al. 2003). In addition, more general issues include

- Can relationships between genetically effective population size, N_e, and census size be predicted across species and conditions? How does the magnitude and duration of genetic diversity reduction affect population vulnerability?
- Can relationships be developed for relative changes in genetic diversity and fitness across species and conditions? How do the magnitude and frequency of population size variations affect population vulnerability to genetic diversity loss?
- Can contemporary and historical population effects be distinguished, for example, through complementary information on processes with varying temporal scales, such as migration and gene flow?

Knowledge of functional variation in genes associated with specific functions linked to fitness or toxic chemical response is useful for diagnosing causative stressors, but examples are limited. Although there are few candidate genes sufficiently characterized in wildlife species to fulfill this role, rapidly advancing genomic technologies hold great promise. Currently, phenotypic measurements of performance or chemical tolerance are being used to identify stressor-adapted populations while genetic mechanisms are under investigation. However, the relevance of adaptation to understanding population and ecosystem condition requires further consideration. Questions concerning the value of adaptation in population-level ERAs include

- Is stressor-specific adaptation (i.e., inherited phenotype) an efficacious way to confirm a stressor as the causative agent of evolutionary change?
- Does the occurrence of adapted species provide a diagnostic signature of a specific stressor as the selective agent?
- Is adaptation as sensitive an indicator of historical stress as is loss in species diversity?

Knowledge of functional variation in genes associated with specific functions linked to fitness or toxic chemical response is useful for diagnosing causative stressors, but examples are limited.

Genetic Contributions to Modeling Assessments

There are few field examples of applications in which genetic information has been integrated into modeling assessments to predict effects of stressors on population dynamics (see Chapter 9). In fact, the integration of population genetics and dynamics would be an important advancement in this area. Although the use of genetic information to define population structure has been explored, linking genetic variation and population fitness is less well developed, reflecting limited information on gene function. However, there are some examples from conservation genetics where the implications of changes in genetic variation on population viability are explored (Reed and Frankham 2003). Modeling mechanisms to incorporate genetic diversity effects have also been demonstrated. For example, inbreeding depression can be included in demographic model projections (Brook et al. 2002). Other approaches with restricted, but potentially useful, application include metapopulation (Hanski and Gilpin 1997) and individual-based modeling (Lomnicki 1988) approaches. Recently, 2 genetically explicit individual-based simulation approaches have been developed with the capacity of predicting genome-wide changes in population genetic structure in response to a spatially and temporally varying environment (Strand 2002; Topping et al. 2003b). Spatially explicit approaches could be integrated into a larger landscape genetics framework. However, the value of specific genetic information in predicting population viability is limited by our knowledge of relationships between genotype and performance at the organism and population aggregate levels. The need to include potential effects on population genetics of stressors has also received little attention. Stressor factors that may be important in evaluating the probability of genetic effects include intensity, duration, multiplicity, spatial extent, and heterogeneity. Despite predictions, a limited number of studies have identified field evidence of genetic effects. Specifically, reduced genetic diversity is not often found, even in extremely degraded sites. These findings may reflect the choice of species for study, and consideration of how attributes like population size should contribute to expectations. The choice of species should reflect the goals of the assessment (i.e., whether the selected species is one of special concern or a representative of ecosystem health). More specific questions include

- How do we best measure "ecologically important" genetic variation, and then translate changes in variation into changes in population persistence? Are those genes whose expression changes upon stressor exposure the targets of selection? Can unique gene patterns be identified that serve as signatures of specific environmental stressors?
- Can we develop useful rules predicting the nature of genetic effects of stressors on populations? For example, is biochemical mechanism of action (i.e., specific versus general target) predictive of rapid adaptation? Are long-term costs of adaptation predictable based on adaptive mechanism (e.g., upregulation of detoxification system)?
- Is genetic diversity loss only of concern for small populations? Or is vulnerability associated with factors including reproductive strategies, generation time, and population size? Is there a predictable relationship between genetic diversity and fitness across species or categories of species that can be included in population modeling projections?
- How can genetic variation among individuals inform us relative to past and current population structure, connectivity, and dynamics? Can we incorporate genetics into models predicting effects of spatial heterogeneity of stressors on gene flow and metapopulation dynamics?

FUTURE RESEARCH NEEDS AND CHALLENGES

Genetic information has important potential to contribute uniquely and effectively to population-level ERAs. Technical limitations associated with obtaining genetic information have been greatly diminished by rapidly developing molecular methods, and the greatest challenges now involve interpreting genetic information. To this end, we define some important research areas where advancements will greatly facilitate the realization of the potential importance of genetic information to population-level ERAs:

- Molecular markers are increasingly accessible because of their short development time, their transferability, and the increased availability of specialized equipment and trained technical personnel. Their application will be facilitated by a theoretical framework and quantitative methods to analyze these new marker types. The application of molecular methods to reveal target gene complexes and population processes will improve diagnostic capabilities for unique stressors.
- Improved understanding of toxicological mechanisms will provide a sound basis for species extrapolation and identification of vulnerable categories of species. Better understanding of stressor characteristics may further specify those likely to produce genetic effects. However, factors related to exposure may also contribute to potential for genetic effects (e.g., mixtures of toxicologically disparate chemicals, cooccurrence with other stressors, duration of exposure and effects via continuing deposition, environmental persistence and biomagnification).
- Theoretical and experimental information using model species provides important advances but species biology must also be considered when predicting population effects. Field and experimental studies should also include species ranging in life history strategies (demographic categories), including those less easily studied (e.g., those with long generation times).
- Platforms for integrating population dynamics and genetics will be important for predicting and testing implications of genetic changes. Current population modeling software seldom account for genetics, excepting VORTEX (http://pw1.netcom.com/~rlacy/vortex.html), which incorporates inbreeding effects on fitness. Various strategies could incorporate genetic variation in neutral markers and markers under selection at the level of the organism, population, and metapopulation (see Chapter 9). A variety of modeling approaches probably are needed, and all require testing.
- Landscape genetics and recently developed geospatial statistical techniques are being applied to define population and metapopulation dynamics, infer demographic history and reveal temporal patterns of population size, growth rate, migration, and subdivision (reviewed in Emerson et al. 2001; Hare 2001; Brumfield et al. 2003; Manel et al. 2003). Theoretical and practical advances are needed to address temporal scale issues, and improve ecological interpretation (e.g., to separate recent genetic change from historical factors).

CONCLUSIONS

Genetics can inform population-level ERAs with respect to issues of population condition, vulnerabilities, and persistence. For example, genetic information can improve characterization of population structure and health, prediction of risks to population viability, and understanding of (long-term) effects of stressors, including chemical pollutants. More specifically, genetic information can be used to infer metapopulation constructs, and structural features that can implicate interactive dynamics (e.g., source-sink relationships). When interpreted with respect to environmental features and stressors on large and small spatial scales, such a "landscape genetics framework" could be extremely useful in revealing how multiple environmental and biological factors can interact to affect populations. Similarly, as mechanisms to interpret genetic histories are improved, environmental changes and population histories may be correlated to reveal patterns of contemporary evolution.

Conservation genetics reflects the recognition that inclusion of genetic variation to describe population condition is critical for small, threatened populations faced with significant genetic erosion. However, reductions in genetic variation may also contribute to population trajectories for populations that are not critically small in size. Improved understanding of gene function will have enormous impacts as genomics provides the mechanism to link target genes, complex genotypes and performance and fitness traits. Finally, genetic information is providing an improved understanding of the nature and magnitude of stressor effects on populations worldwide. Adapted populations serve as diagnostic indicators that chemicals and other stressors are acting as strong selective agents, and provide documentation of biodiversity effects on a local scale.

Research integrating genetics, demography and toxicology can contribute to predictions of vulnerable species and populations within a context of improved understanding of the underlying molecular mechanisms of toxicity and tolerance. To make advances in all of these areas, it will be necessary to collect genetic information synoptically with other measures, extend theoretical frameworks incorporating genetics, and adapt analytical procedures to address these new data. Studies in the laboratory and in the field can provide experimental and long-term observations to test predictions linking genetics and population persistence.

Research integrating genetics, demography, and toxicology can contribute to predictions of vulnerable species and populations within a context of improved understanding of the underlying molecular mechanisms of toxicity and tolerance.

6 The Spatial Structure of Populations and Ecological Risk Assessment

Wayne G. Landis and Andrew Deines

INTRODUCTION

Other chapters in this book deal with the density-dependent regulation of populations (Chapter 4) and population genetics (Chapter 5). However, if we are to perform ecological risk assessment on populations, we need to understand the other basic parameters of these ecological units. Populations do not exist as 2-dimensional graphs, but reside within a complex 3-dimensional heterogeneous and dynamic landscape. Species differentiate the use of this landscape spatially and temporally. The distribution and utilization of this landscape by species has been one of the fundamental forces driving ecological and evolutionary thought.

A SHORT HISTORY

Biogeography is one of the oldest areas in ecology and evolutionary biology. Darwin (1859, reprinted in 1998) spent 2 chapters (XII and XIII) discussing the importance of barriers, means of dispersal, colonization, mainland-island relationships, and the inhabitants of oceanic islands in understanding patterns of species distributions and speciation. Often overlooked, Wallace (1867, 1881) also published extensively on the patterns of distribution of animals and plants, speciation and extinction that are found on islands. MacArthur and Wilson's (1967) classic treatise, *The Theory of Island Biogeography,* set the stage for the quantification of ecology and the modeling of patterns of species in patchy environments. It became clear that biogeography was fundamental to understanding the spatial patterns of populations and species as well as in the evaluation of risks.

In the 1990s, a shift in conceptualization occurred with the recognition of metapopulation biology as a general model for the distribution of species in space (Hanski and Simberloff 1997; Hanski 1999). Part of this shift was due to the realization that many species exhibited a patchy spatial structure and that the theory and data derived from metapopulation investigation could be useful in conservation biology and other applied areas.

As fundamental as biogeography is to the understanding of populations, it is rarely incorporated into ecological risk assessments. Although landscape considerations are discussed in the latest US Environmental Protection Agency guidance (1998) they have proven difficult to implement. Part of the issue is that the fundamental science of understanding the dynamics of populations within landscapes is still under development. The effects of toxicants, barriers to migration, and changes in the patch structure of the landscape are additional factors that must be incorporated into the framework so that space and time can effectively be incorporated into risk assessments. Fortunately, there have been developments that make the application of the landscape features of population in ecological risk assessment become possible.

Introduction to This Chapter

The purpose of this chapter is to set the stage for the incorporation of landscapes and patch dynamics into the risk assessment of populations. First, we list the spatial considerations that should be considered for a population. Second, we briefly review the status of incorporating stressors and especially toxicants in patch and landscape dynamics using McLaughlin and Landis (2000) as a summary and a starting point. Next, we try to describe the potential types of effects that stressors have on populations in a landscape context using current theory and data. Finally, we discuss some of the implications of looking at the landscape scale of population dynamics, including the paradox of the existence of a reference or control site.

POPULATIONS IN SPACE

Many populations use only certain areas of the landscape for spawning, other areas for maturation, and adults may be segregated to an alternate part of the landscape. The distribution of populations of other species can be described as a population of populations, patchy in distribution but connected by migration. Other species can be characterized by using various combinations of these strategies in a continuum. In general, this continuum can be broken into 5 general categories of spatial structure for purposes of investigating the effects of toxicants on populations (McLaughlin and Landis 2000).

1) Isolated populations
2) Classical metapopulations
3) Patchy populations
4) Mainland-island and source-sink metapopulations
5) Continuous populations

The first 4 categories are illustrated in Figure 6.1 and form the basis of the discussions. Each of these systems has discrete habitat patches that supply the resources for survivorship and reproduction. This habitat is distinguished in this discussion from areas that provide corridors for migration between these patches. Both are important for the persistence of individuals and populations.

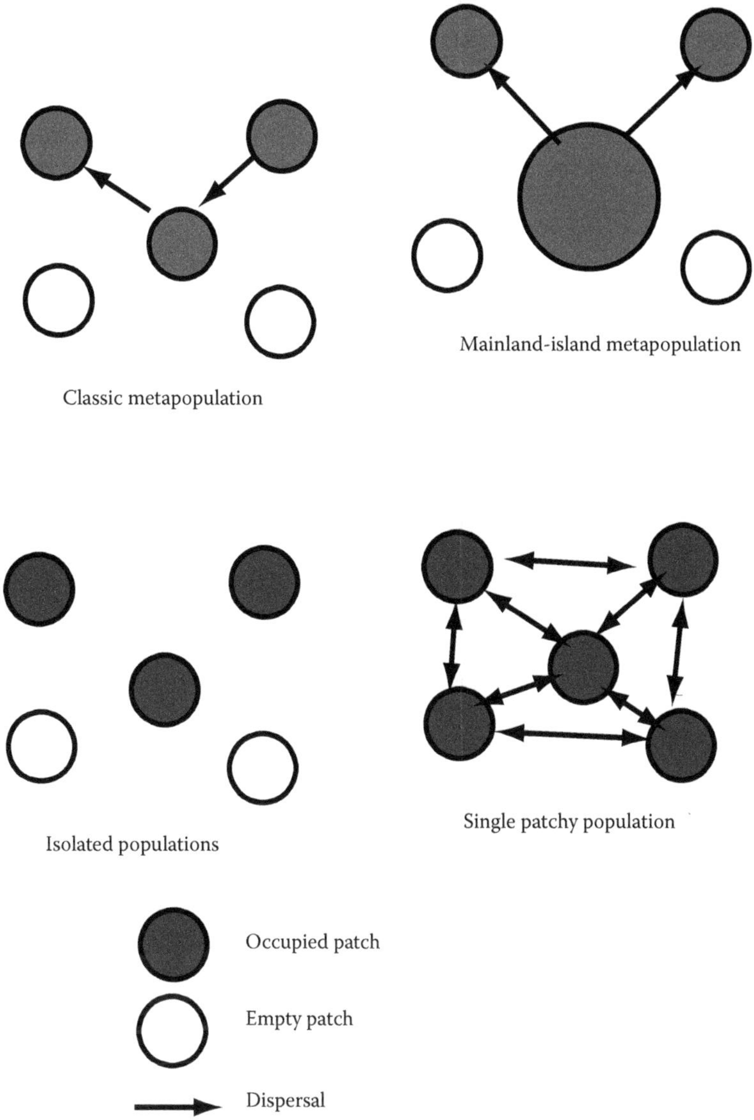

FIGURE 6.1 Spatial structure of patchy populations.

1) Isolated populations are a collection of habitats without migration or dispersal between them. These isolated populations act as if they are self-contained. Contaminants or other stressors in one isolated patch do not affect dynamics in other patches. On the other hand, after extinction has occurred in a patch, recolonization does not occur.
2) Metapopulations result from low to intermediate migration between habitat patches. A metapopulation is a "population of populations" (Levins 1969).

Not all potential habitats necessarily contain populations. Migration between patches affects the dynamics of local populations, even including recolonization after extinction. If sufficient dispersal between patches exists, then a "rescue effect" can prevent local extinctions. Persistence of a metapopulation requires migration rates between patches sufficient to offset local extinction rates. This definition is very specific, but often a broader definition of metapopulation is useful and has been used to estimate environmental impacts. The broader definition states that a metapopulation is a set of geographically or ecologically distinct populations in which migration has a discernible effect on local demography (Stacey et al. 1997; McLaughlin and Landis 2000).

3) Patchy populations are characterized by higher rates of migration between habitat patches compared with a metapopulation. Because of these high rates of migration, the dynamics within the patch may be dominated by the migration instead of the local characteristics of the patch. A characteristic of patchy populations is that 1 organism may spend a part of its lifetime in several patches. In contrast, in a metapopulation the organism will likely spend all of its lifespan within 1 patch unless migration is forced by events.
4) Source-sink and mainland-island metapopulations result when 1 or more of the local populations differ in the probability of local extinction. In a source-sink structure, the source has excess organisms that migrate to other habitat patches. The other habitat patches, sinks, do not contain the resources to maintain growing populations. In contrast to a classic metapopulation, dispersal is not equal between patches but is from the source to the sinks. In a mainland-island metapopulation, the difference is principally the size of the population within the patch and that all patches can support viable populations. Because smaller populations run a greater risk of extinction, the mainland can often provide a source for recolonization and establishment of a new population on that patch. Conversely, islands can also act as refugia when a mainland population becomes extinct (MacArthur and Wilson 1967).
5) Continuous populations have high rates of dispersal. Gaps in the distribution are rapidly filled and individuals appear widespread in the environment. The organisms may vary in their utilization of the landscape, breeding in areas different from where they spend most of the year. Other species may be more evenly spaced in the environment. The environment can be heterogeneous and the organisms may be clumped in space. The behavior of the organisms in the environment may amplify, blur, or reflect the heterogeneity (McLaughlin and Landis 2000).

Examples of Spatial Structure

Organisms vary in the use of a landscape depending on breeding requirements, food resources, and life stages. Gray whales, monarch butterflies, and many bird species are famous for the wide expanse of ecological landscapes used during a year. Conversely, many marine invertebrates take an opposite strategy: the adults are sessile and only the gametes and larvae are free swimming and use currents to disperse. The way that an organism exploits the landscape clearly determines exposure to a variety of stressors, and blockages of this migration can severely alter the viability of the population.

Several animal and plant populations have been examined in the context of understanding the influence of spatial structure on their dynamics (see Hanski and Gilpin 1997; Hanski 1999). Butterfly populations (Thomas and Hanski 1997; McLaughlin et al. 2002) have been the subject of field investigations in Europe and North America. American pika (*Ochotona princeps*) exhibits a population dynamic that is strongly influenced by spatial relationships (Smith and Gilpin 1997). The populations of marine species are also influenced by spatial structure.

Pacific herring (*Clupea pallasi*) of the Cherry Point run spawn along a narrow stretch of intertidal zone off the coast of Washington state during May and early June (EVS 1999). When not spawning, these fish use all of Georgia Strait and even include the West Coast of Vancouver Island in their range. Migrations of hundreds of kilometers take place during the nonspawning months. Just before spawning, the animals congregate just offshore and are segregated by age in their schools.

This broad utilization of the landscape exposes the herring population to a variety of potential stressors (EVS 1999; Landis et al. 2004). Climate change can alter the seawater temperature, altering the presence of prey species and changing the type and number of predator species. Because Pacific herring are predators, they may be exposed to polychlorinated biphenyls (PCBs) and other materials that bioaccumulate over the habitat range that the population uses. Although fishing is currently banned, the Cherry Point run of Pacific herring was fished as it congregated for spawning, and both fish and eggs were taken as they spawned along the beach. Spawning adults, eggs, and larvae could be exposed to contaminants in the near-shore area, spawning substrate could be altered, predatory seabirds that could be attracted by the spawning congregations and storm events could occur. It is not clear if the Cherry Point run exists as its own population or as part of the metapopulation along the British Columbia coast.

Although the same species, the Pacific herring of the British Columbia coast, has been identified as a metapopulation comprised of several patches or subpopulations (Ware et al. 2000). Pacific herring are an important commercial fishery in Canadian waters. Simulation models (Landis et al. 2004) suggest that natural or anthropogenic impacts to 1 part of the metapopulation could have important effects to the apparent numbers in other parts of the range of the fish. The causes would be spatially and likely temporally displaced from the impacts, making attribution of declines or prediction of future numbers problematic without understanding the spatial construction of the population.

It has become increasingly recognized that populations do exist at landscape scales in patchy populations, metapopulations, and other types of spatial relationships. Thorrold et al. (2001) discovered that the fish *Cynoscion regalis* (weakfish) exist along the Atlantic Coast of North America as metapopulations. Examination of the chemical composition of the otoliths demonstrated that juvenile fish could be identified as being from 1 of 5 East Coast estuaries. Examination of spawners in these estuaries demonstrates a similar but not exact correspondence in the composition of the otoliths. Estimates of the straying of adults ranged from 60% to 81%.

Straying was not at random. The strays from 1 estuary would migrate to a proximate spawning area. This spatial structure was not reflected in the genetic analyses, based on allozymes and mitochondrial DNA markers, and that indicated no genetic divergence among the spawning areas. Weakfish appear to exist as a patchy population, with spatial differentiation but with enough migration so that the genetic markers could detect population level differences.

The population structure of Pacific herring in British Columbia has been determined by Beacham et al. (2001) using DNA microsatellite loci and these runs were compared with those of California and Alaska. Variation at 13 loci with a sample of 20000 herring from 78 locations were investigated. A mixture of population structures was observed. No convincing evidence of substructure was observed in the North Coast or Central Coast stocks along the coast of British Columbia. The Straits of Georgia had a more complex pattern. The runs the stocks spawning along the east coast of Vancouver Island demonstrated no significant differentiation. However, stocks spawning at Esquimault Harbor (Vancouver Island near Victoria) and Secret Cove along the mainland were distinct from the Strait of Georgia population. The Cherry Point stock of northwest Washington is also distinct. Some of the *C. pallasi* runs clearly exhibit sufficient mixing to ensure a genetic similarity, yet there are genetically distinct populations within the same geographic area. Beacham et al. attribute the isolated to a temporal isolation with differences in spawning times the culprit although some geographic isolation may occur.

As demonstrated by the extensive review of McElhany et al. (2000), salmon of the west coast of the United States exhibit complex spatial and temporal patterns. In these species, there is leakage of adults from their home spawning site to other rivers and streams in a geographic area. So the spawning runs, although substantially isolated, are still linked by this leakage to other runs. Fish considered the same species can also be reproductively isolated because of different timing of the runs within the same spawning region. To estimate risk to these groups of salmonids National Oceanic and Atmospheric Administration (NOAA) has established grouping by evolutionarily significant units to recognize these important geographic relationships. Evaluation of risks is now conducted at the evolutionarily significant unit (ESU) level.

The previous examples demonstrate that a variety of spatial structures can exist, even in the same species and even in the same geographic areas. The next sections develop the application of patch-dynamic models to investigate the interactions between toxicants and the use of the landscape by the organism in the determination of outcomes at the population level. An example of a laboratory experiment is presented to illustrate these principles. Finally, we discuss the existence of several modeling environments that allow simulations of patch dynamics that could be modified for risk assessment applications.

APPLICATION OF METAPOPULATION AND PATCH-DYNAMIC MODELS TO INVESTIGATE TOXICANT EFFECTS

Metapopulation dynamics is a useful tool in evaluating the consequences of a stress over both time and space. First we need to introduce basic terminology and assumptions. A general assumption is that there is a minimum viable population (MVP) size below which patch extinction will occur in a deterministic model. The carrying capacity is the population size that can just be maintained without a tendency to increase or decrease. A subpopulation serves as a sink if it is below the MVP and is draining immigrants. A subpopulation serves as a source for nearby patches by providing immigrants. Hanski (1991) defined this process as the "rescue effect," a population that is below the minimum viable population can be rescued by organisms from a source.

The arrangement of patches, the size of the population in a patch, and migration paths in a landscape was demonstrated by Wu et al. (1993) to be important factors in the persistence of populations within a landscape. Hanski (1999) summarizes a variety of cases that illustrate the importance of these factors and the life-history characteristics of the species in determining the persistence of species.

The effects of toxicants have also been investigated. Metapopulation models (see Chapter 9) have been used to examine the dynamics of populations resulting from pesticide application. Sherratt and Jepson (1993) investigated the impacts of pesticides to invertebrates using both single species and predator-prey metapopulation models. In the case of the polyphagous predator, persistence of the population in the landscape is enhanced if only a few fields are sprayed, the application rate of the pesticide is low, or the intrinsic toxicity of the pesticide is low. There also appears to be an optimal dispersal rate that maximizes the likelihood of persistence of the predator in a sprayed field. Importantly, there are also patterns of pesticide application that would cause the prey insect population to reach higher densities than would occur otherwise. Dispersal rates of the predator and the prey are important factors determining the prey population densities. The importance of dispersal in the determination of the persistence of a population in a contaminated landscape was discovered in a subsequent study.

Maurer and Holt (1996) have used several types of metapopulation models to investigate the importance of migration and other factors in determining the impacts of pesticides. The exposure to the pesticide was assumed to decline geometrically to simulate degradation. An increase in migration rate among patches was found to decrease the persistence of the population. The more toxic the pesticide, the less persistent the population. An increase in the rate of reproduction improved the persistence of the population in the landscape. Further investigation also demonstrated that as more of the patches became contaminated, the persistence of the population decreased as the number of potential sites for colonization was reduced.

Spromberg et al. (1998) used modified versions of the Wu et al. (1993) model to examine the impact of a contaminated patch on a 3-patch metapopulation of varying arrangements. The simulation models also allow the recognition of a variety of dynamics and rates of population growth. A simple stochastic function for exposure to the toxicant was incorporated. Linear and circular 3-patch models were used and the distance between each patch could be varied. No toxicant was allowed to cross between patches. In some instances, the toxicant was persistent, in other simulations the toxicant was degraded over time. Two important sets of findings were made.

The first finding was that populations in patches removed from the contamination were affected by the presence of the toxicant. In the case of the linear, persistent toxicant model, the effects were the reduction of the population below carrying capacity and fluctuation in population size. The reduction in number and the fluctuations in the nondosed patches resulted in population sizes that were equal to those in the dosed patches, even with a dose equivalent to the effects concentration for 50% of organisms tested (EC_{50}). In the simulations when the dosed patch was at an EC_{100}, organisms could still be found because of immigration from other patches. Because of the stochastic nature of the exposure between the toxicant and the organism, the simulations are repeatable only in type of outcome, not in the specific dynamics. As the number of patches and types of stressors increase, so would the uncertainty in the exact dynamics, even if the types of outcomes were limited to a few types.

The second finding was that in the simulations that incorporated toxicant degradation several discrete outcomes are available from the same set of initial parameters. The range and types of outcomes depend on the specifics of toxicant concentration, initial population size, and distance between patches. The outcomes can be as varied as all 3 populations reaching carrying capacity to all 3 becoming extinct with associated probabilities of occurrence. In this simulation, only 1 patch contained a degradable toxicant that disappeared halfway through the simulation. With an initial population size of 100 in each of the patches, 80% of the simulations resulted in all 3 of the populations in the patches reaching the minimum viable population. At MVP, 1 less organism and the population becomes extinct, at 1 more the population can increase in size. All 3 populations reached carrying capacity in 20% of the simulations. In contrast are the simulations beginning with all 3 of the patches having initial population sizes of 140. In no instance did the populations decline to the MVP. In 82% of the simulations, all 3 populations reached the carrying capacity. However, 18% of the simulations resulted in a stable oscillation, or bifurcation, in the dynamics of all three populations.

Further modeling using the same approach was performed to examine the effect of the placement of the toxicant within the landscape (Landis and McLaughlin 2000). A series of additional simulations with a 3-patch linear arrangement and a degradable toxicant were conducted. The initial population sizes were 200, 50, and 50 in patches 1, 2, and 3, respectively. In some of the simulations, the toxicant was placed at the end of the linear arrangement and in others it was placed in the middle. In every instance the patch dosed was the source patch for the simulated landscape.

When the source patch 1 was dosed the 4 outcomes were 1) In 50% of the simulations, only the population in patch 2 survived at the MVP. In this modeling environment, the MVP is the population size in which the removal of 1 organism results in a negative growth rate for that population. 2) In 26% of the runs, the populations in patches 1 and 2 survived at the minimum viable population (MVP). 3) In 10% of the simulations, all 3 populations survived at the MVP. 4) In only 14% of the runs, all 3 populations reached carrying capacity. Placing the toxicant in nonsource patches 2 and 3 in the middle and at the far end with population densities at 200, 50, and 50 for patches 1, 2, and 3, respectively, resulted in all populations reaching carrying capacity.

Johnson (2002) has also modeled patch dynamics with toxicants using an individual-based approach. The model uses stochastic functions to model birth, death, and dispersal. Two patches are employed — a source patch and a sink. Two toxicant scenarios are modeled. In the first, a spill instantaneously contaminates the patch and then degradation is modeled as a first-order function with a half-life of 1 year. The second scenario is a constant input to the patch with first-order degradation. A physiological toxicokinetic model is used to calculate uptake and concentration of the toxicant by each organism. A dose-response relationship is used to assess the increase in mortality to the adults.

The dynamics of the model without toxicant were determined. If no migration occurs between the patches, then the population in the sink patch declines toward extinction while the population within the source patch persists. When the patches are connected by migration so that each individual has a 20% probability of moving to the other patch, nearly equal populations exist in each patch.

When a 1-time introduction of toxicant occurs to the sink patch a relatively small dip in the population of each patch occurs, followed by an increase that fluctuates around numbers typical for the simulations without toxicant. If the source patch is dosed, a larger dip is observed in the output of the simulation.

The continuous or chronic input of toxicant produces different simulation results. In these simulations the amount of toxicant was set at zero, then toxicant was added until a steady state was reached. The individual-based effect of the toxicant is set to reduce the fecundity of the individual organisms. If the source is dosed, then extinction occurs in both patches relatively rapidly. If the sink is dosed, extinction occurs but over a 100-year time frame.

These simulations confirmed the results of Spromberg et al. (1998), and McLaughlin and Landis (2000) described that the effects of toxicants can occur outside the area of contamination resulting from changes in demography, rates of reproduction, and other population characteristics. These findings are consistent although the modeling strategies are very different. It appears that action at a distance is a characteristic of both modeling approaches.

There are three general types of impacts from toxicants or other stressors that may be considered. Figure 6.2 depicts a 4-patch landscape in which patch 1 is always the source patch. Shading of the patch indicates the input of a toxicant. In case A, the source patch is dosed. Immigrants from patch 2 can migrate into patch 1, along with those from patch 4, and result in the restoration of the population in patch 1. In this instance, patch 3 can be the ultimate loser because additional migration from this patch to the lower densities in patch 2 can now occur. However, patch 3 is not connected to any other patch, so the most persistent impact may occur in the patch furthest from the toxicant input. Such a pattern has been observed in simulations. In case B, a sink is dosed. Migration from patches 2 and 1 will ensure that the effect is short-lived or at least compensated so that a relatively minimal effect is observed. Case C presents a toxicant or other stressor blocking immigration between patches 1 and 4. Because patch 4 is a sink, the population within will go to extinction. The patch relationships between the remaining patches have been altered, changing the dynamics of a stressor being introduced into this novel arrangement. Note that in this case no direct interaction between the toxicant and the organisms of the populations has to occur. All that needs to happen is that the immigration pathway is destroyed by the alteration of habitat type or fragmentation so that movement is impaired.

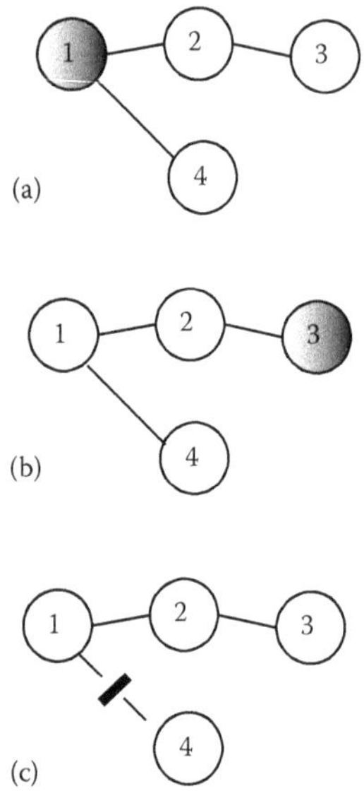

FIGURE 6.2 Effects of toxicant or other stressors on a metapopulation or patchy population.

Patch and metapopulation dynamics have several important implications for predicting the impact of chemical toxicants: There are 6 conclusions that can be derived from the outcomes of these sets of modeling exercises:

1) Effects can be promulgated between patches, even if the toxicant is not transferred. There is action at a distance between populations connected by immigration as a result of the source-sink dynamics established by the toxicant.
2) The effects of the toxicant can persist even after the degradation of the material from the contaminated patch.
3) Multiple discrete outcomes can occur from the same set of initial parameters in certain areas of state space.
4) Small differences in initial population sizes can dramatically alter the frequency of outcomes. It is not only the properties of the chemical and its interaction with an organism, but the status of the population that determines relative risk.
5) Patch arrangement, the relative sizes of the populations and the placement of the stressor are important in determining the outcome.
6) Patches used as a reference cannot be linked by migration to the contaminated patch. If connected, the reference patch can be affected by the toxicant. This finding places severe limitations on the likelihood that reference sites are possible within a landscape. The implications of this finding are further discussed in The Problem of the Reference Site.

Are these results artifacts of the numerical simulations or can they be expressed in simulated populations and ecological systems?

EXPERIMENTAL RESULTS

A Metapopulation Experiment to Simulate Toxicant Effects

Partial experimental confirmation of the impacts of a simulated toxicant on a metapopulation "action at a distance" has been provided by the research of Macovsky (1998). This study created a laboratory patch model based on the simulation framework of Spromberg et al. (1998) using the flour beetle, *Tribolium castaneum*. Vials of flour were connected by string bridges to allow movement of adult beetles. This allowed habitat patches to be linked by density-dependent dispersal of the adult morph. The numbers of adults, pupae, and larvae were assessed in each patch. To simulate a stressor, adults were removed from one of the end patches for either 50% or 100% mortality. Patches were monitored for the indirect effects on population demographics beyond the patch that received a simulated mortality of adults over the period of approximately 1.5 egg-to-adult cycles.

It was demonstrated that indirect effects do occur in patches beyond the patch directly impacted. The indirect effects were dose related and correlated to distance from the directly impacted patch.

APPLICATION OF SPATIAL RELATIONSHIPS INTO THE RISK ASSESSMENT OF POPULATIONS

Recognition of the importance of spatial relationships in determining the types and range of impacts requires that this information is gathered and analyzed in the course of a risk assessment. This same recognition also opens the door for ecological risk assessment to be applied to risks from invasive species, genetically engineered organisms (GMOs), and zoning designation and land use management issues that change the structure of the landscape.

Consideration of Spatial Structure of Populations in Risk Assessment

The potential pattern of propagation of effects from the spatial relationships among populations requires an understanding of these interactions. If the populations are isolated, then the effects can be isolated and populations compared for determining causality. Wide-ranging populations may be subject to multiple stressors that may mislead the attribution of effects to local stressors. If the populations are linked, then effects may be more widely distributed to patches removed from the contaminated site, or dampened by the rescue effect of source patches.

Understanding these possibilities requires that sampling programs be constructed so that the natural history and patch dynamics of the relevant populations are understood. This determination may require tagging studies, genetic analyses, and intense observation. However, we can answer a variety of critical uncertainties in the risk assessment process. How many populations are included in the risk assessment? What are other likely stressors that may confound the risk estimate? What is the geographic reach of the impact? Is recovery of the local population likely, or does extinction require specific mitigation? How should populations be compared in the determination of causality — as a single entity or as an interacting set of subpopulations?

An understanding of the spatial relationship of the populations within the habitat will also determine the type and number of environmental gradients to be considered in the risk assessment. If the individuals of the population move large distances within the region, then a number of broad scale gradients need to be understood and the location of specific stressors mapped over a large area. Patchy populations should also be understood within a landscape context, but also at a very local area. An analysis of connected patches also requires an understanding of the corridors and potential stressors within these areas as well. Small isolated populations require a reduced scale of spatial considerations.

The Problem of the Reference Site

As discussed previously, spatial interactions among connected patches has important implications for the use of reference or control sites. At the spatial scale of a population, the range of gradients of controlling factors, stressors, effects, and the lack of independence of many sites makes the search for a reference or control site essentially irrelevant for studies of risk. The likelihood that another patch with a population of similar genetic structure, spatial relationships to various controlling factors, patch interactions identical or even very similar to another site, and similar evolutionary history is small enough to be considered not practical. Furthermore, reference sites in this sense are not required and the replacement approach is straightforward. A regional and landscape basis for inferring causality coupled with statistical tools such as patch analysis and the weight of evidence approach will allow an understanding of mechanisms and effects that correspond to the scale of the assessment. The effects of the environmental gradients can then be allocated, causal pathways confirmed and the risk calculated. If the gradient is steep and only a few are present, then only a few sample sites close together may suffice. In situations where the slope of the gradients are shallow, multiple gradients are interacting, and when differentiating populations may be difficult, a large number of samples spread along the various gradients at some distance will be required. The goal becomes understanding the landscape of environmental gradients, impacts, and risks, not making a unrealistic 1-to-1 comparison.

Tools for the Analysis of Spatial Relationships

One of the potential criticisms of examining effects from spatial relationships at landscape and regional scales is that the tools and techniques do not currently exist. At the beginning of a research and development cycle, there is a gap between the recognition of the feature and the everyday application. The field is in its early stages, but there are tools available for implementing spatial considerations into the risk assessment process.

1) Geographic information systems (GIS) are now very common in environmental laboratories in academic, governmental, and business settings. GIS is an excellent means of compiling and sorting data, visualizing spatial

relationships, and establishing sampling programs. There are also useful clustering and other analysis tools that can be used.
2) There are programs of varied usefulness and applicability that can model populations in a spatial manner, some of which use GIS data as basis for the model landscape (Table 6.1; see Chapter 9 for description of modeling approaches). Modifications may be necessary to answer specific questions in a risk assessment context, but these programs are good starting points.
3) Site-specific models and programs can be employed. This approach will make necessary the addition of expertise in population and landscape ecology, but those skills are becoming more common. Collaboration between risk assessors and population-landscape ecologists has already begun, and should be encouraged.

When to Ignore Spatial Structure?

Given the pressures governing the risk assessment and decision-making process, there is an impetus to ignore spatial structure. McLaughlin and Landis (2000) have already presented criteria for when spatial structure should be included or ignored in the assessment of populations.

1) Spatial structure can be ignored when little or no migration occurs between isolated habitats or when large, relatively homogenous areas are being considered and organisms move short distances compared to the extent of the habitat.
2) Spatial structure should be considered when there are moderate rates of migration between habitat patches or movement within a moderately heterogeneous but continuous habitat.
3) Spatial structure should be explicitly considered when migration rates between patches are high, or when the dispersal distances are long even within a continuous habitat.

Note that each of these criterion is not set by a physical size limit (such as 10 m^2) but by the use of the environment by the population under consideration. With many populations undergoing a risk assessment, many of these factors will be known and a judgment made about the size of the study and a minimum size limit set for when spatial scale can be safely ignored. If the information to make these judgments is not known, then it is likely that the population-level ecological risk assessment will have an unacceptable level of uncertainty.

Expansion of Risk Assessment beyond Chemical Impacts

The expansion of population-level ecological risk assessment also broadens the types of questions that can be approached. Many of the critical environmental decisions are for items that are at a scale broader than the scale in typical contamination issues.

A critical current issue is the management of invasive species and making risk-based decisions. The introduction, spread, and impacts of invasive species are all dependent on spatial interactions coupled with population dynamics. The location of

TABLE 6.1
Spatially explicit population dynamics models and programs

Name	Citation	Description and notes
CAPS	Plotnick and Gardner (2002)	CAPS was created to simulate sessile species patterns, abundance, and the effects of competition and establishment in heterogeneous landscapes. The program is spatially explicit and uses stochastic features for certain parameters. The required species characteristics inputs include fecundity, dispersal, and habitat preference. Imported geographic information systems data can be used in the analysis to provide information concerning habitat heterogeneity and disturbance.
LANDIS	Mladenoff et al. (1996)	LANDIS was created to predict the changes in forest structure and species composition, particularly age distributions and dominant tree species. LANDIS was programmed to simulate disturbances, succession, and the effect of management practices.
Program to Assist in Tracking Critical Habitat (PATCH)	Schumaker (1998)	PATCH was designed to model the movement of territorial, terrestrial vertebrates through time. Patch is spatially explicit and incorporates habitat data in the form of raster files. However, the model is female-only and is an individual-based single-species model, thus it cannot model interspecies interactions. www.epa.gov/wed/pages/models/patch/patchmain.htm
RAMAS GIS	Akçakaya (1998)	RAMAS GIS is designed to be able to link geographic and demographic data and simulate changes in species demographics in both patches and a metapopulation, though the model does not consider multiple species or interactions. RAMAS offers a variety of similar programs, however RAMAS was the only model considered that was not in open distribution. www.ramas.com
Spatially Explicit Landscape Event Simulator (SELES)	Fall and Fall (2001)	SELES creates a spatial and temporal "state space" to reveal how changes in that space (i.e., by management practices) combined with changes in structure and process generate changes at the landscape scale. SELES incorporates raster images to define a landscape as to its habitats and other features. The accompanying language editor assists in the creation of the model language in a value = function format. Stochastic functions may be applied as well as a Monte Carlo approach. The simulator creates and runs a queue of actions that occur sequentially in the model. The model can be programmed to output multiple forms of demographic and relationship data and to produce a variety of raster images altered from the original based on the model run. www.cs.sfu.ca/research/SEED
Spatial Modeling Environment (SME)	Costanza et al. (1992)	SME allows the construction of sophisticated models of ecosystems in a point-and-click graphical environment. SME is a grid-based tool that allows each grid cell to independently run a STELLA model (High Performance Systems, Inc. www.hps-inc.com). SME is available for UNIX systems. www.uvm.edu/giee/SME3/index.html

the introduction will have a large bearing on the future spread of the invasives. The life-history strategy regarding the use of space will be an important factor in determining the potential rate of spread. The spatial characteristics of the potential assessment endpoints are also critical in estimating the probability of adverse impact. Are the patches linked by migration corridors, are special habitats required for spawning, do the organisms spread by pelagic larvae or do the adults move? These questions are all important in calculating risk in a landscape context. We and other groups are in the process of adapting patch dynamic models for the estimation of risk due to invasives.

The risk assessment of GMOs has many similarities to the risk assessment of invasives. In the risk assessment of GMOs, both the spatial and dynamic aspects of the host and the introduced genetic element need to be considered. If the introduced element is confined to the GMO, then the population and spatial dynamics of the host can be followed in a manner similar to an introduced species. If the introduced element is mobile, then the population genetics of that movable element and its effects on its new host should be considered. These interactions can be modeled using adaptations of basic patch dynamics models (Landis et al. 2000).

Management of fisheries, wildlife, and ecological systems is in need of the framework provided by ecological risk assessment. Many that deal in the decision-making process for these management areas see risk assessment as limited to dealing with toxic materials at contaminated sites or on a spatial and temporal limited scale. The recognition that risk assessment can be expanded to scales familiar to these researchers will enable this tool to be used in the management of these systems.

Environmental decisions at the local government level are often about zoning and other land use issues. Although local at first glance, these collective decisions establish the large-scale regional patterns of urban, industrial, and agricultural patches. Risk assessment applied as part of this critical decision-making process may be able to inform zoning and other land use planning.

SUMMARY

The introduction of spatial considerations to the risk assessment of populations provides a number of insights and opportunities for ecological risk assessment. There are 3 main conclusions:

1) Representation of spatial relationships and dynamics places the population in a more realistic context and leads to the realization that dynamics unanticipated in models without spatial structure can occur.
2) It is possible to observe and model the effect of spatial relationships on the impacts of toxicants at large scales. Although the study of spatial relationships, such as island biogeography, has been in existence for more than 140 years, its application in risk assessment is still in an early phase.
3) The introduction of spatial relationships and dynamics will allow ecological risk assessment to be applied to a variety of other decision-making activities, from introduced species to land use planning. Ten years from now, such applications will be the norm for a new generation of risk assessors.

7 What Conservation Biology and Natural Resource Management Can Offer Population-Level Ecological Risk Assessment

Jennifer A. Gervais and Helen M. Regan

INTRODUCTION

As ecological risk assessment moves to incorporate increasing levels of biological complexity, it is confronted by the same issues of uncertainty and imperfect knowledge that have faced the fields of natural resource management and conservation biology for decades. In a similar vein, conservation biology and natural resource management are only slowly beginning to recognize that chemical impacts on the natural environment go well beyond DDT, and need to be considered as one of many factors that may cause plant and animal population declines. Up until very recently, the disciplines have not often crossed the boundaries to learn from each other's experience and knowledge base. It is encouraging to note that articles dealing with ecological risk assessment are now cropping up in interdisciplinary journals, such as *Ecological Applications*, because this should help increase the flow of ideas. The purpose of this chapter is to highlight some of the areas that have developed within conservation biology and natural resource management that have special relevance for the newly emerging discipline of population-level ecological risk assessment. These developments will also be useful to the development of ecological risk assessments that include community dynamics and ecosystem services.

Population-level ecological risk assessment is emerging as an important approach to assess the impact of contaminants on organisms in the environment. One of its major goals is to quantify the probability of adverse impacts of contaminant exposure on populations of organisms — impacts at this level of biological organization are frequently very different from those that can be measured at the level of the organism (e.g., Forbes and Calow 1999). As explained elsewhere in this book, assessing risks at the population level requires a different approach to the established practices of organism-level risk assessment. In most instances, it is not sufficient to simply scale up from individuals to aggregates of individuals, because of complications such as density-dependent interactions (Chapter 4). Furthermore, population-level ecological risk assessment usually entails a different set of regulatory and management questions which, in turn, require a different approach to assessment (Chapter 3).

The recent move to consider populations and even communities or ecosystem services as appropriate ecological assessment endpoints means that population ecology and dynamics must be routinely considered when evaluating risks, and that population status must be determined at a minimum. In addition, there is a growing recognition that stressors must be considered in total, and that environmental conditions can have a major influence on a stressor's impact (e.g., Hatch and Blaustein 2000; Boone and Semlitsch 2002; Boone and James 2003). Environmental toxicology and chemistry have a long and distinguished record of predicting chemical fate and transport in the environment, and in determining biochemical modes of action within individual organisms. These research traditions translate poorly into the study of natural populations subject to complex biological interactions and exposed to all the vagaries of the environment. Conservation and resource management, on the other hand, have been concerned primarily with the protection and management of natural populations within the ecological community since their inception — the experiences and tools gained over the years have a great deal of relevance for population-level ecological risk assessment.

There are at least 5 major contributions that conservation and resource management have to offer population-level ecological risk assessment. The first is the establishment of environmental and demographic variation, or stochasticity, as important influences on population dynamics and population-level responses to stressors. The second is an enormous and growing toolbox for better empirical estimation of attributes such as abundance, survival, and reproduction in natural populations in the face of environmental variation. The third is the application of statistical approaches to evaluating hypotheses represented by models, beyond the traditional null-hypothesis statistical test. These methods of inference are better suited than traditional statistics to situations where manipulative experimentation is either infeasible or simply not possible. Fourth is modeling techniques to project the possible outcomes of different management or ecological scenarios that are powerful tools for exploring the effects of uncertainty on scenario outcomes and the sensitivity of these outcomes to various model parameters. In particular, the development and use of stochastic models may be one of the most important contributions to the developing field of population-level ecological risk assessment. Fifth, conservation and resource management have a history of methodological development that is relevant to population-level ecological risk assessment, including how models should be developed and changing perceptions of how models should be used in a decision-making context.

ENVIRONMENTAL AND DEMOGRAPHIC VARIATION

The dynamics of natural populations and the behavior of ecosystems are dependent on environmental factors that vary in time and space and that are influenced by changes in other related variables. Variation is due partly to interactions that we can predict, but most often it results from processes that we either cannot foresee or for which we have incomplete knowledge (Regan et al. 2003a). It is an inherent feature of dynamic systems and generally can be classified into 2 broad classes — demographic and environmental stochasticity.

Demographic stochasticity refers to random variation in individuals' fitness that is independent across individuals (Lande 2002). This causes variation in population growth rates, the magnitude of which is inversely proportional to the size of the population, so that for small populations the effects of demographic stochasticity are greater than for large populations. Environmental stochasticity refers to random fluctuations in biotic or abiotic factors that do not originate from individual characteristics. In this case, variation in growth rates as a result of environmental stochasticity does not depend on population size. Catastrophes are a special case of environmental variation. They are low-probability, high-consequence events that dramatically reduce population size (Shaffer 1987). Examples of catastrophes that have population-level responses are floods, fires, epidemics, oil spills, volcanic eruptions, and hurricanes. Catastrophes can be naturally occurring or human-induced.

The effects of variation on population growth rates are important to consider because they directly influence relevant endpoints for population-level ecological risk assessment, namely probabilities of extinction or decline. In fact, it could be argued that variation enables us to consider the risk of population extinction or decline, because risk refers to the probability of adverse effects given variation in the system. Small populations are particularly prone to the adverse effects of demographic and environmental stochasticity. Demographic stochasticity can lead to the onset of inbreeding depression and Allee effects, and environmental stochasticity can further reduce small populations through a random reduction in population growth rate. This makes small populations particularly at risk of extinction. Variation also has important implications for metapopulation persistence (Hanski and Gilpin 1997; Lande et al. 1998).

Conservation biology has long recognized the need to explicitly consider variation in risk assessments of populations. Population viability analyses that incorporate stochasticity are essentially population-level ecological risk assessments. Although stochastic population models have been used in ecology since the 1970s (see Goel and Richter-Dyn 1974; May 1974; Poole 1974; Pielou 1977; Nisbet and Gurney 1982 for notable examples), Ginzburg et al. (1982) were the first to treat the problem from a risk-analytic perspective. These efforts have been extended to incorporate increasing detail and relevance for conservation and resource management contexts (for reviews see Lande 2002 and Ferson 2002). As a result, there is a burgeoning literature on the implications of stochasticity on natural populations under stress. Some models have extended these considerations to populations impacted by chemical stressors (Spencer and Ferson 1997; Hakoyama and Iwasa 2000; Tanaka 1998) but this is still far from common.

The impact of demographic and environmental stochasticity on populations is perhaps the most important concept that conservation and wildlife management bring to population-level ecological risk assessment. It is clear that consideration of the effect of demographic and environmental variation on the risk of extinction or decline of populations is a necessary component of any reliable population-level ecological risk assessment.

TOOLS FOR PARAMETER ESTIMATION

Conservation and resource management are primarily concerned with the status of populations of organisms in natural communities. However, determining population status is difficult at best, requiring that inferences must be made on the basis of estimates of the quantities of interest. Field sampling methodology and associated parameter estimation is a large and growing subdiscipline, attesting to the importance of these issues in population-level management. Although this chapter is not by any means a comprehensive review of the topic and further discussion can be found in Chapter 8, some examples follow.

DEMOGRAPHIC PARAMETER ESTIMATION TECHNIQUES

For animals, mark-recapture methodology has been highly developed to estimate survival (Lebreton et al. 1992; Nichols 1992; Kendall et al. 1995; Schwarz and Arnason 1996; Seber 2001). These methods not only allow the estimation of survival rates, but can be used in the direct estimation of population growth rate as well (Pradel 1996; Nichols and Hines 2002). In addition, mark-recapture methodology can be used to estimate movements (Hestbeck et al. 1991; Schwarz et al. 1993; Joe and Pollock 2002), with obvious implications for estimating metapopulation dynamics. These techniques have also been expanded to examine transitions among life history stages or phenotypic states (Williams et al. 2002). Density dependence can even be explored with mark-recapture data (Barker et al. 2002). Standardized software packages are available for the analysis of these data (e.g., program MARK, White and Burnham 1999; Program RELEASE, Burnham et al. 1987; Program JOLLY, Pollock et al. 1990; Program POPAN, Arnason and Schwarz 1999). Although mark-recapture studies can be time-consuming and require substantial field effort, they are based on well-developed theory and are well documented for supporting empirical methods for risk assessment and for estimating some of the parameters critical to population modeling. This will be critical in cases in which demographic rates are close to the values determined to be thresholds for contrasting management or remediation strategies. Population-level ecological risk assessment will lead to this scenario on occasion, and using the best available techniques that minimize bias and variance in estimates will help alleviate ambiguity in interpretation.

Estimates of reproduction are more species-specific, but also must account for the fact that exact numbers of recruits to adult stages can rarely be known in the field. For free-ranging populations of animals, estimates based on direct counts of young are common. For birds, for example, the classic method is determining nest survival using the Mayfield estimator (Mayfield 1961; Williams et al. 2002). Other methods of estimating reproduction include counting ovulation scars in mammals, radio-tagging samples of young animals, or using nest boxes so that direct counts can be accurately obtained. Fish stock recruitment may be estimated from electrofishing, redd surveys, or mark-recapture of released smolts (Hilborn and Walters 1992). Methods for invertebrates include sampling in the field for young or eggs. Plants are perhaps the most difficult because the seed bank allows reproduction to be staggered through time as well as dispersed through space, but transect sampling or quadrat sampling for seedlings, seed traps, and examination of samples of soil are possible methods for estimating recruitment from the seed bank (Auld and Denham 1999; Auld et al. 2000). Regardless of the purpose of the study, researching and using the best available techniques will lead to the most robust estimations of the parameters of interest, and aid in the decision-making process that follows the initial assessment.

Population Size and Density Estimation Techniques

Population size can be determined from mark-recapture studies, although they are less commonly used for this purpose. The earliest uses of mark-recapture methodologies were actually for estimating population size, such as the Lincoln-Peterson estimator (Williams et al. 2002), and more sophisticated estimators that allow recapture probabilities to vary (Program CAPTURE, White et al. 1982).

Distance sampling methodologies that do not require marking individuals have been developed to determine quantities such as density and abundances in both plants and animals (Bonham 1989; Buckland et al. 1993; Ralph et al. 1995). Specialized software is available for analysis of the resulting data (Program DISTANCE, Thomas et al. 2002). A population assessment can also be carried out in the absence of abundance estimators using changes in sex ratios or numbers of individuals in age or size classes (Cooper et al. 2003). This may be particularly useful in ecotoxicological applications, as baseline data are frequently lacking. More detail on this topic can be found in Chapter 8.

Minimizing Sampling and Parameter Estimation Error

Methods have been developed to remove sampling error from variance estimates entirely if enough sampling events exist (Burnham et al. 1987; Dennis et al. 1991; White 2000; Morris and Doak 2002), leading to more precise confidence intervals on the parameter estimates. Variation is usually confounded between sampling variation and process variation even in the best of field studies. Patterns of interest can be obscured by sampling error that is an artifact of study design. Removing sampling error thus allows for better inference, and makes the most of available data. Sampling error removal was first applied to fisheries data (Burnham et al. 1987), but the techniques are suitable for any organisms if there are enough sampling events in the data set. A discussion of the major techniques, examples, and recent advances can be found in Morris and Doak (2002). There are obvious applications in population-level ecological risk assessment, in which parameter estimates are used to determine a course of action or remediation; the more accurate the estimate, the more likely that the appropriate choice will be made. In other words, we will be more likely to correctly identify impacts when they are present, and avoid situations when unnecessary remediation is undertaken.

One of the tools to help with sampling design is the use of power analysis to determine whether the proposed study and sampling methods are likely to detect changes considered biologically important (Cohen 1988). This technique can be used to compare various sampling strategies and potential effect size combinations to ensure adequate field sampling — it may even indicate when it may not be possible to obtain enough data to document an effect. A prospective analysis examining what levels of declines were reasonable monitoring targets for seabirds is one example of an application, found in Hatch (2003). This article also compares 2 commercially available software packages for power analysis, and discusses issues such as the effects of variance structure and model choice on analysis outcomes. Detection of the effects of habitat alteration on survival estimates of Chinook salmon (*Oncorhynchus tshawytscha*) is another applied example (Paulsen and Fisher 2003). Power analysis should not be used in a post-hoc fashion in an attempt to justify a nonsignificant statistical result (Steidl et al. 1997).

Sampling design, data collection protocol, and parameter estimation are areas in which uncertainty can be minimized using the techniques that have been developed for this purpose and that are constantly being refined. Great care should be taken to collect data in a way that allows the most efficient estimation of the quantities of interest, and the tools for estimation should be selected to match the data collection limitations and logistical challenges presented by the species under study. Ideally, sampling strategies should be set up with both logistics and a plan for data analysis already in mind.

The estimation problem is not trivial; its importance is illustrated by the fact that no fewer than 11 articles on the topic appeared in 2002 in *Journal of Wildlife Management* alone. Conferences have been devoted to new and better ways of estimating demographic parameters and population growth rate based on mark-recapture data (Kanji 2002), and there are numerous books dealing with the subject (e.g., McCallum 2000; Williams et al. 2002). This is an enormous and well-developed knowledge base for use in population-level ecological risk assessment.

METHODS OF INFERENCE

The classic research paradigm in the biological sciences and management fields involves testing statistical hypotheses, by comparing a null hypothesis and a research hypothesis. The objective is to determine the likelihood of the observed data given that the null hypothesis is in fact true. If the observed data are significantly different (usually at a $p = 0.05$ level), then the null hypothesis is rejected in favor of the alternative research hypothesis. This is a time-tested tool for designed experiments, and has certainly contributed substantially to science's knowledge base in general (Platt 1964), but it often works poorly when it is used on observational data that may have been generated by multiple processes (Johnson 1999). How do we choose the best research hypothesis, however we define it, when we may have more than one that are plausible for the data? There are two developments worth noting that are applicable to population-level ecological risk assessment.

INFORMATION–THEORETIC APPROACHES

The first is hardly a new idea. T.C. Chamberlin (1890) first warned of the dangers of considering only one research hypothesis at a time. Consideration of a suite of hypotheses, each tested with an experiment that could potentially rule out one hypothesis over another has also been strongly advocated by Platt (1964). However, traditional statistical techniques make it difficult to evaluate relative levels of support for multiple working hypotheses simultaneously, and conservation biology, natural resource management, and ecological risk assessment are frequently forced to draw inference from unreplicated, nonexperimental systems. Information–theoretic techniques such as Akaike's Information Criterion (AIC), Bayesian Information Criterion (BIC), and Takeuchi's Information Criterion (TIC) are powerful methods for ranking various hypotheses that are represented by statistical models. The 3 techniques have different strengths relative to one another, and an accessible comparison and discussion can be found in Burnham and Anderson (2002). The use of AIC in particular has rapidly expanded in the field of wildlife management in recent years (Burnham and Anderson 1998, 2002; Franklin et al. 2000, 2001), and the following discussion will focus on this particular method.

The general method involves building a set of candidate models based on our best understanding of the system under study. AIC is based on likelihood theory and the theory of parsimony. It ranks the candidate models according to how well they fit the data as a compromise between model variance and bias. The greater the number of parameters, the less the bias in parameter estimates but the variance in the estimates will also be greater — the model with the most parameters is not necessarily the most appropriate one (Franklin et al. 2001; Burnham and Anderson 2002). The hypotheses and the models that represent them do not have to be nested, allowing greater consideration of alternatives. An excellent illustration of the use of AIC can be found in Franklin et al. (2000), where various hypotheses that included habitat and climate factors were compared to estimate their effects on survival and reproduction of northern spotted owls (*Strix occidentalis caurina*). AIC has been used in a population-level ecological risk assessment to evaluate the relative effects of fluctuating food supply, nesting population density, habitat features, and p,p′-DDE on the dynamics of a burrowing owl population (Gervais and Anthony 2003).

Although these methods are not by any means a replacement for traditional statistics (Guthery et al. 2001; Eberhardt 2003), they will be extremely useful to population-level ecological risk assessment. For example, information-theoretic methods are very powerful tools for retrospective studies that are observational as opposed to experimental manipulations. They can also be used as a method to better estimate parameters by computing weighted averages over the models from which they were derived (Burnham and Anderson 1998, 2002). Although this technique has not been as widely used as the model selection applications, it may have even greater utility in population-level ecological risk assessment, where threshold parameter values may be the trigger for very different management actions. In this case, an estimate that incorporates the information provided by a suite of hypotheses may be better suited for inference than simply choosing the parameter estimate for the highest ranking model, particularly if data are sparse and no one model is clearly better than any of the others in the set.

Bayesian Techniques

Bayesian methods also allow the simultaneous consideration of multiple hypotheses (Hilborn and Mangel 1997; Brooks et al. 2002). However, their particular distinction is the use of information already available to inform the degree of belief the analyst has in each of the alternate hypotheses under consideration — these degrees of belief are then updated using the new data set (Gelman et al. 1995; Berry 1996; Hilborn and Mangel 1997). These methods are also not new, although recent theoretical advances have greatly facilitated their use (e.g., Link et al. 2002). In addition, increasing power in desktop computers makes the techniques more accessible.

The essential difference between traditional statistical ("frequentist") and Bayesian approaches is in how the parameter that governs the distribution of the data is viewed. For example, sample data can be regarded as coming from a distribution $f(Y|\theta)$ governed by the parameter θ (Link et al. 2002). Frequentist approaches view θ as a fixed, if unknown, parameter. The expected value for this parameter is the mean of an infinite number of repeated random samples (Ellison 1996). However, in Bayesian analyses, this parameter is a random variable.

An excellent introduction to the general Bayes approach can be found in Hilborn and Mangel (1997). Although more thorough treatments exist (e.g., Gelman et al. 1995; Berry 1996), Hilborn and Mangel (1997) describe the utility of the Bayesian approach using several clear and simple examples. They then give a detailed example analyzing catch per unit effort data from hake fisheries in Namibia (Hilborn and Mangel 1997). Their emphasis is on conceptual application throughout.

Bayesian approaches to model fitting, parameter estimation, and decision-making have all been advocated in the ecological and conservation literature (e.g., Ellison 1996; Wade 2001; Dorazio and Johnson 2003). Interestingly, although they have not yet been widely used in conservation or wildlife management, fisheries management has employed Bayesian techniques routinely for years, particularly for stock assessment (e.g., Hilborn and Walters 1992). This has particular relevance for population-level ecological risk assessment in that data about the populations of interest are often sparse and the relationships between anthropogenic pressures and ecological response are poorly understood, although some prior information from similar systems or populations may be available.

Fisheries stock assessments using Bayesian approaches continue to be refined (Meyer and Millar 1999), and these refinements are likely to be directly applicable to population-level ecological risk assessment. Specifically, Bayesian methods allowed the consideration of nonlinear state-space modeling of population dynamics (Meyer and Millar 1999). Much more complex model structures than can traditionally be fitted may be warranted based on the biology of the species of interest, and Bayesian methods can accommodate a greater range of model structures than was previously possible using traditional statistical methods.

Another difficulty with traditional frequentist statistical approaches is that of incorporating hierarchical scales in analyses. In particular, individual heterogeneity has been shown to substantially influence population characteristics such as persistence time (White 2000). Bayesian methodologies allow the fitting of such models that would be difficult if not impossible to fit otherwise. An example is the analysis of black-legged kittiwake (*Rissa tridactyla*) data to explore the effects of latent factors on survival and reproduction (Link et al. 2002). Discussion of various aspects of Bayesian methodology is also included.

Model averaging is a technique that can be implemented within the Bayesian framework as well. Habitat preferences and spatial distributions of greater gliders

(*Petauroides volans*) were examined using a Bayesian model averaging approach (Wintle et al. 2003). Regardless of the statistical paradigm, model averaging has a great deal of relevance for population-level ecological risk assessment and incorporation of uncertainty.

Bayesian methods are not without their limitations; in particular, there has been a good deal of controversy over the effects of the choice of prior probability distributions on outcomes (Gelman et al. 1995; Punt and Hilborn 1997). A reasonable solution appears to be that of performing sensitivity analyses on the priors themselves to evaluate their impact on the posterior degrees of belief assigned to the analysis. When prior information of some reasonable quality is available, however, Bayesian estimators appear to outperform traditional frequentist estimators in the sense that the estimated parameter value is nearer the true value (Samaniego and Renau 1994). Bayesian methods offer useful approaches for addressing the inherent complexity, variability, and uncertainty associated with ecological systems, and as software and training make Bayesian approaches more accessible, they will undoubtedly be very useful in ecological risk assessment.

THE USE OF POPULATION MODELS

Population models are useful tools available to risk assessors to investigate population-level effects of contaminant exposure of organisms. They have a long history of development and successful application in theoretical ecology and have subsequently played an important role in conservation and natural resource management. Such models attempt to incorporate life history traits of populations into a theoretical framework, and are used to assist in predicting likely future outcomes or the projection of possible scenarios (Starfield and Bleloch 1991; Hilborn and Walters 1992; Shenk and Franklin 2001). They have been used routinely in ecology to study the dynamics of populations, to provide explanations for observations, to make predictions and to explore different scenarios on possible population outcomes. In natural resource management, population models are frequently the basis for setting harvest limits, implementing control strategies, or for evaluating various management alternatives. A detailed discussion of the types of models available to risk assessors can be found in Chapter 9, Burgman et al. (1993), Caswell (2001), Maltby et al. (2001), Ferson and Burgman (2002), and Pastorok et al. (2002). Although models are used in an incredibly broad array of applications in conservation and resource management, there are several broad dichotomies worth noting before exploring some of the applications with particular relevance to population-level ecological risk assessment. A hard-earned lesson that should be immediately applied across discipline boundaries is the need to ensure that all parties agree on the intended purpose of a model and the types of questions it is used to answer (Bunnell 1989).

RETROSPECTIVE VERSUS PROSPECTIVE MODELING

The empirical methods discussed previously and in Chapter 8 are concerned with describing attributes of the population that existed when it was sampled. Modeling methods can also be used to explore the relative impacts of toxicants on specified population parameters (e.g., Caswell 1996; see also Chapter 9). These approaches are concerned with actual, measured effects that have already occurred, and are considered retrospective analyses. In other cases, investigators will have measured population attributes or decomposed toxicant impacts on population growth rate, but are also interested in comparing the relative effectiveness of remediation strategies or potential future population size with a given level of toxicant exposure. These analyses, focused on the future, are prospective in nature. Prospective analyses can be further classified into predictions and projections of population outcomes under given assumptions.

Prediction versus Projection

Prediction usually refers to instances when a single best answer is required. A frequent use of predictive modeling is that of setting harvest limits, which are point estimates of sustainable take from managed populations. Projection, on the other hand, is usually the comparison of multiple scenarios as a way of exploring the impact of various model parameters or estimates on future states. Projection analyses are frequently used in conservation and resource management to assess the relative effectiveness of 2 or more courses of action on some objective, such as population persistence. Such a technique could be extremely useful in applications such as comparing remediation strategies.

In addition, projections can be used to explore the relative importance of various factors or the precision of estimates on the chosen model outcome. This is known as a sensitivity analysis. Sensitivity analyses can be used to determine which model input variables have the greatest influence on model output, and therefore should be estimated as precisely as possible. This application is particularly relevant for predictive modeling. Sensitivity analyses can also help determine the relative importance of various factors in the model, as a technique to understand system dynamics. In this case, comparing specific scenarios is secondary to exploring the interplay of factors included within the model.

For example, a prospective analysis was used in a population-level ecological risk assessment for burrowing owls (*Athene cunicularia*) living in an agricultural site in California. Evidence of an interactive effect between food scarcity and moderate levels of p,p′-DDE in the eggs of the owls was found, which led to lowered reproductive success (Gervais and Anthony 2003). Food scarcity was a function of California vole (*Microtus californicus*) population dynamics. Although high-latitude vole populations are known to exhibit cyclic patterns, there were few data available for this mid-latitude species in highly disturbed landscapes. The relative effects of the frequency of vole population fluctuations and varying levels of p,p′-DDE contamination within the owl population was explored by projecting the results of differing frequencies of vole cycles and reproductive impairment on the growth rate of the burrowing owl population. From those analyses, it was clear that voles and their dynamics had far greater impact on the burrowing owl population than did the p,p′-DDE when vole population peaks were frequent, but the overall impact of the contaminants was mediated by the voles' dynamics (Gervais et al. 2006).

Because the role of such an analysis is to compare relative outcomes and explore parameter influence, not to determine an absolute outcome, these approaches can be extremely valuable when necessary data are lacking, conjectural, or based on other species or locations than the one of interest. They can be helpful even when important ecological processes such as density dependence and environmental stochasticity are so poorly understood that they cannot be modeled with any degree of confidence. Elasticity analyses are also used to explore the relative importance of model parameters on output, but they are nothing more than sensitivity analyses on a proportional scale (Caswell 2001).

Predictive models can be further delineated from projection exercises in that their primary purpose is to determine a solution to a problem (although see Recognizing Model Limitations and Appropriate Usage for limitations of this use). This is in contrast to the more exploratory nature of many prospective analyses, which primarily seek to increase understanding of a system. These distinctions can also be referred to as applied versus heuristic outcomes.

Heuristic versus Applied Models

Models can be generally classified into 2 general types — those developed for specific applications, where some level of prediction is required, and those whose major purpose is to increase understanding (Bunnell 1989; Starfield 1997). Both types have many features in common, the primary one being that they strive to simplify complex ecological systems. Although the dichotomy is not absolute, it is helpful to look at these 2 general goals separately to appreciate how model use in conservation and resource management can be applied to population-level ecological risk assessment.

Models have been used widely in the resource management fields to aid in decision-making involving specific management targets. These models strive to give empirical estimates with confidence intervals sufficiently narrow to be "useful," depending on the accuracy required of the model's intended use. There do not necessarily need to be any theoretical underpinnings for this purpose, although including relevant biological processes may improve the model's output. A model designed for a practical application may not specifically be intended to shed light on biological or population-level processes, but evaluating the performance of model output and modifying the model to improve performance can be useful in learning about ecological processes (Bunnell 1989; Starfield and Bleloch 1991).

Models can also be used to formally state the best understanding of a system or process, and can be used to increase understanding of the system. In this application, a model is considered successful if it leads to improved understanding, or if it highlights an issue that had not been previously considered. In this case, models that fail to produce the desired output may be very valuable, by indicating that understanding of the system is insufficient to project its dynamics. It is vitally important to determine the objectives of the modeling exercise clearly before starting, as the goals of predictive and exploratory modeling are very different (Bunnell 1989).

Models and the Inescapable Uncertainty

After demographic rates have been estimated, they themselves may be used as a basis for decision-making, but they are frequently used in models that are built to estimate quantities that are very difficult or impossible to measure directly. Such quantities that have been used in the conservation and resource management fields include population growth rate, time to extinction or quasiextinction, probability of extinction or decline, habitat suitability indices, and elasticities of population models (see Ginzburg et al. 1982; Lande and Orzack 1988; Caswell 2001; Maltby et al. 2001; Morris and Doak 2002; Chapter 9). Although population models are now a standard component of conservation and resource management decision-making, it is important to remember that models are not replicas of the truth, are not intended to be such, and therefore are themselves subject to uncertainty beyond the uncertainty in the input values.

Model uncertainty can arise from 2 major sources. The first is through simplification of ecological reality. Usually, only variables and processes thought to be important to estimate the endpoint of interest are included in the model. The set of variables deemed to be important to the system is determined through observations, the management context of the modeling exercise, available data, exploration of the impact of alternative combinations of variables on the model output, sensitivity analyses, and on theoretical grounds. In most ecological applications, models should be a compromise between the level of understanding of the system, and the kinds of questions necessary to answer (Levins 1966; Burgman et al. 1993).

The second source of model uncertainty is the mathematical abstraction of processes that are essentially ecological. The Ricker equation (Ricker 1975) is a mathematical construction that describes how a population changes in time under density dependence. It assumes that the rate of change of a population at time *t* depends on the population size at time *t*. The use of the first-order derivative, or difference equation, to describe how populations change in time is a mathematical construct based on an underlying theory about population growth rates. Nevertheless, it is still a representation of a natural process; individuals die, reproduce, ingest chemicals, and compete with other individuals within the system for resources. None of these activities is fundamentally mathematical and yet each has an impact on population abundance and ecological interactions, and may be represented in a variety of mathematical forms (Regan et al. 2002b, 2003a). The simplification and mathematical abstraction of ecological processes will always result in model uncertainty. This uncertainty is insidious and impossible to eliminate entirely, hence all models will be uncertain. It is clear then, that models need to be chosen in such a way as to reflect the level of understanding of the system, the ecological theory underpinning population models, and the management context the model is intended to support.

After a model has been selected, parameter uncertainty can also render population model results unreliable. The iterative nature of population models simulated through time leads to the propagation of uncertainty. This is exacerbated with more complex models that rely on many parameters and assumptions. Greater complexity in a model (i.e., more parameters) does not necessarily lead to convergence to the precise population-level endpoint. The greater the number of uncertain parameters there are in a model, the greater the uncertainty in the result. Reliable population-level ecological risk assessment needs to be a balance between model complexity and parameter uncertainty. This is a matter of addressing model uncertainty in conjunction with parameter uncertainty. Parsimonious models should be preferred over more complex models when there is uncertainty in the input parameters (i.e., a model should be employed that is representative of the system and entails a minimum number of parameter estimates). Monitoring and data collection efforts aimed toward reducing overall uncertainty can be better directed when there are fewer parameters to address. However, care needs to be taken to ensure that an appropriate level of theoretical foundation and realism is not sacrificed for simplicity as a result of parameter uncertainty. For instance, the absence of detailed data on density dependence should not rule out a model incorporating density dependence if there is good reason to believe that this might be an important process for the assessment population. In such cases, even though there may be substantial uncertainty in the type of density dependence or in carrying capacity estimates, the exploration of scenarios to determine the importance of these features can be extremely useful in guiding further data collection and honing in on the most appropriate model for the available data set. Appropriate treatment of parameter uncertainty, through Monte Carlo simulations and sensitivity analyses, goes some way toward addressing the inherent problems of making predictions and projections with population models with uncertain input parameters.

Coping with Uncertainty: Two Approaches

Perhaps one of the most concise statements about models is the quote, "all models are wrong, but some are useful" (Box 1976). A model is an abstract representation of reality and, at best, incorporates only the elements necessary to capture the dynamic of interest. The real standard is whether the model is appropriate and useful for the problem it was developed to address, recognizing that frequently decisions must be made based on limited information. In fact, some argue that the results of population models are so uncertain that they should never be used to predict the actual fate of a population (Beissinger and Westphal 1998; White 2000). However, what information we do have can still be used in a decision-making context, and in this too conservation biology and natural resource management have a good deal to offer the new field of population-level ecological risk assessment.

Conservation biology, by definition, concerns itself with species and populations that are rare, small, declining, or any combination of the three. Rarely are sufficient data available for robust estimation of parameters or model outcomes. Natural resource management, particularly of populations that are very difficult to sample, has a long history of decision-making based on data that are frequently less than optimal. It is unusual for a resource dispute to wait for appropriate data to be collected before decisions are made. Instead, approaches have been developed to make the most of whatever data are available, frequently while also highlighting further data needs. Although the 2 general approaches discussed are not mutually exclusive, and should often be used together, they still represent different philosophical solutions.

Solution 1: Simplify the Questions to Fit the Available Data

One approach to a lack of information in modeling is to simplify both the goals of the modeling exercise and the model's data requirements. If only rough estimates of demographic data are available, for example, some idea of the relative susceptibility of the life history stages of an organism to a toxicant, or responsiveness to a remediation scheme, can be gained from an elasticity or sensitivity analysis (Caswell 2001). Although this level of simplification prevents the exploration of complex hypotheses, management and conservation questions can be addressed well enough to identify a management target or justify a conservation decision. One of the best-known examples involves the loggerhead sea turtle (*Caretta caretta*) — elasticity analyses revealed the importance of the survival of large subadult turtles to population persistence. This in turn was used to justify the implementation of turtle excluder devices to prevent turtle bycatch by the shrimp fishery in the southeastern United States (Crouse et al. 1987; Crowder et al. 1994). Other approaches that require minimal data and caveats to sensitivity techniques are briefly discussed in Chapter 9, but particularly when few data are available, they have proven very useful in conservation and resource management.

Solution 2: Consider Multiple Working Hypotheses and Seek Relative Answers

This solution rests on the suggestions that Chamberlin made well over a century ago, and that were discussed earlier in this chapter (Chamberlin 1890). Rather than seeking a single solution, resource managers and conservation biologists have employed modeling as a comparative tool: Which of a given set of management options has the greatest likelihood of returning the desired result? What factors appear to be most important in determining the endangerment or recovery of a population?

The approach relies on prospective data analysis, in which projection of multiple scenarios allows comparisons of model outcomes. Although one of the most frequently used model outputs is time to quasiextinction (defined as the time taken for a population to decline to some predetermined critically small size), other outcomes are also used, such as risk of extinction or decline in a management time frame, or long-term population growth rate in deterministic models. Models can range from quite simple, deterministic forms to highly complex individual-based models incorporating a full range of environmental and demographic stochasticity. They are similar in their application, however. Further discussion of the various possible models can be found elsewhere (Maltby et al. 2001; Pastorok et al. 2002; Chapter 9). To illustrate the approach, we will concentrate here on one particularly powerful class of models that have had enormous impact on conservation biology — stochastic population models.

The role stochastic population models can play in decision-making is often overlooked in risk assessment, but may be one of the most important contributions of conservation and natural resource management to population-level ecological risk assessment. The extra flexibility gained through stochastic population models allows assessors and managers to investigate and rank management strategies to reduce risk of decline or extinction of populations. Hence, in addition to assessing the impact of stressors on a system, stochastic population models can assist in identifying the strategies most likely to succeed in reducing the long-term effects of those stressors. Stochastic population models can also assist in planning research, data collection, and monitoring programs to improve knowledge and management of the system of interest and to guide resources in fruitful directions. While there is a necessary cost for this added flexibility in the form of extra data requirements, the added flexibility and information gained from population models is yet another reason to consider an ecological risk assessment at a more biologically complex scale than the organism level.

Population viability analysis (PVA) uses stochastic demographic population simulation models and other methods to assess the risk of quasiextinction of a species, metapopulation, or population, over a specified time frame, in the face of demographic, environmental, and genetic stochasticity (reviewed by Possingham et al. 2001; Beissinger 2002; Beissinger and McCollough 2002). A set of possible management or conservation scenarios is created. Each hypothesis is projected in the form of a PVA model, and the outcomes of the models compared across the set. Monte Carlo simulations are employed to incorporate variation and uncertainty in demographic and environmental parameters to provide a suite of population trajectories, from which the probability of extinction or decline is calculated (Regan et al. 2003a). By incorporating important life history traits, such as survival rates, fecundity, extent and rate of dispersal, density dependence, and demographic and environmental stochasticity into population models, the chance of persistence or decline of a population can be quantified over a relevant time frame. The likely impacts of disturbance, stressors and catastrophes can be explored and the effects of management actions can be incorporated to provide a ranking of viable strategies to assist the persistence of the species or population. PVA provides quantitative measures for the likely fate of populations contingent upon underlying assumptions. Furthermore, the importance of spatial arrangement on the persistence of metapopulations can be explicitly investigated to identify which subpopulations are likely to become locally extirpated in the future (Akçakaya 2000a,b). There are many examples in the literature of the application of PVA to species management. For example, Regan et al. (2003b) used a stochastic individual-based population model to rank fire and predation management options for an endangered plant threatened by adverse fire regimes and massive seed predation in years directly after a fire. They estimated the optimal combination of fire interval and seed predation reduction necessary to decrease risk of extinction. They showed that there was a tradeoff between fire interval length and seed production and used this tradeoff to rank a range of fire and seed predation management options. For a compendium of management applications of stochastic population models to a variety of taxa, refer to Akçakaya et al. (2004).

The application of PVA and the general comparative approach has become much easier because of the availability of a number of software packages such as VORTEX (Lacy 1993), ALEX (Possingham and Davies 1995), and RAMAS GIS (Akçakaya 2002). Comparisons of some of these packages can be found in Lindenmayer et al. (1995), Mills et al. (1996), Brook et al. (2000), Pastorok et al. (2002), and Regan et al. (2003a). It is important to note, however, that comparative analyses may be carried out in general software packages such as SAS (SAS Institute, Cary, NC) or MATLAB (The MathWorks, Natick, MA), and even spreadsheet programs allow quite complex modeling as a first step in exploration. Whatever the software used, however, the modeler must still deal with uncertainty, and ensure that the model is interpreted and applied correctly.

APPLICATION, INTERPRETATION, AND COMMUNICATION OF MODEL RESULTS

Although practical tools and methods are of great value, perhaps the most important lesson to be gained from the fields of conservation biology and natural resource management is in the appropriate use and interpretation of models. First, the realization that models are not truth, and are in fact not meant to be truth, is a critical step to successfully using them in research and management. Second, model output used for decision-making or other applied purposes is only as good as the quality of input data — reliance on model output should be modified accordingly. Third, models are only one part of the decision-support process, and should be updated as more information becomes available. The use of adaptive management techniques has great potential for ecological risk assessment. Finally, the most successful modeling work will be achieved as a collaborative effort among the various stakeholders.

A Realistic View of the Role of Models

Too frequently, models are seen as "the" final answer, rather than as a tool and an aid to decision-making. Models are nothing more than abstractions of reality that attempt to simplify the system enough so that defensible decisions can be made. Assumptions should be made explicit, and every effort made to incorporate understanding of uncertainty into the modeling process. The importance of ensuring that all parties have a realistic understanding of the goal of the modeling exercise can hardly be overemphasized (Bunnell 1989).

Recognizing Model Limitations and Appropriate Usage

An illustrative example of unrealistic interpretation can be found in the history of the use of PVA. Although early uses of these analyses emphasized prediction of absolute risk of extinction to a population, it became clear that this was not an appropriate use for these models (Gilpin 1996). Rarely are there adequate data available to parameterize the complex models that are frequently built to assess population viability (Beissinger and Westphal 1998; White 2000), yet small changes in parameters may have dramatic impact on the model output (Taylor 1995; Pascual et al. 1997). Biological processes that are intuitively important may be too poorly understood to estimate, such as dispersal or density dependence. An additional problem is that it is simply not feasible to test the accuracy of most models predicting time to extinction; the point of the exercise is generally to prevent extinction, although this is not always true (for example, the extirpation of a pest species). In addition, stochastic models are examined over long spans that prevent validation in many instances (Beissinger and Westphal 1998; but see also Brook et al. 2000).

Although the original modeling format has remained much the same in the last 30 years, the application of PVA has undergone a dramatic shift away from attempting to accurately model one scenario, to comparing relative outcomes of a set of scenarios. This approach better addresses some of the major criticisms, while still making the most of available data and information about the system.

Fisheries management has long dealt with the need to make reliable forecasts of sustainable harvests (Hilborn and Walters 1992; Rose and Cowan 2003). The goal of setting sustainable harvest can be extremely difficult, plagued by inadequate data and conflicting interpretations. For example, red grouper modeling conclusions on stock status ranged from severely overfished to allowing for slightly increased harvests over the course of just a few years, which led to vastly different projected harvests. The difference in the results appeared to be primarily because of the incorporation of new data and reinterpretation of previous data (Rose and Cowan 2003). Based on this and other experiences, Rose and Cowan (2003) warned of several "portable lessons" for terrestrial ecologists, and these apply equally well to population-level ecological risk assessment. First, forecasting will need to be done under oversimplified conditions, and results will be presented without all of their uncertainties and caveats. Results will be closely scrutinized, but often will fail to change anyone's mind. Second, there will almost never be enough data, and calls for more data will be ignored until there is a crisis. Modelers and decision-makers will have to make do with what is available. Third, tools in modeling, as with everything else, will run in cycles of popularity and calls to use the model du jour can easily further confuse the issues. Fourth, ecological understanding will eventually improve as management issues feed back onto the science, demanding better methods and models. Multidisciplinary work will be crucial. The final 2 points relate to the interface of science and policy: there will always be challenges in explaining model outcomes and the inherent uncertainties to stakeholders, and doing so will require time, patience, and absolute honesty (Rose and Cowan 2003). Models are a crucial part of management decisions, but they cannot be substituted for the decisions themselves.

Models as One Step in the Decision-Making Process

Population viability analysis or other modeling approaches are only one component of the larger decision-making process. Figure 7.1 displays a general framework for PVA and how and where population models fit into the overall process (Akçakaya et al. 1999). The first step in the process is to identify and frame the problem to be addressed. This will involve collating data and information on the population and specifying the management context, including feasible management alternatives (Step 1). The model chosen to investigate projected population scenarios will depend on the available data and the types of questions managers need to answer (Step 2). The model parameters are then estimated and key data gaps are identified (Step 3). The estimation of parameters will depend on the model structure and the data already available. If necessary and feasible, further field studies may be performed to obtain missing data. In many circumstances, it may not be possible to delay assessment for further data collection, or additional field studies may be unfeasible because of resource and time constraints. In such cases, it may be necessary to incorporate subjective judgment from a species expert or gather collateral data for related species and habitats (Andelman et al. 2001; Johnson and Gillingham 2004; Martin et al. 2005). Building a model requires the incorporation of existing information into projections of population decline or persistence under different scenarios, different

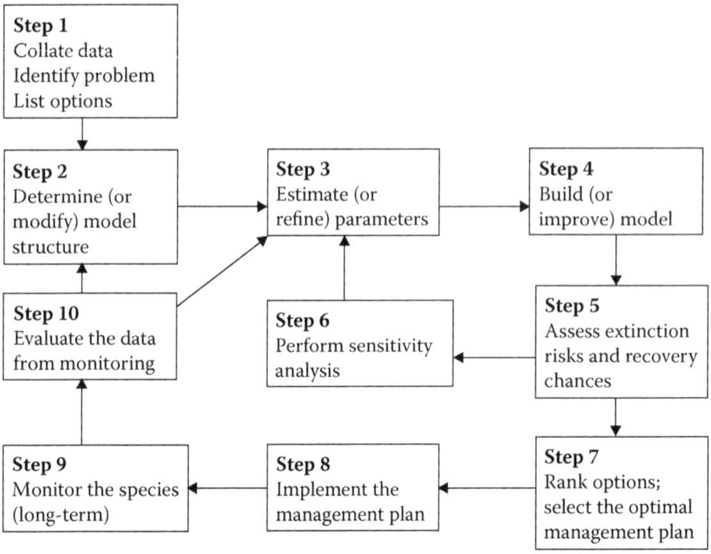

FIGURE 7.1 The main components of a population viability analysis (from Akçakaya et al. 1999). Reprinted with the permission of Applied Biomathematics®.

assumptions about environmental conditions, and different management alternatives (Steps 4 and 5). The extent of uncertainty and variation in the model assumptions and parameters will affect the conclusions made about possible population outcomes and the most appropriate management alternatives. Hence, it is crucial that a sensitivity analysis be performed to determine the assumptions to which the model results are most sensitive (Step 6). A sensitivity analysis can assist in determining the parameters that require more careful estimation. At this stage in the process, Steps 3 to 6 in Figure 7.1 may be repeated for multiple iterations until the assessor or manager is satisfied that the available data and knowledge of the system will not yield further improvements to the model input and output.

Ranking and implementing management activities, monitoring the effects of those activities on populations, and using monitoring data and analyses to assess conservation or natural resource management in the light of social and resource needs are part of the ongoing management process (Steps 7 to 10). After a decision has been made (Step 7) and the selected alternative implemented (Step 8), ongoing monitoring (Step 9) must include sampling to check the effects of management on species of concern. Using ongoing monitoring to gauge population responses, estimate population parameters, and verify biological relationships incorporated into population models will incrementally improve both data and models and, ultimately, the management decisions that rely on them (Akçakaya et al. 1999; Lindenmeyer 1999; Noon 2003).

Understanding the effects of ongoing management actions is a necessary component of a monitoring program. Gathering information that can lead to improved management requires active experimentation to determine the effects of potential management actions that differ from the ongoing actions. This is referred to as active adaptive management (Bormann et al. 1999; Committee of Scientists 1999) and it provides an opportunity to explore the possible effectiveness of management plans that are not yet implemented but may need to be implemented in the future. In this way, population-level ecological risk assessment should be seen as an ongoing and iterative process, with management plans adapted and revised according to the new information gained throughout and feedback loops leading to improved decisions.

A Precautionary Tale

As the issues with PVA illustrate, information frequently is lacking that might be extremely helpful in modeling a system and making a decision. One aspect of this is that 2 models with very different outcomes may fit the available data equally well, leaving the decision-making process without clear guidance. How should models be evaluated in that case? Again, conservation and natural resource management have grappled with these issues, and although frameworks for decision-making in the presence of uncertainty are quite well developed, frequently it seems that such situations wind up as the basis of litigation. An example of inconclusive modeling and the subsequent fallout is provided by one of the many court battles involving the conservation of the northern spotted owl (*Strix occidentalis caurina*) in the Pacific Northwest of the United States (*United States of America versus West Coast Forest Resources Limited Partnership and Dean Mt. Logging Co., Civil No. 96-1575-HO*).

A pair of federally endangered spotted owls nesting in western Oregon on private land became the focal point of a legal battle between the US Fish and Wildlife Service and a consortium of timber companies. At issue was the fate of a stand of valuable old-growth forest. The spotted owls did not nest within the stand, but arguments about the importance of the stand to the maintenance of the pair became the defining issue in the case. The judge, deciding that there were insufficient data available, ordered that a radio telemetry study be done to determine whether the owls used the contested stand. Over the next year, the owls were detected in or adjacent to the stand on numerous occasions. Both sides then modeled habitat selection by the owls based on the radio telemetry data to assess the stand's importance to the owls. The US Fish and Wildlife Service biologists created a model incorporating central-place foraging behavior by the owls. Their model suggested that the owls used the stand far more often than could be expected by chance, and therefore was important in the owls' continued survival and reproduction. The consultants hired by the timber company presented a second model, which included different assumptions on spotted owl foraging behavior. Their results suggested that the stand was not selected by the owls out of proportion to its availability and therefore could be harvested without negative impacts on the owls. Both modeling techniques were subsequently published in the peer-reviewed literature (Rosenberg and McKelvey 1999; McDonald and McDonald 2002).

The judge listened to the arguments, asked questions, and examined various graphs and charts. Then the hearing was interrupted by the arrival of a local law school class that had come to watch the case. The judge called a halt to the proceedings, stripped off his robe, and went to talk to the law students about the case before him. "I'm told that these 2 models both accurately depict spotted owl behavior," he told them. "But I don't understand why this model, this third-order polynomial, is any improvement over this simpler model over here." He pointed in turn to the US Fish and Wildlife Service model and the timber consortium model (J.A. Gervais, personal observation). Ultimately, the judge decided in favor of the timber consortium. The old-growth stand was harvested in 1999. Interestingly, the pair of owls that occupies the area around the cutover area have produced only 2 young in the 8 years since they have either not nested or have failed when they attempted to nest (E. Forsman, USDA Forest Service, personal communication).

This particular example illustrates several issues relevant to the use of models in population-level ecological risk assessment, and highlights the need for a more holistic solution to resource conflicts. First, it is entirely possible for more than one model to fit the system's dynamics adequately, but for each to lead to very different conclusions. In this case, the effort to obtain more data might have been expanded, or the judge might have made a very explicit argument about the tie-breaking value system he relied on in his judgment, whether economic concerns and private landowner rights or the principle of parsimony. Models cannot remove value judgments from decisions. If there are no conclusive outcomes, either more information must be obtained to discriminate among the models, or decisions must be made using other criteria. In this case, that the timber industry model did not adequately capture the details of the system should be used in future decision-making contexts, even though it is too late for this particular instance. Such efforts to learn from previous actions are at the core of adaptive management.

The judge had made it clear to the law students that he himself did not understand the difference between the 2 models, and in fact disliked the complex mathematics used in one of them, because he did not see why they were necessary. Models built for applied purposes must be accessible to the end users, and ideally these users should be involved in the identification of assessment approaches (Bunnell 1989). In any practical model-building exercise, it is important to be aware of how models fit into the overall management context and the role models play in informing management action. For many managers and practitioners, models form only one part of the risk assessment process. When the output of models is used to guide management, it is crucial that stakeholders have some sort of input to give the models credibility. It is important not to lose sight of the fact that many managers, stakeholders, and members of the legal profession have very little knowledge of the theory, structure, and interpretation of models and of their benefits and limitations. For many managers and stakeholders, the modeling process is a black art and model output is treated with the same skepticism as an oracle's portent. If models are to have any practical import, it is necessary that the people who will be guided by their output have some understanding and ownership of the modeling process and the ultimate results. This may require enlisting an agency expert to provide life history information for the species of interest, agency employees to provide a management context, and other stakeholder input to provide alternative scenarios and points of view for exploration in the model. In extreme cases, involvement of managers and stakeholders may require workshops to elicit information from groups and education of the modeling process and the limitations and benefits of model output. Although these activities can seem time consuming and perhaps unnecessarily exhaustive at face value, they are invaluable investments and can ensure that model results will not be dismissed without good reason. The close involvement of agencies or clients in the modeling process can also help prevent development of redundant and unrealistic models; the modeler is rarely as informed of the biological system and management context as are the stakeholders, managers, and species experts.

More Constructive Approaches

Although the previous example highlights an insidious impediment to the use of population models in guiding policy, the modeling process can prove to be a useful and fruitful tool in guiding management decisions when appropriate communication and involvement of relevant parties occurs. For instance, in 1999, Forestry Tasmania in Australia requested a population viability analysis for a rare land snail occurring on private lands managed for timber harvest, but administered and regulated at a state government level. The population model was constructed by academics with close involvement of an agency expert on the life history of the snail and the agency conservation scientist. All aspects of the model were planned and discussed with agency workers and the relevant Forestry Tasmania managers. The model was constructed in a software package that was delivered to the agency and explained through a series of workshops. This enabled some flexibility and a sense of ownership of the model and its results by the agency. Furthermore, agency workers were instructed on how to run the model and change scenarios as needed. Two workshops were held for the purpose of gathering information on the forest management context and the species' life history. Additional workshops were held to explain the model structure and results and to elicit the opinion of managers and forest workers on relevant scenarios for exploration with the model. As a result of the intense collaboration of the agency with the model builders, the results of the model made a definite contribution to the subsequent decisions on harvesting strategy. This is a clear sign that the involvement of representatives of the relevant parties in a management context can result in the successful incorporation of a model into the ultimate management plan (Regan et al. 2001; Taylor et al. 2003). Again, fisheries science and policy has long recognized the need to involve stakeholders and openly acknowledge and incorporate uncertainty, and many approaches to both have been widely applied in fisheries management (e.g., Berkson et al. 2002; Rose and Cowan 2003). Both terrestrial resource management and population-level ecological risk assessment should be aware of the precedents.

The 2 examples described previously clearly highlight the importance of interpretation and communication of model results to a range of interested parties. The perhaps justifiable perception that many scientists are unable to effectively communicate technical results to nonscientific stakeholders and the general public is a thorn in the side of any endeavor to inform policy with science. This is one of the major lessons to be learned from conservation and natural resource management. Because conservation management is a highly volatile and sensitive issue that has social, political, and economic implications, it has been necessary to involve a vast array of stakeholders and contexts into the PVA and subsequent decision-making process. Although this tends to complicate the analysis greatly, delay recommendations and outcomes, and increase the costs of the analysis, it may eliminate potential controversies and legal action that would also delay outcomes and increase costs substantially. Ecological risk assessment is as equally volatile and sensitive an issue to stakeholders and society at large and one would expect that scientific methodologies and results need to be interpreted and communicated accordingly. That societal risk perception has received much attention in the risk assessment literature identifies communication of risk as an important area for improvement (Johnson and Slovic 1995; Slovic et al. 1997; Pidgeon 1998; Bickerstaff 2004; Mullet et al. 2004).

CONCLUSIONS

As population-level ecological risk assessment moves into the complex and uncertain realities of ecological systems, there is much to be gained from examining the experiences and tools of the fields of conservation biology and natural resource management. It is unrealistic to suppose that all risk assessors will develop the knowledge and skills necessary for conservation management, or that conservation and resource management professionals will develop expertise in ecotoxicology. However, collaborations among the professions will greatly enhance their ability to meet their respective goals. As both sides emphasize the analysis and management of populations of organisms, the goals are essentially the same. Further, the data requirements are increasingly similar. Perhaps this can be the motivation not only for greater collaborative efforts among the disciplines, but also for greater emphasis on monitoring and data collection on numerous species that could be used by all parties in population management. Sharing of knowledge, skills, and data across disciplines may be the only way to adequately address the many challenges facing conservation, resource management and population-level ecological risk assessment.

This discussion has been limited to that of population-level ecological risk assessment, but the need to keep the larger ecological context in mind cannot be overemphasized. The next step in ecological risk assessment will be incorporating formal consideration of the many interactions inherent among the components of natural systems (e.g., Landis et al. 2000; Clements and Newman 2002; Pastorok et al. 2002). This will require an even greater level of cross-discipline training and collaboration.

8 Empirical Approaches to Population-Level Ecological Risk Assessment

Tina M. Carlsen, S. Jannicke Moe, Sandra Brasfield, Peter F. Chapman, Ary Hoffmann, Wayne G. Landis, Diane E. Nacci, Helen Noel, and Julann A. Spromberg

INTRODUCTION

Empirical studies and modeling efforts interact intimately. Models need to be grounded in empirical observations, and the model predictions need to be tested against reality. Conversely, field observations and observational studies need to be guided by the insights provided by models describing ecological relationships and dynamics. By empirical studies we mean the collection of empirical laboratory or field data and its evaluation with respect to an attribute of the assessment population to characterize an effect of a stressor. The evaluation of the empirical data can involve some mathematical manipulation, but would not typically be considered modeling.

Modeling, as used here, refers to a mathematical construct to explore the possible effects of a stressor exposure on a specific population attribute, and does not include statistical models for parameter estimation.

In this chapter, we examine empirical methods that can be used in evaluating population-level ecological impacts, both as stand-alone assessment techniques and for use in models. We begin by further describing the empirical data useful in population-level ERAs, initially introduced in Chapter 3. We next describe methods available for collecting empirical data and evaluating that data for evidence of population-level impacts. We follow this with examples of how these methods can be used in addressing 2 of the management scenarios in Appendix 1 — pesticide registration and hazardous waste sites. We conclude with a discussion of how empirical methods could be more fully used in population-level ERAs.

The roles of empirical data fall into 3 basic categories:

- field observations and experiments to directly measure population effects,
- field observations and experiments to estimate parameters for models, and
- field observations and experiments to validate or confirm model results.

Empirical studies that directly measure population effects are used to quantify the causal relationships linking the stressor to the observed impact. These types of studies are very diverse and include, for example, conventional toxicity concentration-response research, field studies describing the impacts of fire on forest species diversity, and the linkage between landscape fragmentation and population abundance. These relationships can represent a stand-alone risk assessment or be used directly in a model, such as models exploring how a stressor influences the likelihood of (quasi)extinction. Empirical data from these studies can also be used to derive specific parameters that might be necessary for risk models. Estimations of population numbers, age structure, the distribution of a toxicant, and the construction of survivorship tables are all examples.

Figure 8.1 illustrates the linkages between modeling efforts and empirical studies in the risk assessment process for the prospective (e.g., new product registration, as described in Pesticide Registration and Appendix 1) and retrospective cases (e.g., hazardous waste sites, as described in Terrestrial Wildlife Populations Inhabiting Contaminated Sites and Appendix 1). There are subtle but important differences in form. In the prospective case there is a candidate material, land use or action that is being evaluated. Information is collected on the assessment species that can include parameters examining the mode of action, the geographic range of the population, habitat requirements, and other parameters. Modeling is then typically conducted to describe the relationship between the prospective action and the impact on the assessment population. Thus modeling has a central role in the prospective case. The model is evaluated and additional empirical data are collected as necessary to ensure accuracy. Finally, predictions are made that can be tested using a variety of empirical techniques.

In a scientific process, empirical data are used to validate or confirm (*sensu* Oreskes et al. 1994) the predictions made by ecological models. In many cases, it might not be possible because of resources, time or other constraints to directly test the model predictions for the entire risk assessment. However, tests of predictions of the model may be possible for localized regions or in microcosm or field tests. If the model is not confirmed (this is the term of Oreskes et al.) then further refinement or the use of alternative models can be applied.

The retrospective case implies an impact might have occurred. Empirical studies are conducted to determine the level of impact. These studies use available data on potential stressors, their location, types and distribution of habitat, and other ecological parameters. If correctly designed and of sufficient detail, empirical studies alone can, in some cases, be sufficient to determine the level of impact to an assessment population from a stressor. Thus empirical studies have a central role in the retrospective case. However, modeling is often useful in further evaluating possible linkages between proposed stressors and the impacts identified in the population. These models can inform lines of evidence in a weight-of-evidence decision

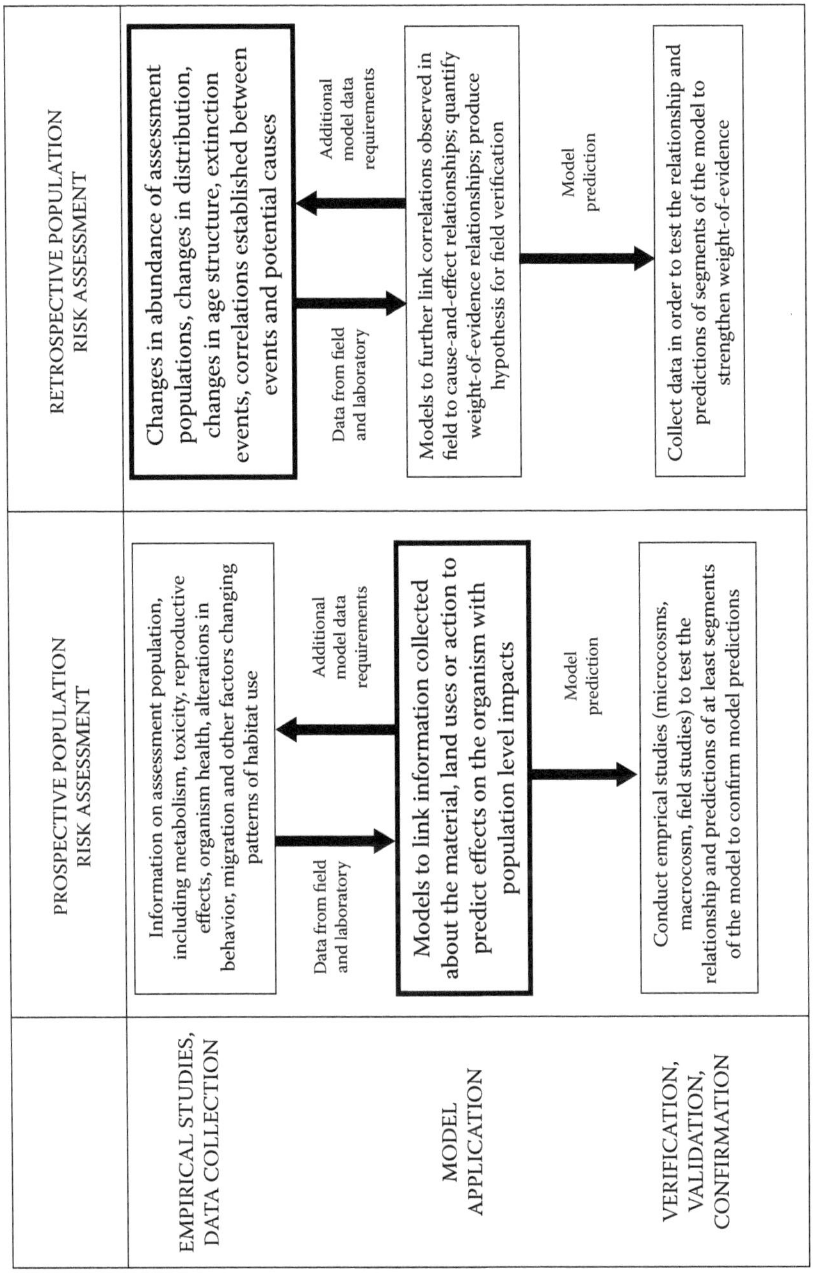

FIGURE 8.1 The link between empirical studies and modeling for population-level ecological risk assessment.

process. Outputs from the models can be used to assess the probability of a stressor causing the observed impact. In many cases, additional data might be required to reduce uncertainty or to establish a better linkage between cause and effect.

The combination of data and population models has been demonstrated to be useful in generating predictions and in estimating risks from contaminants. Models can also be useful in discovering causal relationships in a retrospective risk assessment or weight-of-evidence application.

Ideally, the predictions of the models should be compared to the observed patterns of impacts. In some cases, mechanistic studies might be required to provide evidence linking stressors to observed impacts. In other instances the predictions can involve parameters not previously observed in the environment. For instance, the model predicts that an unnoticed habitat parameter will change in concert with the observed impact. Further examination of the habitat confirms that the parameter did change as predicted. This interaction between model and empirical data bolsters the weight of evidence. In some situations, 1 line of evidence will be supported, in other situations several potential causes will be identified with no means of allocating relative contributions among stressors.

Empirical data and modeling interact closely. Observations of the ecological system inform the modeling process and test the modeling assumptions and predictions. Models can supply a framework for collecting data and distinguishing between candidate causes. In the risk assessment of populations there is a strong link between these 2 processes.

EMPIRICAL DATA USEFUL IN POPULATION-LEVEL ERAS

Depending on the type and scale of the risk assessment being conducted, a wide range of empirical data might be required. However, all risk assessments will require 3 basic types of data (the level of resolution required depending on the type of risk assessment being conducted):

1) data describing the attributes of the site, region, or ecosystem type under consideration,
2) data describing the biological attributes of the assessment population, and
3) data describing the population stressor attributes under consideration.

Data can be obtained from published literature and available databases, and from field or laboratory studies. The method used will likely be determined by the scope and detail of the assessment. The primary scientific literature provides a valuable resource for this information, and can be searched effectively using a variety of commercial databases. In addition, many government agencies have supported the development of biological, toxicological and geographic data resources. A review of these sources should be done during the scoping phase of the assessment and before the development of field studies. However, a set of data quality goals should be set a priori for the inclusion of data from any source. Often the data about the data (metadata) — what measurement methods were used, when the data were collected, and other factors — are not included. Tracking and compiling the uncertainties of the data are also critical to the documentation and estimation of the uncertainty that will be part of the final risk calculation.

This section emphasizes field-based population studies, although we recognize that laboratory-based population studies are widely performed and used in ecological risk assessments. The use of laboratory-based population studies in agrochemical risk assessments are summarized in Pesticide Registration in this chapter.

Site Attributes

Site attributes are important to population-level ERAs for a number of reasons. They characterize the habitat in which the assessment population is found. They can be used to estimate the likelihood that an exposure to a stressor might occur. Data on site attributes can be used in purely empirical population-level ERAs, as inputs into models, or a combination of both. The degree of detail and resolution required is determined by the type of assessment being performed. For prospective risk assessments such as pesticide registration or manufactured chemical regulation, all that might be required is a set of generic environmental characteristics that can be used to define exposure scenarios broadly. For retrospective assessments such as contaminated site assessments, detailed descriptions of site attributes might be needed to identify populations at risk and to develop parameters for site-specific assessment models. This section emphasizes site-specific attributes because the generic scenarios and site characteristics used in prospective assessments are usually specified in guidance documents.

Table 8.1 lists example site attributes that can be used in a population-level ERA. It is unlikely that data describing all the attributes listed in Table 8.1 will be required for every population-level ERA. Rather, the type and scale of the assessment will drive the selection of site attribute data to be collected.

Stressor distribution refers to the distribution of the particular stressor under evaluation across the landscape, or in time. In the case of a chemical stressor in a retrospective risk assessment, chemical distributions would be measured across the site or landscape. It could also be the current distribution of an invasive species, or maps on the distribution of past flooding or fire disturbance.

The distribution of a stressor across the landscape is an important variable in selecting the assessment population.

TABLE 8.1
Potentially useful site attributes for population-level ecological risk assessments

Site attribute
Scale for consideration
Topography
Bedrock geology
Hydrology
Hydrogeology
Soil types
General water chemistry
Stressor distribution
Productivity
Distribution of vegetation and animal communities
Current and future land use
Climate patterns
Adjacent land use (i.e., industrial development)

The section on Habitat Characterization describes some of the methods for collecting and analyzing site or habitat data. In a risk assessment conducted at a hazardous waste site in the United States, habitat characterization revealed that deer, while present at the site, did not use regions of the site containing cadmium-contaminated soil (Ferry et al. 1999). In this assessment, the site attributes of plant (appropriate deer forage), animal (locations of deer within appropriate habitat), and stressor (locations of cadmium-contaminated soil) distributions were the only attributes used to conclude cadmium in soil was not a risk to deer populations.

Biological Attributes of Assessment Population

As discussed in Chapter 3, the assessment population can range in composition from a number of individuals that are a subset of the actual local biological population, to the biological population, to a metapopulation. Thus, empirical data needs can span the range from data on individual organisms to aggregate data on populations.

In general, empirical biological data are collected primarily on individual organisms, with population attributes estimated from the data collected on those organisms. Chapter 3 gives examples of important biological attributes used in population-level ERAs. In this chapter, we make a further distinction between primary and secondary population attributes. A primary population attribute is directly estimated from organism-level attributes, whereas a secondary population attribute is estimated from primary population attributes. For example, 1 measure of abundance is the number of individual organisms within an area. The empirical data collected focuses on organisms (individual counts), and the primary population attribute is the sum total of counts. If repeated counts are performed over time, a secondary population attribute, population growth rate, can be estimated. Biomass is also a commonly measured primary population attribute, especially in aquatic and vegetation studies.

TABLE 8.2
Potentially useful biological attributes for population-level ecological risk assessments

Attributes of organisms	Primary population attributes (estimated from organism attributes)	Secondary population attributes (estimated from primary population attributes)
Demographic characteristics	**Demographic characteristics**	
Presence	Abundance	Population growth rate
Age and stage	Population size	Variance of abundance
Age and stage at death	Number	Carrying capacity
Sex	Biomass	Density dependence
Size	Population Density	Probability of (quasi)extinction
Individual weight	Age and stage structure	Time to recovery or extinction
Individual length	Sex ratio	
Somatic growth rate	Recruitment	
Reproduction	Survivorship	
Fecundity		
Number viable offspring		
Development		
Physiological characteristics		
Vital physiological rates		
Respiration rate		
Food intake rate		
Metabolism		
Diet		
Landscape/habitat use	**Landscape/habitat use**	
Movement/dispersal ability	Habitat preference	Dispersal
Home range	Critical patch size	
Location (specific time)	Spatial distribution	
Genetic characteristics	**Genetic characteristics**	
Individual genotype	Measure of genetic diversity	

Table 8.2 summarizes the biological attributes of the assessment population that can be used in a population-level ERA. Again, it is not expected that data on all the attributes listed in Table 8.2 will be used in every population-level ERA.

Empirical data on biological attributes will be collected for 2 primary purposes. One is to parameterize or test conceptual or mathematical population models for use in either retrospective or prospective risk assessments. The second purpose is for use as direct indicators of exposure to evaluate the potential effects of the stressor under consideration.

Attributes of Organisms

Measurable attributes of organisms within the assessment population are typically used to estimate population parameters. These attributes are also widely used as indicators of stress imposed on both organisms and populations. Changes in somatic growth rates and other organism-level performance measures can often be detected before changes in recruitment or abundance are noticed. Key performance measures include somatic growth, reproduction, and survival. For example, a decrease in fecundity translated across a significant proportion of the population can cause a decrease in the population growth rate. Populations and communities can exhibit a graded response to stressors, and a properly selected suite of organism-level performance measures can contribute to detecting and diagnosing impacts on these higher levels of organization (Munkittrick and McCarty 1995; Munkittrick and McMaster 2000).

Attributes of organisms can roughly be grouped as demographic attributes, physiological attributes, landscape or habitat use attributes and genetic attributes. These attributes are briefly described in Chapter 3. Demographic studies are probably the most common empirical method used to evaluate populations. There are several studies at the US national level designed to develop demographic data of specific taxonomic groups. Some examples include the Monitoring Avian Productivity and Survival (DeSante et al. 1995) and the North American Waterfowl Banding Program. Physiological attributes for organisms in the assessment population are most commonly obtained from literature sources. Published laboratory studies on the species of interest (or a related species) should be reviewed. The Wildlife Exposure Factors Handbook (US Environmental Protection Agency [USEPA] 1993b) and the Wildlife Contaminants Exposure Model are also good sources for species natural history information and literature. The use of genetic attributes is a reasonably new addition to population-level ERAs (see Chapter 5). Genetic attributes are important because adverse changes in the genetic structure of a population can lead to increased risk of extinction, and because selection for genetic traits conferring resistance to a stressor can reduce extinction risk. Reduction of genetic variation can reduce the population's potential for adapting to the stressor. The section on Methods for Measuring Genetic Variation describes methods for measuring genetic variation in wild populations.

Primary Population Attributes

Population attributes include abundance, population structure, fecundity, and survivorship. Population abundance can be characterized in terms of total numbers, biomass, density (numbers per unit area or volume), or size relative to some critical resource. Population structure includes numbers or proportions of individuals in categories such as age, life stage, sex, or genotype. To properly characterize the temporal dynamics of populations, information is also needed concerning rates of reproduction (numbers of offspring per female per year or some other relevant measure of time) and survival. Collectively, these characteristics are termed "vital rates" or "demographic parameters." Survivorship and reproduction are often estimated from changes in abundance and age structure through time. Age is difficult to assess for many species, so life stage is often used to characterize population structure. In other species, particularly plants, fungi, and algae, biomass is a more useful measure than numerical abundance, and populations can be considered to consist of identical organisms without any explicit age, stage, or sex structure.

Attributes of organisms within the assessment population can be integrated to compute overall attributes of the population under study. These population attributes directly determined from measured data of individuals within that population are well suited for use as direct indicators of stress to the population. They can be evaluated for perturbations from expected values. They can also be used to parameterize models. Population attributes can be roughly categorized as demographic, landscape use, and genetic attributes.

For terrestrial wildlife, the spatial distribution of a population together with information on its dispersal ability provide information on migration rates and possible metapopulation dynamics (see Chapter 6), as well as help to identify potential stressors and the proportion of the population that might be exposed. Critical patch size is the quantity of area required to maintain a population. Although this concept seems straightforward, it is quite difficult to parameterize. Most data on critical patch size come from studies on different patches of habitat resulting from fragmentation. In these studies, critical patch size has been interpreted as 1) the minimum patch size below which the species is never found, 2) the minimum patch size below which the species is not present in 100% of the patches, 3) the minimum patch size that can sustain a viable population, or 4) the percentage of habitat that must be remaining for the species to be found in the landscape (Carlsen et al. 2004).

Population genetic markers are used to characterize genetic variation for a number of purposes. Selection of specific approaches and markers depends on the goals of the study (Sunnucks 2000; van Tienderen et al. 2002) as well as technological constraints of measurement and availability of markers for the species of concern. As long as markers are neutral or near neutral, they can be used to assess population processes such as gene flow, genetic drift, population bottlenecks, and mutation rate (Chapter 5). Population parameters typically estimated to describe population health include effective population size, allelic diversity, mean population heterozygosity, and spatial structuring.

Secondary Population Attributes

Several important secondary population attributes are typically estimated from primary population attributes. These include population growth rate, carrying capacity, variance in abundance, probability of extinction, time to extinction or recovery, and density dependence. These attributes relate to the dynamics of the population over time. All of them are model dependent, meaning that they are defined within the context of specific models, and different choices of models can lead to different estimates of attribute values. Model-dependent attributes are further described in Chapters 3, 4, and 9.

For example, the carrying capacity of a population, defined as the long-term average abundance of the population under relatively constant conditions, is an attribute of interest for risk assessment. The concept of carrying capacity ultimately derives from the parameter K of the logistic population growth model (see Chapters 4 and 9), which defines the equilibrium population size reached in a constant environment. Carrying capacity values are sometimes estimated using simple closed laboratory or long-term averages from field data. However, many factors related to population growth rate, resource availability, and stressor pressures interact to control the abundance of natural populations. Conceptually, it is often more realistic to think of population abundance as fluctuating around an average value that can itself change through time as a function of long-term changes in environmental factors such as climate or nutrient inputs. Rather than attempt to define a carrying capacity strictly in the traditional manner, a dynamic carrying capacity could be estimated by calculating a moving average of abundance over long periods. Changes or trends in the moving average could then be used to estimate dynamic carrying capacity within the context of a risk assessment model.

Life History

Life-history information can be very useful in the problem formulation stages of population-level ERAs. This information can be extremely important in developing conceptual and mathematical models for the risk assessment. Life-history information includes reproductive behaviors, average life span, migratory patterns, population age structure, ecological niche, habitat requirements, and life-history strategies.

Life-history information gathered from the assessment population can be compared to the standard or historical natural history information (or alternatively regional data on populations of the same species in a similar habitat) to identify potential differences that might have developed after an impact or that can indicate increased susceptibility to potential impacts. Thus, the incorporation of life history information into population-level ecological risk assessments can be very valuable.

Life-history strategies evolve that tend to maximize the fitness of individuals in their environment. Resource availability and survival potential influence the time and energy invested in each life stage (i.e., for fish the embryo, larva, juvenile, and adult stages). We use the term "strategy" to denote the way individual organisms allocate resources to reproduction under different circumstances (Ware 1984). This allocation is a heritable trait, which is subject to natural selection in the same manner as any other characteristic determining the fitness of an organism. The life history strategy develops as a set of tradeoffs in allocation of energy and resources between growth, current reproduction, future reproduction, and survival. Energy allocation tradeoffs between growth and reproduction evolve to maximize the individual's total reproductive value, which is composed of both organism survival and the individual's reproductive contribution to the population (Roff 1984). For organisms that continue to grow after reproductive maturity (fish, reptiles, amphibians, mollusks, and annelids) there is a negative correlation between instantaneous mortality rate and age at maturity (Koztowski 1996), which implies a tradeoff between resource allocation to survival versus reproduction.

Adaptations to local environmental factors such as seasonal conditions and severity of stochastic impacts occur over many generations, leading to the evolution of variation in life-history strategies. Some strategies can confer greater or lesser susceptibility to anthropogenic perturbations that fall outside of the species' adaptive history (Stark 2003). Life-history strategy can affect a population's resistance (i.e., the amount of perturbation that it can sustain without impact) and its resilience (i.e., the time required to recover from the impact of a perturbation). For example, populations of organisms that reproduce multiple times each year can be more resilient to the effects of a 1-time impact than are those displaying strategies with less frequent reproduction (Spromberg and Birge 2005).

Species with certain suites of life-history traits can be especially vulnerable to effects of environmental stressors, including nonchemical stressors. For example, Williams and Hero (1998) classified rainforest frogs of the wet tropics in Australia into 9 guilds based on reproduction, habitat use, temporal activity, body size, and microhabitat in an attempt to determine patterns of species loss. They found that there was no single feature shared by the declining species, but found a trend combining the characteristics of low fecundity and high habitat specificity (Williams and Hero 1998). These characteristics might render the populations more susceptible to extinction risk, and are similar to characteristics in fish listed under the Endangered Species Act in the United States. Many listed species, including the pupfishes and spinedace, are short-lived specialists inhabiting small ranges, which make them highly susceptible to habitat destruction or the introduction of exotic species (Trautman 1957; Carlander 1969a,b; Miller 1972; Morrow 1980; Sigler and Sigler 1987; Crossman 1991). As this implies, susceptibilities to anthropogenic impacts that are influenced by life-history strategy are important considerations in population-level ERAs.

EMPIRICAL METHODS FOR CHARACTERIZING POPULATIONS

This section provides a brief introduction to methods available for characterizing the attributes discussed in Empirical Data Useful in Population-Level ERAs. It is not intended to be exhaustive, but rather to identify important information resources, experimental methods, survey techniques, and software that can be used to support empirically based population-level ERAs.

HABITAT CHARACTERIZATION

Habitat characterization should be the first step in an empirical population-level ERA. It is conducted to identify the list of potential stressors and populations potentially at risk and forms the basis of subsequent analysis. Carefully conducted habitat characterization and evaluation can, in some instances, substitute for a more comprehensive population-level ERA.

Geographic information systems (GIS) provide ready frameworks for organizing spatial habitat data and tying toxicological, physical, ecological, and biological information together using a geographic reference. ArcInfo (www.esri.com/software/arcgis/arcinfo/) and ArcView (www.esri.com/software/arcview/) are 2 of the more commonly used programs, but other programs such as GRASS provide similar capabilities. GIS packages also possess analytical capability, for example, to calculate the area of coverage by a particular land use or habitat type, or to delineate the environment by watersheds. Clustering algorithms such as Jenk's optimization can be used to explore patterns in a spatial context. The ability of GIS to organize and analyze spatial data can inform empirical and modeling approaches to assessing risk. Data can be portrayed in a spatial manner, and many modeling programs, such as PATCH (Schumaker 1998) and RAMAS GIS (http://www.ramas.com/ramas.htm), take GIS input as the basis of the modeling process.

A large portion of the data on the physical characteristics of United States and other countries worldwide is now available through government organizations as layers (shape, raster, or vector files) that can be used in GIS software (Hall et al. 2001). Available data can range from very fine grain to very coarse. In the United States, for example, the STATSGO (State Soil Geographic) database provides coverage of the coterminous United States at a scale of 1:250 000, and a minimum area of delineation of approximately 625 hectares (1544 acres). Delineations depict the dominant soils that make up the landscape. For finer scale, SSURGO (Soil Survey Geographic) provides a county-level soils database.

The US Geological Survey (USGS) has layers available on land use, vegetation, topography, hydrologic units (watersheds), ecoregions, stream flow, and hypsography at a variety of scales and resolution, depending on the region. Digital elevation models are available at scales of 1:24 000 and 1:250 000. The 1:24 000 digital elevation models cover a smaller area of the Earth's surface and are available in 10- and 30-m resolution. Thirty-meter coverage is available for the coterminous Unites States, whereas 10-m coverage is relatively rare. Also available are digital orthophoto quarter quads, which are georeferenced, fully orthorectified, digital aerial photographs. Each dataset represents one-quarter of a 24K quad sheet. They are extremely useful as an overlay for verifying, revising, and supplementing data from other layers, and are an invaluable tool to aid environmental mapping. The USGS also provides water resources data pertaining to levels, flows, and quality for surface and ground water at 1.5 million sites in the United States. Meteorologic data typically are readily available for most sites from nearby weather stations and can be obtained for the United States from sources like the National Oceanic and Atmospheric Administration. The National Climatic Data Center provides an archive of world climate conditions. The National Ocean Service distributes bathymetric maps, bathymetric and fishing maps, regional maps, and geophysical maps for the US coasts and waterways. Similar data sources exist for other regions of the globe.

For biological characteristics of a site, state natural diversity databases, National Biological Information Infrastructure databases, heritage, and gap analysis programs

(GAP) programs are probably the best sources of information for the United States, in addition to the USGS land cover data. Vegetation maps are typically available for most regions. The spatial distribution of animals is somewhat more difficult to obtain, although known occurrences of some species are usually available in various natural diversity databases. In the United States, the Bird Breeding Survey is a good beginning source for the distribution of avian species. State and county departments of fish and wildlife collect distribution data for species considered threatened or endangered, as well as for commercially important species. GIS layers are also available from National Oceanic and Atmospheric Administration (NOAA) for environmental sensitivity index maps documenting information on shoreline classification (specifically, sensitivity to oiling), human-use resources, and biological resources.

Remote sensing has become an important method for acquiring site data. Aerial photographs have long been used for obtaining geophysical data, but the use of color and sensors that use light outside of the visible spectrum can supply information on the type of plant cover, the extent of macrophytes, and evidence of currents. Larger landscapes can be sampled from spacecraft as well, providing a very large-scale picture of the region.

Existing datasets and information from remote sensing can assist in placing site or region in context, but surveying and sampling on location can markedly reduce the uncertainty of a site-specific ecological risk assessment. Given the rapid change in landscapes because of the development of housing subdivisions, commercial sites, and changes in industrial processes, a ground survey can greatly reduce uncertainty in this area. Repeated visits to the site and comparisons with remote sensing data can also identify trends in the region.

Biological Surveys

For biological data, the initial survey will typically be very general, and is essentially a ground-truth of biological information collected from other sources. This fieldwork can include surveys to verify species present in the area. Occasionally, species not identified as part of the site in large databases can in fact occur in the study area.

The next major step in site-specific, empirical population-level ERAs is estimating abundance, densities, or other demographic parameters of the selected assessment populations. The method of estimating abundance or density will depend on the assessment population. It could range from actually counting the number of individuals per unit area (for example, plants), to catch and release studies (many animals), to the use of surrogates for more cryptic animals (scat, prints, etc.). The section on Demographic Studies describes demographic methods in more detail.

Depending on the type of assessment, information on the temporal dynamics of abundance or density of the assessment population can be important. These types of data require multiple surveys over time that are representative of the suspected fluctuations in the population. If the population is known to have the potential to fluctuate over large numbers very rapidly, then frequent samples taken over short time periods will be necessary. If the population is characterized by fluctuations that are of many decades in length, then samples might not need to be as often. The danger is in selecting a sampling period that does not support adequate estimation

TABLE 8.3
Demographic parameter estimation software

Estimation software	Examples
Abundance and density	CAPTURE (Rexstad and Burnham 1991) http://www.mbr-pwrc.usgs.gov/software.html
Abundance (mark–recapture)	MARK (White and Burnham 1999) http://www.cnr.colostate.edu/~gwhite/software.html
Survival, movement	SURVIV (White 1983); SURGE (Lebreton and Colbert 1986; Pradel and Lebreton 1991) http://www.phidot.org/software

of the extremes and potential rate of change in the numbers of individuals. Too low a sampling frequency will likely underestimate the extremes and the rate of change.

Field research can also place the population in ecological context within its community. What factors limit the size of the population in its natural environment? Is the population regulated by density-dependent processes (Chapter 4)? Is there intense competition for resources with closely related species? Are predators prevalent? Are there sufficient resources available for a population to increase or are resources in sharp decline? Answers to these types of questions obtained by field research should reduce the uncertainty inherent in the empirical risk assessment.

Demographic Studies

Demographic studies play a central role in population-level ERAs. These studies can provide much of the data described in Biologic Attributes of Assessment Population that are specific to the assessment population. They frequently are used as the primary field-based assessment technique, and also provide the primary data for many types of modeling efforts. Because of their central role in population-level ERAs, we describe the methods for conducting demographic studies in some detail.

A formal parameter estimate requires description of both central tendency and variation. In addition, methods and studies designed to estimate population parameters should be based on sound statistical methods, require minimal assumptions, and be robust to assumption violations. More generally, information about natural populations is often available opportunistically and therefore imperfectly collected for intended uses. To maximize their utility for risk assessment, it is important to state assumptions related to data collection and use clearly, and to rely on valid statistical techniques to account for variation. Studies that include consideration of appropriate estimation methods and statistical models permit valid inference about population parameters (for examples and reference to specific methods, see Williams et al. 2002). Several programs are available to estimate population parameters, including MATLAB (Catchpole 1995) and publicly available software with more specialized functions (Table 8.3).

For some types of wildlife, population abundance (and by extension, population density) can be determined by census (counted completely). In most cases, abundance must be estimated from samples. Traditional methods to estimate abundance have been described in detail (e.g., Seber 1982). Typical methods to estimate abundance include population surveys, radio telemetry, banding, and tag-recapture procedures. More extensive descriptions of these methods and the statistical analyses used to translate count data into useful estimators are provided elsewhere (e.g., Williams et al. 2002).

Demographic parameters often must be estimated as input to population dynamic models. Methods of estimation and brief descriptions are presented here; for a more extensive discussion, see Williams et al. (2002). Population growth rate can be estimated simply as the ratio of abundance in 2 sequential years, and therefore, its estimation from observational data is dependent on abundance estimates and their statistical treatments. Reproduction rates are often defined as the ratio of young per mature individuals, adjusting for probabilities of detection for each group. Estimates of survival rates and movement probabilities can be developed from statistical evaluations of data from tag-release studies. For example, annual survival probability can be estimated as the ratio of individuals marked in 1 year that are observed again the following year, accounting for the probability of recapturing the individuals. Similarly, annual movement probabilities from one to another site can be determined as the ratio of individuals observed at a second site after 1 year (e.g., accounting for similar probabilities of resighting because of mortalities).

Population models that are structured to incorporate organism age (or stage) can be used to project population structure as well as total abundance (e.g., Wood 1994; Manly 1997a/b), and to estimate intrinsic population growth rates (r, Chapter 9). Models of this type require estimates of survival and reproduction parameters for each age group. For many population-level ERAs, values obtained from published literature can be sufficiently accurate. The data requirements for estimating age-specific mortality rates from field data are much more extensive than are requirements for estimating abundance, age structure, or reproduction. If site-specific mortality rates are required, they must be obtained either by 1) repeated censuses of individual cohorts (groups of organisms born at the same time) or 2) measurements of the relative abundances of differently aged animals at a single point in time. Methods for implementing both approaches can be found in general treatises such as Seber (1982) for wildlife and Ricker (1975) for fish. Both methods are subject to numerous potential biases and uncertainties, which should be considered in selecting a method for use in any given application (see also Chapter 4).

Field Manipulation

Experimental manipulation of ecological systems can be a powerful tool in providing information for a population-level ERA. Model ecological systems can be created in which to examine, for example, the effects of toxicants, nutrient additions, different combinations of plant diversity, or the introduction of an invasive species. These systems have the advantage that replication is possible and therefore are amenable to a number of experimental designs. However, for practical reasons, they will often be beyond the scope and resources of a risk assessment. One disadvantage is that they typically include only a portion of an ecological system, so extrapolations have to be made to components not included in the test system.

Field manipulations can be useful tools for large-scale assessments. Large fields and even forests can be subdivided and different treatments applied to designated plots. Logistically, these experiments are difficult and often need to be conducted over long periods. Because it is never possible to ensure uniformity of experimental plots, textbooks on experimental statistics generally recommend a random-block design. Experimental manipulations at such scales ease many of the issues of extrapolation. However, data analysis can be quite complex, owing in part to the large number of variables left uncontrolled. Multivariate statistics are advisable because of the numerous factors present even in a simple field study.

With proper experimental design, the effects of spatial heterogeneity, differences in the species number and proportion in the habitat, and a variety of other factors can be examined using field manipulations. However, the utility of these systems has to be balanced with the cost necessary to conduct successful experiments at these scales. Experimental systems at smaller scales can be used to control the costs of data collection.

Mesocosms and microcosms are smaller systems that can be used to evaluate stressor-effect relationships. They are important tools for assessing environmental fate and effects of stressors on communities of plants and animals because results are considered applicable to natural systems (Giddings et al. 1997). These types of studies enclose part of an ecological system so that properties important to answering specific risk assessment questions are included in the design. Typically, a smaller set of species is incorporated in these types of systems, although the number of repeated treatments can be economically increased. Mesocosms and microcosms can be conducted in ambient conditions or inside a laboratory or greenhouse for a tighter control of the conditions. Because these experiments typically include multiple species and factors, the use of appropriate multivariate tools is recommended.

Compared with the field manipulations, microcosm and mesocosm studies are not as realistic ecologically, so extrapolations should be carefully considered. Conversely, statistical power is often enhanced as a result of the greater replication of treatments that can be accommodated. Careful design of experiment to answer specific risk questions can reduce extrapolation error.

For example, artificial streams can be defined as a constructed channel having controlled flow of water, and can be used to study some physical, chemical, or biological property of natural streams. The strength of laboratory stream research is not in the outcome of any particular experiment, but in a logical sequence of experiments, the results of which hook together into an integrated view of how the ecosystem works. This means that investigators need a strategy for synthesizing the results of a meaningful series of experiments, and for the integration of the experimental work with field observations to inform the risk assessment.

Methods for Measuring Genetic Variation in Wild Populations

Population genetic markers are used to characterize genetic variation for a number of purposes. As described previously, selection of specific approaches and markers depends on the goals of the study (Sunnucks 2000; van Tienderen et al. 2002) and technological constraints of measurement and availability of markers for the assessment population. As long as markers are selectively neutral or near neutral, they can be used to assess population processes such as gene flow, genetic drift, population bottlenecks, and mutation rate.

Neutral approaches use protein, nuclear and mitochondrial genetic markers that are chosen on the basis of goal and technology (e.g., method feasibility, species-specific development and capacity). Other features of neutral markers that contribute to their suitability to different goals include their variability (precision in differentiating groups), extent to which the genome is represented (comprehensiveness regarding genomic diversity), and degree of prior usage (well-developed theoretical basis and analytical approach, comparability among studies). Traditionally, neutral approaches have employed allozymes, which are expressed proteins of common metabolic pathways whose allelic variants can be visualized using low-technology apparatus. With this approach, little species-specific methodological development is required, there is a well-developed theoretical basis for interpretation and well-documented statistical analytical methods, and results are often directly comparable to other studies. Advances in nucleic acid technologies have meant that molecular approaches now often replace this strategy. Common molecular markers include microsatellites, amplified fragment length polymorphisms, and single nucleotide polymorphisms. Molecular markers can be scored with small amounts of biological material and many molecular markers are more variable than allozymes. The advantages of various molecular marker types are described more fully in Chapter 5 and reviewed elsewhere (e.g., van Tienderen et al. 2002).

Whatever neutral markers are chosen, the main purpose of studying them is to describe population processes. Population genetic structure is described by partitioning variation among individual organisms, populations, and geographic regions (Hartl and Clark 1997). This partitioning involves the calculation of F-statistics and similar measures. Other parameters that are estimated frequently include connectedness or migration rate within and among populations. In addition to population discrimination, statistical tests are used often to determine whether population genetic structure reflects geographic distance among units. Population parameters typically estimated to describe population health include effective population size, allelic diversity, and mean population heterozygosity.

More specialized analyses use genetic information to infer historical information about populations. For example, historical demography uses genetic information from extant populations to infer demographic history and reveal temporal patterns of population size, growth rate, migration, and subdivision (reviewed in Avise 2000; Emerson et al. 2001; Hare 2001; Brumfield et al. 2003). The field of landscape genetics has developed taking advantage of the availability of molecular genetic information, attempts to describe patterns of gene flow and local adaptation related to how landscape structures influence genetic variation within and between populations (reviewed by Manel et al. 2003). For these purposes, a strategy might be chosen that includes the simultaneous collection of data from multiple marker systems (i.e., neutral and nonneutral markers). Complementary marker strategies have been advocated to take advantage of complementary information provided by marker types. In addition, a combined analysis of neutral markers and patterns of quantitative genetic information can provide evidence of selection on quantitative traits and the strength of selection acting along geographic gradients.

Nonneutral genetic markers represent loci of known and important function (i.e., fitness-associated), making them useful for investigating selection. Currently, there are a limited number of loci whose functions are known and important for fitness, especially for nonmodel species. Therefore, screening for targeted loci will be useful in circumstances where stressors are suspected and candidate loci for fitness or stressor-associated markers have been characterized. Candidates for this type of analysis include metallothionein for metal contamination, the major histocompatability complex for pathogens and parasites, and genes of the p450 family, esterases and glutathione transferases for organic contaminants and pesticides (see Chapter 5). Advances associated with species genome sequencing and microarray technologies might soon make it more feasible to apply genetic approaches to characterize the selective effects of environmental agents on populations of species that are not well known genetically.

STATISTICAL METHODS

Statistical methods are a key component of empirical work in ecology and in empirical population-level ERAs as well. The principles of experimental and survey design ensure that study resources are used optimally to obtain unbiased results of maximum precision (or minimum variance). Design issues include the number of treatments, the structure of treatments (e.g., factorial arrangements) and the nature of treatments (e.g., selection of concentrations to be used), the form of replication and the number of replicates per treatment. So-called power analyses can help determine the number samples needed to detect a given effect magnitude with a given probability, based on assumptions about the variables' statistical distribution. Randomization procedures can be used to remove bias by ensuring that treatments are allocated to experimental units (e.g., field plots) in a way that does not correlate with environmental gradients, and that the experimental procedure does not introduce systematic errors.

Several statistical methods are available for analyzing observations. Some methods are more appropriate for designed experiments, others for observational studies. A main distinction is between univariate and multivariate methods (for 1 or more response variables, respectively). Univariate methods such as linear regression and analysis of variance are based on normally distributed data, but generalized linear models with link functions are available for a wide range of distributions (e.g., binomial, Poisson, gamma). Nonparametric methods make no assumptions about the data distribution. Multivariate methods can be extremely useful for explorative analyses of main patterns in population studies that monitor many species.

Clear and unambiguous objectives (or questions to be addressed) are necessary for successful empirical work. The objective naturally points to an appropriate form of analysis, which in turn points to an appropriate design. For example, if we wish to estimate the environmental concentration of some toxicant that corresponds to a specific effect on a population attribute, then the response should be measured along a toxicant concentration gradient, to support description of a concentration-response curve that can be used to interpolate the critical concentration.

EXAMPLES

This section offers detailed examples of empirical methods for predictive and retrospective population-level ecological risk assessments. Pesticide registration provides the context for the predictive example, because pesticide risk assessments typically involve much more extensive and complex data collection and assessment methods than do other predictive assessments for industrial chemicals. Contaminated sites provide the context for the example of a retrospective assessment, because these assessments are necessarily site specific and usually focus on questions of existing impacts.

PESTICIDE REGISTRATION

To register a new pesticide under most regulatory authorities worldwide, the registrant must submit a risk assessment and supporting data to the appropriate regulatory authority. The various risk assessment tiers and the purpose of each are described in the Agricultural Pesticide Registration Scenario (Appendix 1), which presents in broad outline the process for registering an agricultural pesticide. This section discusses in more detail the approaches for collection of empirical data, with an emphasis on population-level studies.

Standard Laboratory Studies

Early tier assessments focus on direct acute and subchronic toxicological effects on individual organisms. Initially, the registrant carries out a number of laboratory studies to evaluate the hazard associated with the compound.

Early tests are basic laboratory tests using a single species and can be classified broadly under the 2 headings of acute and chronic. The precise list of tests required varies between regulatory authorities, although there is considerable similarity and overlap. The list of species includes mammals, birds (mallard and quail), bees, other beneficial arthropods, earthworms, soil microorganisms, non-target plants, fish, aquatic invertebrates, aquatic plants and amphibians. Studies are usually carried out to standards laid down in guidelines issued by organizations such as the Organisation for European Cooperation and Development, International Organisation for Biological and Integrated Control, or the International Standards Organisation. Use of standard guidelines obviates the need to repeat tests under different guidelines for different regulators.

For European Union registration, a toxicity exposure ratio (TER) is computed by dividing a measure of toxicity (e.g., a NOEC or LD_{50}) by a measure of exposure. A TER value greater than 10 is taken to indicate low risk, whilst a lower TER is a potential cause for concern. US registration uses a very similar procedure based on a risk quotient in which a measure of exposure is divided by a measure of toxicity (the inverse of the TER).

Laboratory studies as described here are not always directly relevant for measurement of population effects, although some endpoints are (e.g., population growth rate), and some can be useful as inputs for population models (e.g., somatic growth, mortality and reproduction). Endpoints measured at different doses can be used to model populations under different levels of pesticide stress. Standard tests, however, tend to use more sensitive life stages, so they can overestimate potential for impacts of pesticides.

If a lower tier risk assessment indicates cause for concern, then additional studies and more refined assessments will usually be carried out. Initially this can mean higher tier laboratory studies designed to address specific issues and not conforming to a standard approach. Study designs tend to be unique and risk assessments quite specific.

Empirical Measurement of Population Effects — Aquatic Species

Measurement of aquatic population-level effects has received considerable attention and been the subject of many publications. The HARAP workshop (Campbell et al. 1998) resulted in a guidance document for higher tier aquatic risk assessment. The CLASSIC workshop (Giddings et al. 2002) looked at aquatic systems for community-level risk assessments, but also made recommendations about population-level assessments.

Population laboratory studies on single species are possible for a number of plant and invertebrate species. They can be relatively simple in involving only a single life stage, or made more complex by including multiple life stage studies. Taylor et al. (1993), Maund et al. (1992), and Blockwell et al. (1999) describe studies with the midge *Chironomus riparius*, the amphipod *Gammarus pulex* and the amphipod *Hyalla azteca*, respectively. Van Straalen and Kammenga (1998) present methods for assessing effects on population growth rate from such experiments.

A wide variety of laboratory multispecies studies have been carried out. Studies are performed in bounded systems, varying in size from small flasks to large aquaria. Designs vary with objectives. Multispecies test systems contain a community-like aggregation of species, and include trophic interactions among those species. Larger test systems include populations of microorganisms, planktonic, periphytic and benthic algae, zooplankton, meiofauna, macroinvertebrates, and macrophytes. Examples of studies with very simple test systems can be found in Day and Kaushik (1987), Klüttgen et al. (1996), Gomez et al. (1997), and Hamers and Krough (1997). Examples with larger test systems can be found in Taub (1969, 1974), Leffler (1981), and Kersting (1991). Examples with even larger and more complex systems can be found in Breneman and Pontasch (1994), Fliedner et al. (1997), and van den Brink et al. (1997).

Multispecies aquatic field studies should ideally be carried out if on completion of laboratory work outstanding concerns remain. The objective is usually to measure, under environmentally realistic exposure conditions, any impact of the pesticide relative to an untreated control, on abundance. Time to recovery should also be measured where impacts are observed. Each study is designed with a specific purpose determined from results of previous laboratory studies, and so is unique in at least some respects. The objective and resulting design determine the species to include, level of taxonomic analysis required, exposure regimes, sampling methods and timing, and so forth. Both direct and indirect impacts are measured — direct impacts usually take the form of a decline in abundance, but indirect effects can be observed either as increases (when the direct effect is on a predator) or as decreases (when the direct effect is on a prey) in abundance.

Guidance for conducting aquatic field studies can be found in Monks Wood (SETAC-Europe, 1991), Wintergreen (SETAC.RESOLVE 1991), and Organisation for European Cooperation and Development (OECD 1998). Studies vary enormously in terms of test system size, number of species sampled, and duration. Several containers are used, 1 for each replicate of each treatment to be used. For a very large study, the container could be 0.1 ha in surface, or larger. A depth of natural sediment is placed in the bottom of each and they are filled with water from an appropriate source. They are then stocked with a range of organisms and left to settle for an appropriate period of time. Organisms are sampled for a period leading up to treatment and at regular intervals after treatment — the precise method of sampling depending on the organism concerned. Temperature, pH, and a whole host of abiotic variables are also sampled both before and after treatment. Experimental treatments usually comprise an untreated control, a number of different doses of a pesticide, and a toxic standard. The method of applying the pesticide usually attempts to mimic a realistic route of exposure, such as spray drift or runoff after heavy rain.

Interpreting the data from a multispecies mesocosm study presents a challenge. In keeping with tradition, NOECs are usually computed for each species at each sampling occasion, but the large number of resulting NOECs often serves to confuse rather than elucidate. The CLASSIC workshop (Giddings 1999) did, however, recommend using univariate statistical methods for evaluating population-level effects.

Multivariate methods have been proposed and successfully used for evaluating community (or assemblage) effects. In particular, the method of principal response curves has become a standard tool. Originally known as reduced "rank regression" and included in the CANOCO software under the name "redundancy analysis" (ter Braak et al. 1996), it was developed specifically for use in pesticide multispecies studies. It is most easily described as a 2-stage process. In the first stage, a regression procedure is used to remove all variation not directly attributable to treatment (i.e., random variation and variation because of time). In the second stage, a principal component analysis is carried out and the number of components needed to adequately summarize the data is determined. Each component is a linearly weighted sum of the original species and so can be likened to a population- or community-level attribute. Whether or not populations (or communities) recover and time to recovery are determined by comparing component values for the treatments through time with component values for the untreated control.

Many other methods, collectively called ordination techniques, such as principal component analysis, correspondence analysis, canonical correspondence analysis, canonical correlation analysis, and canonical variate analysis are also frequently used in ecology. For a very accessible introduction to all of these methods in an ecological context see Legendre and Legendre (1998). For a somewhat more challenging discussion on the methods see Jongman et al. (1995) and ter Braak et al. (1996). Van den Brink et al. (2003) also compares and contrasts the different methods. Nonmetric ordination and permutation methodology makes few assumptions and is commonly used to analyze multispecies experiments. The methodology is implemented within the PRIMER software (Carr 1996). Examples of the use of the methodology can be found in Clarke and Warwick (1994).

Empirical Measurement of Population Effects — Terrestrial Species

Multispecies field terrestrial experiments are used extensively to measure effects of pesticides on earthworm and nontarget arthropod populations.

Experiments on earthworms are relatively straightforward. Plots tend to be fairly small and subplots within the main plot area are used to sample on different occasions. Sampling is carried out by soaking the sublot with formalin, thereby inducing worms to come to the surface, or by physically removing the soil from the sublot and sifting out the worms. Sampling is carried out both before and after application of a pesticide. Sampled worms are partitioned into different species and different life stages, and then counted. The purpose is usually to determine whether populations from treated plots recover as compared with an untreated control and the time they take to recover.

Guidance on higher tier testing for nontarget arthropods can be found in the report of the ESCORT 2 Workshop (Candolfi et al. 2001). Candolfi et al. (2000) discuss the interpretation of field trials in the context of regulatory testing. This document discusses laboratory testing, semifield testing in which single species are held in cages outside and subjected to more realistic environments than in the laboratory, and multispecies field studies. Population-level endpoints are not discussed in the context of laboratory and semifield studies and, in contrast to aquatic work, there does not seem to be a history of population-level experimental laboratory work for terrestrial species. However, field studies are recommended specifically to evaluate population-level endpoints in real agricultural environments.

In concept, field studies for NTAs are very similar to mesocosm studies. Plots size varies with objective, but tends to be large. Many species are measured by a variety of sampling methods both before and after treatment. Both direct and indirect effects are observed. The multivariate methods developed for analyzing aquatic studies are equally suitable for terrestrial studies.

Terrestrial Wildlife Populations Inhabiting Contaminated Sites

Many countries have legislation that requires ecological risk assessments to be conducted on hazardous waste sites (Appendix 1). In the United States, the USEPA requires the assessment of risk to ecological receptors during the site assessment and characterization phase for contaminated sites placed on the National Priorities List under the Comprehensive Environmental Response, Compensation, and Liability Act. The USEPA specifically calls for the protection of ecological receptors at the population level, except in the case of rare and endangered species, where protection of individual organisms is required. Even so, the early ERAs at Comprehensive Environmental Response, Compensation, and Liability Act sites followed the human health model, in which hazard indices were calculated for individuals of selected species. However, to address the risk of contaminants to ecological populations directly, the risk assessor will need to turn to many of the methods outlined in Table 8.4 and described in Empiric Methods for Characterizing Populations and Statistical Methods. This section describes in detail the ways in which these methods can be used to assess risks to terrestrial wildlife populations inhabiting contaminated sites.

TABLE 8.4
Empirical methods for use in evaluation of contaminated sites

Assessment method	Use and information
Geographical information systems mapping and analysis	Delineation of habitat types, sources of stressors, location of assessment species, migration patterns, land uses. Excellent for tracking data and metadata, and for visualization of spatial relationships.
Mining of existing databases	Identification of political boundaries, existing and planned land uses. Summaries of biological and physical surveys of the region, recorded species distributions and behaviors, and fisheries and wildlife trends. Summary of existing information.
Remote sensing aircraft and satellite	Determination of the location of habitat types, structure of the habitat, trends of habitat, and land use change over time.
Biological field surveys	Surveys of species distributions, visual observations, and ground-truthing GIS and remote sensing datasets, trapping and identification of assessment species and the other biotic components of the ecological system. Assessment of trends of the biological features over time. Assessment of the utilization of habitat by assessment species and other critical populations.
Surveys of chemical concentrations and other physical characteristics of the location	Sampling of the spatial distributions of chemical concentrations, sediment and soil types, water chemistry, organic content of water and soil, etc. Examination of the trends within the region over time.
Ground surveys of land uses, property boundaries, pipelines, industrial sites, and processes	Identification and ground-truthing of the information available from governmental agencies and written reports. Identification of how land use alters the habitat and structure of the location under consideration. Tracking of changes and trends in land use and anthropogenic activities over time.
Manipulation experiments	Experimental manipulation of the environment followed by observation to examine the hypothesized stressor-response relationships, assessment of competitive interactions and life history characteristics, impacts of spatial heterogeneity, etc.
Laboratory experiments	Laboratory determination of stressor-response relationships to include competition experiments, response of assessment organisms to changes in physical characteristics, laboratory measurements of life history characteristics.

Habitat Evaluation

The goal of a habitat evaluation is to evaluate the contribution of contaminated habitat to available habitat. The general approach would be similar for various species, but would differ in the details. The assessment methods used in this evaluation include GIS mapping and analysis, mining of existing databases, and limited biological field surveys. These methods require specific data on habitat attributes, biological or species attributes, and chemical stressor attributes.

GIS mapping and analysis of habitat information both at the local and regional scale allows the site to be placed in context. A basic geographical information system using readily available information should be constructed, as outlined in Habitat Characterization. Assembling this information will assist in placing the hazardous waste site in a regional context. Land use is particularly important. It is critical to know if the site is within a region of basically intact, undisturbed habitat, or if habitat is restricted to small, isolated patches.

The biological attributes of most interest for this approach are life history attributes and species distribution data. These include the types of habitats preferred by the species of interest, the spatial distribution of these habitats at the regional and local scale, home range and critical patch size requirements for the species, and survey data of the target species in these habitats.

Small mammals and amphibians are often considered in hazardous waste site assessments. A large and varied literature exists on small mammals, their distribution, and life history (e.g., Stenseth et al. 1998). Although less studied, literature is becoming increasingly available on salamanders. Amphibians in general are experiencing a worldwide decline, which is the focus of much research (Kiesecker et al. 2001), thus life history information is becoming increasingly available.

Mining of existing databases and limited biological field surveys are the assessment methods used to generate these data. Every effort must be made to identify available species distribution data. In the absence of actual survey data, life-history and natural-history information from the literature should be used to help identify likely areas at the regional and local scale where the species could reasonably be expected. If life-history data are not available for the specific assessment population, available data for the closest related species can be used as proxies. Regional and local habitat can be ranked for habitat quality or suitability for the assessment population with the number of ranking levels determined by available data and the degree of scale desired. These rankings can then be interpreted as the likelihood that the species will be found in a given area or habitat. Using the available data, a map should be constructed of expected species distribution at the regional and local scale.

The stressor attributes of interest for this approach are spatial distribution of contaminants through the habitat, and, if available, toxicity information expressed as concentration in environmental media. Assessment methods used to generate these data include surveys of field chemical concentrations and mining of existing databases. Preferably, a detailed contaminant distribution map should be created. This will show concentrations of contaminants in surface soil, subsurface soil down to some depth (e.g., 10 m), sediments and surface water. The literature should be evaluated for toxicity information. However, species-specific toxicity data for most chemicals are not currently available, requiring interspecies extrapolation (Suter 2000/2007). The toxicity data can be used to describe "toxicity contours" on the contaminant distribution map.

Measures of potential toxicity, such as NOECs or LD50s, can be used to construct these contours. The contours provide approximate boundaries for areas where contaminant concentration might be expected to have a negative impact, or at least significant exposure, to individuals of the assessment population.

Because of their terrestrial, burrowing nature, the exposure of small mammals to soil contamination has long been studied. Laboratory toxicity data using mice and rats are also relevant. Many aquatic species have also been extensively studied, making species-specific or close-relative data fairly available. Amphibians have been less studied, but some data do exist, and is becoming increasingly available. Mole salamanders from the Ambystomidae family are long-lived (>10 y) and spend much of their adult life in burrows. They have a very permeable integument (skin) used for osmoregulation. Thus their aquatic larval and terrestrial adult life history make them potentially susceptible to chemicals found in their environment and has been the focus of some research.

Once the maps describing species distribution and contaminant distribution are constructed, the actual analysis can take place. The analysis seeks to answer the following 2 questions:

- How important is the local population to the regional population?
- How important is the contaminated site to the regional habitat?

The analysis determines the percentage of available suitable habitat that is contaminated above some given level. Although this sounds straightforward, there are several inherent difficulties. One difficulty is defining the "regional" and "local" scales. Logical geographic boundaries, such as watersheds and valleys, can be used to define the regional landscape. Another method is to delineate the regional scale based on where migration between populations or subpopulations can be expected.

A similar approach is to evaluate home range and critical patch size requirements of the species with respect to the contaminated range. In this approach, home range and critical patch size requirements are derived from the literature. Home range at a minimum is required. If such data are not available in the literature, field studies might be necessary to determine them. Techniques for determining home range are well developed (Carlsen et al. 2004). Less developed, and more difficult to determine, is critical patch size, which requires habitat of different sizes to be available. This is not likely to be available at the typical hazardous waste site even at the regional scale, and thus obtaining these data is not suggested. The amount of available regional or local habitat is then compared with these requirements to first determine if sufficient area is available to meet organism (home range) and population (critical patch size) needs. Next, the amount of contaminated range is subtracted from available habitat and the comparison is performed again. Negative impact is suggested if the contaminated area results in the overall habitat being too small to meet these needs (Carlsen et al. 2004).

Some jurisdictions have attempted to set de minimis areas, below which remediation is not required (Sorensen et al. 2004). The literature on the effects of fragmentation and habitat loss can help define this value. Theoretical studies suggest that significant impact is not expected until habitat loss exceeds 50%. Some empirical studies also support this value. However, blind use of such numbers without placing the site in the context of the surrounding landscape is dangerous (Carlsen et al. 2004). Thus the actual determination of a de minimis area will likely be done with a group of stakeholders interested in that particular site. If the site is very small, and the contaminated habitat a tiny portion of the overall habitat used by the species, then the stakeholders might agree that no further action is necessary. If, on the other hand, there are limited populations throughout the region, and many of the animals reside in the contaminated area, then further assessment is probably warranted. For cases in the middle, more than likely stakeholder interests will require additional analysis.

Population Evaluation

Additional analysis on the assessment population might be necessary should the habitat evaluation suggest that a considerable percentage of the habitat is potentially affected, or if the evaluation is inconclusive. Assessment methods that can be used at this point include laboratory toxicity experiments and additional biological field surveys (see Field Manipulation). If the habitat evaluation reveals that a significant amount of high-quality habitat is contaminated, or if the target species has actually been observed in areas of high contamination, it can be worthwhile to conduct species-specific laboratory toxicity studies. If the species is rare, it might be necessary to use a closely related, but more common, species to conduct the tests. For population-level ecological risk assessments, endpoints for laboratory toxicity tests should include growth, survival and reproductive endpoints that can be interpreted with models that extrapolate organism-level effects to population response (Chapter 9).

Biological field surveys include collecting tissue samples for chemical analysis, conducting field manipulation experiments, and collecting demographic data. Tissue samples can be used to determine if actual exposure is occurring. The degree of exposure to organisms, and the percentage of the population exposed can be estimated. The use of tissue sampling will depend on the target species and properties of the contaminant.

Even if considerable exposure to a contaminant is verified, that in and of itself does not provide information concerning the degree to which the population might be impaired. Modeling the effects of such exposure is one method to address this question (discussed in Chapter 9). The primary empirical approach involves collecting demographic data so that a demographic evaluation of the population can be conducted at the local and/or regional scale (Demographic Studies).

If a substantial portion of a population resides in the contaminated area, demographic data can be collected on this exposed population. If the species is well studied, it often is possible to compare the resulting demographic attributes to those reported in the literature. If sufficient literature is available, confidence intervals can be calculated for each demographic parameter, and the field data compared with literature derived values using these confidence intervals. If literature data are not available, it will be necessary to collect similar demographic data from a reference site and conduct a statistical comparison between the sites. If the species is transient across the contaminated site, or the site only comprises a small portion of the population's range, an evaluation of the regional or biological population can be useful. Again, mark and recapture studies can be used across the range and evaluated for perturbations. Here, a reference site is not useful, as the assessment population represents the entire biological population. Instead, changes in population characteristics along a gradient of contaminant concentrations can be examined.

Selecting a reference site is not trivial. The reference site should be as similar as possible to the contaminated site with the exception of the contamination. This is often not possible. If the site is large enough, demographic data can be collected across a gradient of contamination. However, unless all other aspects of the gradient except contaminant concentration remain constant, the analysis will again be confounded by these additional factors. Thus it is necessary to ensure a sufficient number of replicate sample sites are included in the survey to provide sufficient power in the subsequent analysis. Each time a covariate is added to a correlation analysis, a degree of freedom is lost.

RECOMMENDATIONS

Empirical methods for characterizing populations and for estimating parameters required by population models are well-developed in the scientific literature. However, because most current ecological risk assessments still focus on organism-level effects, risk assessors might be unaware of the existence, value, advantages, and disadvantages of the available empirical methods. Moreover, different methods are appropriate for different decision contexts. The following recommendations are intended to increase the access of ecological risk assessors to accurate information about these methods.

- Guidance is needed on parameter estimation methods for population models used in ecological risk assessments. Because aquatic and terrestrial ecologists often employ substantially different study methods, separate guidance for aquatic and terrestrial populations might be needed.
- Additional decision-specific guidance is needed to aid ecological risk assessors in defining the types of empirical studies appropriate to support a given risk assessment. This guidance is especially important for studies of contaminated sites, because agency guidance for such sites emphasizes site-specific determination of assessment approaches and data needs.
- Because methods for spatially structured population studies, including GIS-based applications, have developed rapidly over the past decade, training courses on the use of these approaches are needed to ensure that ecological risk assessors are knowledgeable concerning the latest developments in this field.

9 Modeling Approaches to Population-Level Ecological Risk Assessment

Wayne R. Munns, Jr., Jennifer Gervais, Ary A. Hoffman, Udo Hommen, Diane E. Nacci, Mayuko Nakamaru, Richard Sibly, and Chris J. Topping

INTRODUCTION

Models support several aspects of chemical risk assessment, including characterizing exposures (existing or predicted) and predicting (or diagnosing) the effects of exposures on assessment endpoints. A decade ago, McCarty and Mackay (1993) suggested that science's ability to model and predict the fate of chemicals was markedly improved. Similar arguments about the acceptance and use of exposure models were offered by Clark et al. in 1988. Although still imperfect, our ability to model the temporal and spatial distributions of chemicals originating from a variety of sources and in a variety of environmental media (including multiple media) is reasonably well developed. Arguably less so is our ability to predict the effects of exposures on assessment endpoints, particularly those at the population level of biological organization. Due in part to the practical difficulties and resource requirements associated both with assessing and predicting impacts on populations empirically, models play a key role in population-level ecological risk assessment. This chapter focuses on the use of models as tools to characterize effects at the population level, and is intended to be a companion to Chapter 8.

In his broad discussion of ecological theory and models, Levins (1966) describes a triangular scheme for ordinating ecological models that has the qualities of generality, realism, and accuracy (originally precision[1]) as its apices. In a simple interpretation of this scheme, general models are those that tend to simplify biological processes and

[1] Use of the terms "accuracy" and "precision" in this chapter follows their connotations in the field of inferential statistics. Accuracy refers to how well an estimate matches the true value of a particular parameter or value being estimated (for example, population abundance), and typically is quantified using some measure of bias. Precision refers to the amount of variation among multiple estimates made of the parameter, and usually is expressed using some measure of scatter.

relationships, and therefore apply to a broad range of situations and are appropriate for exploring relationships among model parameters and outputs. Realistic models attempt to account for known processes and relationships in ecological systems, and as a result can be relatively complex. Accurate models are constructed with an objective to minimize numerical differences between model outputs and actual ecological dynamics. Their case-specific nature often limits their use in broader applications. Levins (1966) suggested that it is not possible to maximize all three model attributes within a single model. Typically, models developed for use in applied situations (e.g., conservation biology) are intended to give realistic, accurate answers. Model parameterization depends on the actual conditions of the situation being modeled (e.g., the specific life history and demographic characteristics of the species of interest). Increased generality can be achieved, for example, by expanding the range of values assumed for particular model coefficients, or by assuming broad functional relationships among parameters, but such actions necessarily reduce the accuracy achievable in any particular application. Model accuracy can be enhanced by increasing the specificity of model construction and parameterization relative to a particular species or environmental situation, but typically at the cost of model generality.

Population models can be used in ecological risk assessment for at least 3 different purposes: detection (and diagnosis), projection, and prediction (or forecasting).

With this ordination scheme in mind, population models can be used in ecological risk assessment for at least 3 different purposes. The first is to "detect" (and perhaps "diagnose" the causes of) previous or ongoing adverse effects on assessment populations. Such uses typically require sufficient high-quality data to be able to detect changes in population abundance or other attributes, and to relate those changes to variation in chemical exposure, habitat quality, or other forms of disturbance. The second purpose is to "project" the consequences of a given set of environmental conditions (or changes in conditions) on the dynamics of a population. Here the intent may be to evaluate the ramifications of particular environmental management decisions as determined by trends in population numbers or changes in extinction probabilities. As such, projection provides information to the decision-maker about the vitality, well-being, or fitness of a population under different environmental scenarios. The final purpose is to "predict" or "forecast" the future behavior of the population based on an understanding of environmental variability and the dynamic interactions of density and biological processes (e.g., births and deaths). (The distinction between projection and forecasting used here follows that given by Caswell 1986, 2001.) This last use of population models may produce the most accurate results, although the generality of the analysis may suffer. It requires knowledge of how the major environmental and biological determinants of population dynamics (including such things as chemical exposures, climate, habitat quality, prey availability) will themselves change over space and through time (or, in a probabilistic sense, how their distributions change), and sufficiently detailed understanding of the mechanisms through which these changes affect the population. Also required is understanding of how population density influences births, deaths, and migration through the density dependence of these rates. Except in rare situations, it is a challenge to obtain and use knowledge of density dependence with confidence (Chapter 4). The particular use—detection, projection, prediction—of population models in ecological risk assessment depends on the management goals and decisions the assessment is intended to support.

We follow this introduction with a general discussion of modeling approaches that support population-level ecological risk assessment, with a particular emphasis on the problem of environmental chemicals. We limit this discussion to mathematical representations of population endpoints that yield explicit descriptions of the dynamics of those endpoints in time and space, as opposed to empirical procedures that attempt to relate measures of population endpoints to environmental conditions statistically (see Chapter 8). In this section, we offer a simplifying categorization or taxonomy of models that supports model selection and interpretation. Also offered are considerations of key assumptions, basic data requirements, and model strengths and limitations relative to their use in risk assessment. A subsequent section discusses incorporation of genetic information into population dynamics models to enhance their realism and utility to address certain risk management questions. This leads to a discussion of the factors affecting selection among models to support risk management and decision-making. Finally, unresolved issues and recommendations relative to modeling approaches are offered to improve the states of application and science within the context of population-level ecological risk assessment. Recommendations for modeling approaches for population-level ecological risk assessments supporting the risk management decisions scenarios described in Appendix 1 are presented with examples throughout the chapter. An illustrative modeling risk assessment for the agricultural pesticide registration scenario is provided in more detail in Appendix 2.

MODELING POPULATIONS IN ECOLOGICAL RISK ASSESSMENT

Population-level ecological risk assessment requires that the composite characteristics of groups of individual organisms be considered rather than (or in addition to) the more traditional approach of using organism-level attributes, such as the chemical concentration causing death, in the risk assessment. Several modeling approaches have been developed in population biology that can be adapted for use in population-level ecological risk assessment. These models range from extremely simple to highly complex. The taxonomy illustrated in Figure 9.1 organizes population models into 5 general classes based on consideration of the level of biological and spatial aggregation reflected in the models as they are typically constructed and used: unstructured (scalar), biologically structured, individual based, metapopulation, and spatially explicit. It should be noted that these classes are not mutually exclusive; for example, one can create a spatially explicit individual based model. Rather, individual model formulations fall along continua of biological and spatial aggregation instead of into discrete classes. This taxonomy is meant to provide a general idea of the potential applications and issues associated with broad model classes.

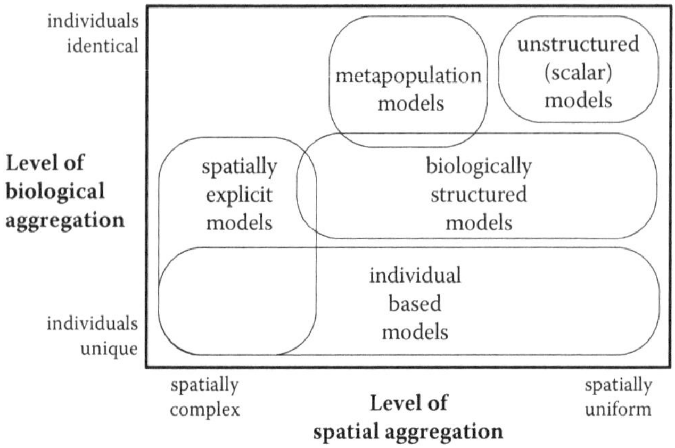

FIGURE 9.1 Taxonomy of population models for population-level ecological risk assessment.

Population models range from extremely simple to highly complex, and can be organized into 5 overlapping classes based on their level of biological and spatial aggregation: unstructured (scalar), biologically structured, individual based, metapopulation, and spatially explicit.

A fair amount has been written elsewhere about population models, and certainly with respect to their use in risk assessment (e.g., Barnthouse et al. 1986a; Emlen 1989; Emlen and Pikitch 1989; Barnthouse 1993, 1996; Maltby et al. 2001a; Pastorok et al. 2002, 2003; Munns and Mitro 2004; also see chapters in Kendall and Lacher 1994). Because of this, we focus our considerations primarily on issues relevant to practical model application to support risk management decisions. For each class of model in Fig. 9.1, we describe modeling formulations in general terms, identifying key model assumptions and describing their qualities relative to Levins' (1966) ordinating scheme. Examples are offered to illustrate their use. We also consider application of each class of model in population-level ecological risk assessment by describing the population attributes (see Chapter 3) they evaluate and the information they require. This information is summarized in Table 9.1.

Unstructured (Scalar) Models

Unstructured population models are perhaps the simplest form for assessing population-level risk. They reflect a single uniform population with no demographic or environmental structure, and usually aggregate species-specific properties into a small number of variables. Individuals within the population are treated identically (in terms of their demographic rates and responses to environmental conditions), and population dynamics are determined by a few controlling processes (typically births and deaths only). Unstructured models can be deterministic, or demographic

TABLE 9.1
Classes of population models for ecological risk assessment

Model class	Key assumptions	Generality	Model qualities realism	Accuracy	Attributes modeled	Data requirements	Recommended uses in risk assessment
Unstructured (scalar)	No biological or spatial structure Density-independent or dependent Deterministic or stochastic	High	Low	Low	Abundance pgr[1] Extinction/recovery	Low	Screening tiers
Biologically structured	Biological structure No spatial structure Density-independent or dependent Deterministic or stochastic	Medium/high	Medium	Medium/high	Abundance pgr Extinction or recovery Biologic structure	Medium	Screening or higher tiers
Individual based	Individuals modeled separately Deterministic or stochastic	Low	High	High	Abundance pgr Extinction or recovery Biological structure	High	Higher tiers
Metapopulation	Local dynamics of subpopulations and migration determines population dynamics Deterministic or stochastic	Medium/high	Medium	Low	Abundance pgr Extinction/recovery Spatial structure	Low/medium	Screening or higher tiers
Spatially explicit	Spatial structure continuous on a landscape Deterministic or stochastic	Low	High	High	Abundance pgr Extinction/recovery Spatial structure	High	Higher tiers

[1] pgr = population growth rate.

and environmental stochasticity can be reflected as variability in aggregated rates. Because of their uncomplicated and aggregated nature, unstructured models emphasize generality at the expense of realism and accuracy.

Unstructured population models are perhaps the simplest form for assessing population-level risk.

Perhaps the simplest (and oldest) unstructured population model is the exponential growth model, explored by Malthus (1798) some 200 years ago. This model has but one state variable, total population size (N), which is assumed to be a function of time (t) and population growth rate, r (= birth rate – death rate), taking the continuous form:

$$dN/dt = rN \qquad \text{Eq. 1}$$

Population dynamics described by this model are exponential, and with fixed growth rate populations increase indefinitely when $r > 0$, decrease to 0 when $r < 0$, or remain constant in abundance when $r = 0$.

Exponential models assume no effects of population density on birth and death rates; that is, they are density independent. They may adequately describe the dynamics of populations newly introduced into an ecosystem, or when environmental conditions have changed substantially in ways that promote rapid changes in abundance. They also can be used to compare responses of populations with different maximum potential growth rates. In most instances, however, population growth likely is dependent on the density of the population itself (see Chapter 4), with population abundance providing a negative feedback on the processes of births and deaths. To reflect this dependence, the basic exponential model has been modified through addition of a feedback term that assumes each individual of the population to require $1/K$ of environmental resources, such that:

$$dN/dt = rN\,(1 - N/K) \qquad \text{Eq. 2}$$

where K is the so-called carrying capacity of the environment. As population abundance increases from low levels, the change in abundance (dN/dt) slows asymptotically to 0 as N approaches K. Thus the trajectory of population abundance assumes a logistic or "S"-shaped curve when initial abundances are low, decreases exponentially to K at starting abundances above K, or remains flat when population size is at K.

Both Eq. 1 and Eq. 2 represent time as a continuous variable; formulations exist for both that treat time incrementally (that is, discretely), and discrete unstructured models with density-dependence have been applied in fisheries management for half a century (e.g., Ricker 1954; Schaefer 1954, 1957; Beverton and Holt 1957; Quinn and Deriso 1999). Environmental and demographic stochasticity can also be introduced into either model (e.g., Hakoyama and Iwasa 2000; Tanaka and Nakanishi

2000), sometimes producing chaotic behavior in modeled dynamics. Pastorok et al. (2002) and Munns and Mitro (2004) describe additional unstructured population models used for fisheries management and other applications.

As recent examples of the use of unstructured models to investigate adverse effects of chemicals, Hendriks and Enserink (1996) estimated the change in abundance of waterflea (*Daphnia magna*) and great cormorant (*Phalacrocorax carbo*) populations as functions of polychlorinated biphenyl (PCB) exposure. Tanaka and Nakanishi (2000) estimated extinction risk of plankton using the logistic population model with environmental fluctuation, building on a model originally developed by Lande (1993). Nakamaru et al. (2002, 2003) applied unstructured models to herring gull and sparrowhawk populations using laboratory and field data. Recently, Barnthouse (2004) used the logistic model to compare recovery rates of different types of aquatic biota to simulated pesticide exposures.

Despite (or maybe because of) their apparent simplicity, unstructured models can be useful tools in some kinds of population-level ecological risk assessment. They can be used to project population abundance and growth rate, offering insights to the potential effects of environmental stressors as they affect the aggregate demographic rates of births and deaths. They also can be used to help identify and diagnose population-level effects that may be associated with, for example, hazardous waste sites when appropriate demographic information, or data describing temporal changes in population size, are available. Further, if environmental and demographic stochasticity are introduced, extinction (recovery) time or extinction (recovery) probability can be estimated, and the results can be translated to other measures of population-level effect. For example, Hakoyama and Iwasa (2000) developed a method to calculate "equivalent habitat loss" from changes in mean extinction time. Equivalent habitat loss may be more readily understandable because the damage caused by chemical stressors can be imagined intuitively.

Because of their assumptions and limited data requirements, unstructured models may be most useful in early tiers of an assessment when understanding of general patterns of population risk is sufficient to inform decisions or when data are limited (see Sibly 1999 for a discussion of efficient experimental designs for estimating density effects). However, these same characteristics may limit their value in more definitive assessments when biological realism is needed to capture important aspects of the assessment population's ecology or the risk problem.

Because of the inclusion of characteristics more accurately reflecting the biology of the species, biologically structured models tend to reflect a higher degree of realism when biological structure is important, although usually at the expense of generality.

Biologically Structured Models

This group of models is distinguished from those of the previous class by their incorporation of biological structure. Typically, this structure is reflected through assignment of demographic characteristics or vital rates to distinct classes of individuals within the population, with all individuals within a class assuming identical vital rates. In the absence of demographic stochasticity, these rates assume the average values of each class. Classes can be identified on the basis of age, developmental stage, size, genetic makeup, or any other biologically meaningful characteristic. A significant consideration in development of biologically structured models is the number of classes modeled. The difficulty can be in determining what number of classes adequately captures the complexity of the species' life history. Often this must be balanced with the number of classes for which estimates of vital rate can be obtained. A discussion of this issue, and a method of determining how many stages are adequate, is given by Caswell (2001); also see Easterling et al. (2000). These models frequently are density-independent and deterministic, but many (particularly projection matrix models) can also incorporate density dependence (e.g., Grant 1998) as well as environmental and demographic stochasticity (e.g., Laskowski 2000). In and of themselves, they do not reflect spatial structure, although several variants have been used in metapopulation and spatially explicit model formulations (see Metapopulation Models and Spatially Explicit Models). Because of the inclusion of characteristics more accurately reflecting the biology of the species, biologically structured models tend to reflect a higher degree of realism when biological structure is important (but see Berryman 2004), although usually at the expense of generality.

A relatively simple formulation that has been popular in the investigation of contaminant effects on population growth rate (e.g., Daniels and Allan 1981; Allan and Daniels 1982; Gentile et al. 1982, 1983) is the Euler-Lotka stable age equation (Lotka 1925), often used in the analysis of life-table response experiments (LTRE) (e.g., Caswell 2000). Age-specific schedules of fecundity and survivorship are the sole data requirements for this model, information that is fairly easily obtained in laboratory experiments or field studies. The discrete version of this model (Leslie 1948) takes the form:

$$1 = \Sigma\lambda^{-(x+1)}\, l_x m_x \qquad \text{Eq. 3}$$

where λ is the geometric rate of population increase, l_x is the probability of an individual surviving to age x, and m_x is the fecundity of an individual of age x. Equation 3 is solved for λ either in an iterative fashion or through algebraic manipulation of the sum of $l_x\, m_x$. Lambda is related to the more commonly used (incorrectly so in these applications; see Michod and Anderson 1980) population growth rate (r) by:

$$\lambda = e^r \qquad \text{Eq. 4}$$

Some important assumptions of this model include that age-specific schedules of fecundity and survivorship are independent of density (i.e., the population grows exponentially) and do not change with time. With the additional assumption of no net migration into or out of the population,

$$N_t = \lambda N_{t-1} \qquad \text{Eq. 5}$$

where N is as defined previously. Thus, although zero population growth occurs when $r = 0$ in Eq. 1, no growth occurs in Eq. 3 when $\lambda = 1$.

Another biologically structured model formulation used frequently in ecotoxicology, conservation biology, and resource management applications is the projection matrix model (Lewis 1942; Leslie 1945, 1948; Caswell 2001). As with the stable age equation, these models incorporate age-specific schedules of fecundity and survivorship to make projections of population dynamics given specified environmental conditions. Unlike that model, however, discrete time steps are incorporated explicitly into projection matrix models, allowing simulation of population dynamics through time and time-dependent variation in vital rates. For populations structured by age (age-structured), the sizes, n, of m age classes at time t are described by a series of difference equations as:

$$n_{0,t} = \Sigma(n_{x,t-1}\, f_x)$$

$$n_{1,t} = n_{0,t-1}\, P_1 \qquad \text{Eq. 6}$$

$$n_{m-1,t} = n_{m-2,t-1}\, P_{m-1}$$

where f_x is the fecundity of age class x and P_x is the probability of survival of females from age class $x - 1$ to class x. In matrix notation:

$$\boldsymbol{n}_t = \mathbf{A}\, \boldsymbol{n}_{t-1} \qquad \text{Eq. 7}$$

where $\boldsymbol{n}_t$ and $\boldsymbol{n}_{t-1}$ are vectors of age class sizes, and $\mathbf{A}$ is the projection matrix consisting of fecundities across the top row, survivorship probabilities down the first subdiagonal, and zeros elsewhere. In this form, Eq. 7 represents the so-called Leslie model. In addition to describing total population size and the distribution of individual ages through time, Eq. 7 can be solved directly for its characteristic root, λ, the geometric rate of population increase:

$$|\boldsymbol{A} - \lambda \boldsymbol{I}| = 0 \qquad \text{Eq. 8}$$

where $\boldsymbol{I}$ is the identity matrix of $\boldsymbol{A}$. To solve analytically, schedules of fecundity and survivorship are assumed to be constant and independent of density and time, with the population experiencing no net migration. With these assumptions, projection matrix models behave similarly to Equation 1: that is, growth is geometric (after a stable age distribution has been reached). The effects of environmental stressors can be projected in a manner similar to that used in the stable age equation: modifications can occur to the original fecundity and survivorship rates in the projection matrix. Additionally, effects on population dynamics of changes in the relative abundances of age classes (n_x), such as what might occur if a one-time exposure to the stressor affects individuals differentially among classes, can be evaluated through modifications of the size vector, $\boldsymbol{n}$, in Equation 7. The projection matrix modeling approach can be generalized such that stages are defined as developmental stages, length classes, weight classes, or any other stage appropriate to a particular species (Lefkovitch 1965). Changes from one stage to another are accomplished by non-zero transition or growth probabilities in A, allowing individuals to remain in their current class between time steps (i.e., no individual growth), or to move to other classes as dictated by individual growth and development rates instead of time alone (see Werner and Caswell (1977) and Caswell (1982, 2001) for additional discussion of stage-classified projection matrix models).

Modeling for Risk Management Scenarios: Consequences of Exceeding Water and Sediment Quality Criteria and Standards

Current methods for deriving chemical criteria and standards protective of aquatic life depend on acute and chronic toxicity test results involving several species. These results are analyzed statistically to identify chemical concentrations that protect the majority of aquatic species a majority of the time. An assumption underlying this approach is that criteria that minimize adverse effects (mortality, reproduction and individual growth) on organism-level attributes will also prevent effects on populations and communities.

An approach to evaluate both the level of protection afforded by existing criteria and standards, and the consequences of exceeding such criteria, involves "extrapolation" of organism-level responses to population response (see Maltby et al. 2001; Munns 2002). For example, Kuhn et al. (2000, 2001) used an age-structured projection matrix model and the results of standard 28-d toxicity tests to explore whether current US water quality criteria were protective against population-level effects. For 20 chemicals evaluated in this manner, criterion concentrations were at or below levels predicted to affect population dynamics of the mysid shrimp, *Americamysis bahia.* However, this presumption of protectiveness assumes the general absence of additive or synergistic interactions of the chemicals with other environmental stressors and biological processes (predator–prey relationships, stochastic events) in the environment. Given the assumptions of the analysis, however, US water quality criteria seem reasonably protective. After parameterized with the appropriate toxicity relationships, extrapolation models can also be used to evaluate the population-level consequences of exceeding criteria. Extrapolation models have been developed for a benthic amphipod (*Ampelisca abdita*) to support similar evaluations of sediment quality criteria (Kuhn et al. 2002).

The Pellston Workshop on *Reevaluation of the State of the Science for Water-Quality Criteria Development* (Reiley et al. 2003) explored other risk-based approaches that focus on population-level response, and efforts are underway to incorporate population modeling into derivation of US water quality criteria.

Other biologically structured model formulations include the z-transform lifecycle graph, and partial differential, delay-differential, and integro-difference equation models. More complete treatments of these models can be found in Tuljapurkar and Caswell (1997), Caswell (2001), Maltby et al. (2001), and Pastorok et al. (2002).

The use of biologically structured models, particularly the stable-age equation and projection matrix model, to characterize the adverse effects of chemical stressors on population dynamics has increased over the last decade or so. Some examples illustrate use of such models within risk assessment. In a prospective evaluation of potential adverse impacts to populations of marine copepod associated with offshore disposal of municipal sewage sludge, Munns et al. (1996) conducted a prospective assessment of risk that used an age-classified projection matrix model, the results of standard toxicity tests of the sewage sludge, and exposure models developed to describe sludge concentrations in the water column around and downstream of the disposal site. They simulated passive advection of the population with prevailing currents through the sludge exposure field, adjusting vital rates on a daily time step in accordance with the concentration of sludge encountered by the population on each day. Environmental stochasticity was simulated by allowing concentrations in the exposure field to vary lognormally. Munns et al. (1996) expressed risk estimates as a 3-dimensional response surface defined by axes of sludge loading rate, environmental stochasticity, and population growth rate with the intent of providing risk managers with the information needed to select acceptable levels of disposal activity.

In support of water quality criteria development (see Appendix 1), Kuhn et al. (2000) used data from lifecycle chronic tests to project the effects of 20 chemicals on the mysid shrimp, *Americamysis* (formerly *Mysidopsis*) *bahia*, population dynamics. They then compared the concentration-based toxicity test statistics derived from standard 96-h acute tests, 7-d rapid chronic tests, and the full lifecycle (28-d chronic) tests for each of these chemicals to the chemical concentration projected by the model to represent the threshold of adverse population effects to evaluate the predictive power of those tests. Kuhn et al. (2001) subsequently demonstrated that the age-classified model, parameterized for the chemical nonylphenol using standard toxicity test data, projected the dynamics of *A. bahia* reasonably well in a 55-d multigenerational experiment conducted in the laboratory. In a final analysis of different biologically structured modeling approaches, data from the 28-d chronic test were reevaluated using 3 additional formulations: a delay-differential equation model, an age-truncated delay-differential equation model that assumed that individuals died after age 28, and a partial-differential equation model (Maltby et al. 2001). Comparisons of model predictions to the 55-d data set suggested model outcomes to be relatively robust to the details of model formulation, but that certain assumptions were important and the veracity of all models suffered by not accounting for density effects (Maltby et al. 2001).

In other examples, Salice and Miller (2003) examined the population growth rate of 2 strains of the freshwater snail, *Biomphalaria glabrata*, chronically exposed to various concentrations of cadmium. They compared the demographic performance of the 2 strains using a life-table response experiment and z-transform lifecycle graph analysis, but also examined the output of 2 different matrix models, and discussed the relative merits of all 3 models. Marshall and Crowder (1996) modeled population dynamics of brook trout (*Salvelinus frontalis*) as a function of various anthropogenic factors, including reduced water pH resulting from acid deposition. They used a size-structured projection matrix model to compare the relative impacts of the stressors on population size and structure in a prospective analysis intended to identify which factors or combinations of factors had the greatest impacts on trout population dynamics. Additional examples of the use of biologically structured models in conservation biology and resource management are described in Chapter 7.

In population-level ecological risk assessment, biologically structured models can be used to evaluate effects on population attributes of abundance, population growth rate, extinction and recovery (e.g., Tanaka and Nakanishi 2000; Ellner and Fieberg 2003), and of course, biological structure (e.g., Stark and Banken 1999; Fox and Gurevitch 2000). In addition, the sensitivity (on an absolute scale) or elasticity (on a proportional scale) of population growth rate to variation in individual model parameters has been widely used in conservation and resource management to identify critical life stages or to prioritize data collection and parameter estimation efforts (e.g., Doak et al. 1994; Heppell et al. 1996; for ecotoxicology applications, see Jensen et al. 2001; Kammenga et al. 2001). These analyses should be used with caution, however, because they are sensitive to the details of model construction (e.g., the number and duration of classes; Gleason et al. 1999) and the state of the population (as reflected in realized vital rates) for which they are generated (Mills et al. 1999). A more thorough discussion of attributes modeled using biologically structured models is given by Maltby et al. (2001).

Models in this class can support both screening and higher tiers of population-level ecological risk assessments. With simplifying assumptions about (the lack of) environmental stochasticity, demographic stochasticity and density dependence, and with the condition of a stable abundance structure among classes, these models behave much like exponential unstructured models and require only estimates of survival and reproduction for each class as input to understand the population's asymptotic behavior. Projections based on these assumptions can be generalized to a variety of situations and perhaps species. Indeed, generalized life history strategies or vital rates drawn from similar species might be used in biologically structured models when sufficient data for the assessment population are not available (Calow et al. 1997; Heppell et al. 2000). However, such approaches introduce additional uncertainty that may be unacceptable in higher tiers of a risk assessment. With increasing sophistication in model assumptions and concomitant increases in data requirements, biologically structured models can be made to be very realistic of the risk problem at hand. In addition to using vital rate information specific to the population, assumptions about demographic stochasticity (see Kaye and Pike (2003) for a comparison of different probability distributions describing demographic stochasticity and their impact on model output) and density-dependence relationships can be included that enhance model realism. With respect to the latter, the effects of density can be incorporated relatively easily in some structured models. However, the choice of a mathematical function describing the relationship between population size or density and vital rates usually must be done almost entirely on theoretical grounds, as rarely are there sufficient empirical data available to help guide modeling of this relationship. Assumptions regarding the form of density dependence can have substantial impacts on model output (Mills et al. 1996; Laskowski 2000). This presents a dilemma for population-level risk assessment and management, because density often influences the response of populations to chemical exposure (e.g., Linke-Gamenick et al. 1999; Sibly et al. 2000; see Chapter 4, but see Sabo et al. 2004). However, for instances in which model outputs are relatively insensitive to the forms and strengths of density dependence, or when conservatism is desired in risk estimates, density dependence can be ignored in the modeling formulation (Ginzberg et al. 1990; Morris and Doak 2002).

Modeling for Risk Management Scenarios: Hazardous Waste Sites

Given the ease with which the responses of organisms to chemical exposure, as measured in toxicity tests, can be linked to population-level attributes, and their relative flexibility to accommodate a wide variety of populations and environmental situations, projection matrix models appear to hold much promise for risk assessments of hazardous waste sites (Appendix 1). With simplifying assumptions, such models can be used in screening tiers with minimal data requirements. Their use in higher tiers requires additional information and incorporation of density dependence in vital rates and immigration. Care is needed when defining the assessment population (see Chapter 3) relative to the size and spatial context of the waste site to help ensure that the assessment supports site management decisions.

Further, analytic methods have been developed to evaluate systematically the relative sensitivity of λ to proportional changes in the transition probabilities (vital rates) of the projection matrix. Called "elasticity analysis" (Caswell et al. 1984; de Kroon et al. 1986), this technique can be used to identify which vital rates influence population growth rate the most when changed, thereby focusing attention on those parameters (say, through targeted toxicity tests; Hansen et al. 1999) in population-level ecological risk assessments.

Individual-Based Models

Formulations in the previous 2 classes model the population as a whole or as interacting collections of subgroups (e.g., age or stage classes). At the opposite end of the biological aggregation spectrum (Fig. 9.1) from unstructured models are individual-based models (IBMs) that, in contrast to biologically aggregated formulations, focus on the individual as the basic element of a population. Often called agent-based models or entity-based models, IBMs can track the characteristics of each individual through time. A key assumption in these models is that individuals can differ with respect to their behavioral and physiological responses to the environment and their chances of being subjected to random events: they can grow at different rates even at the same level of food, they can produce different numbers of offspring under near-identical conditions, and they have different survival probabilities that result in actual (modeled) deaths. The ability to model mechanistically the effects of chemical (and other) stressors on physiological processes and individual behavior distinguishes IBMs from the more aggregated models described earlier, which generally are limited to indirect reflections of these effects on survival, fecundity, and development rates. Individual variability is often modeled as probability distributions from which individual events and realizations are drawn (although deterministic relationships can also be reflected; see the following section). Because the assumptions made in IBMs and their data needs are very specific to the species being modeled, IBMs tend toward a high degree of realism and (ideally) accuracy at the expense of generality.

Assumptions made in individual-based models (IBMs) and their data needs are very specific to the species being modeled. IBMs therefore tend toward a high degree of realism and (ideally) accuracy at the expense of generality.

As a class, IBMs cover an enormous range of specific model formulations (de Roos et al. 1991); Grimm (1999) reviews some 50 IBMs developed for animal populations alone. Thus description of standard mathematical forms is difficult. However, some general comments can be made with respect to their construction and use. IBMs usually consider a specific situation or environment in which individual organisms or entities reside and interact. Procedural rules determine the behaviors of individuals in this environment, which in turn affects their characteristics or attributes. IBMs take 2 broad forms: *i*-state distribution models and *i*-state configuration models (Caswell and John 1992). The *i*-state distribution models are based on the extended McKendrick von-Foerster partial differential equation (Metz and Diekmann 1986) that takes into account the physiological states of individuals in addition to biological structure:

$$\partial\rho/\partial t + \partial\rho/\partial a + \Sigma_i\, \partial(\rho g_i)/\partial m_i = -\mu\rho \qquad \text{Eq. 9}$$

where ρ is the density of the population that is dependent on time t, age of each individual a, and the growth rates g_i of the ith physiologic variable m_i, and where μ is a mortality rate that itself can be a function of individual age, size and population density (Hallam and Lassiter 1994; Hallam 1998; see Hallam et al. 1992 for detailed explanation). The physiological variables, m_i, can represent such factors as lipid utilization, storage of structural components such as proteins, and other parameters that can be modeled in individual organisms (Hallam 1998). These models rely on equations that can be solved analytically, and provide explicit mathematical descriptions of the relationships between organism-level attributes and population-level attributes (see Chapter 3). A limitation to this approach is that many types of attributes, and in particular attributes that depend on the spatial relationships between individuals, cannot be modeled. Conversely, the *i*-state configuration approach can use any type of equation and is analyzed by simulating the activities of many individuals as they eat, grow, reproduce, move, interact with predators and prey, die, and so on. It can be applied to any attribute, but the relationships between organism-level attributes and population-level attributes can only be defined by performing and summarizing repeated simulation experiments—population-level attributes are characterized as the sums of organism-level attributes. More detailed discussion of IBMs are given in DeAngelis and Gross (1992) and Pastorok et al. (2002).

Examples of *i*-state distribution IBMs include Nisbet et al. (1989) and McCauley et al. (1990), modeling daphnid populations as a function of bioenergetics using laboratory data. Kooijman and Metz (1984) and Klok and de Roos (1996) used IBMs to evaluate the effects of chemicals on *Daphnia* and the oligochaete, *Lumbricus rubellus*, respectively. Some examples of *i*-state configuration models include those developed for field populations of fish to investigate trophic interactions and population structure (Adams and DeAngelis 1987), recruitment, population dynamics, and stock management (Madenjian and Carpenter 1991; Rose and Cowan 1993; van Winkle et al. 1993; Sutton et al. 2000), and the effects of environmental stressors such as climate change (Clark et al. 2001) and chemicals (Jaworska et al. 1997). Some of the earliest IBM applications were developed to model tree dynamics (e.g., Botkin et al. 1972; Shugart et al. 1992; Botkin 1993). Topping (1999) describes development of an *i*-state configuration spatially explicit (see Spatially Explicit Models) IBM to evaluate the effects of complex agricultural landscape structure on population-level effects of dispersive linyphiid spiders. Using descriptors based on Danish agriculture settings, Topping (1999) simulated landscapes with varying habitat patch type, area, and size frequency distribution. Juvenile spider dispersal and mortality were modeled as functions of landscape characteristics. Although unaffected by the spatial arrangement of habitat types, spider population size increased with increasing habitat patch size. IBM applications such as this can be used to evaluate environmental factors such as habitat quality and quantity on population dynamics. Appendix 2 describes a case study application of a spatially explicit IBM to support pesticide registration.

In addition to the attributes used in the more aggregated models described earlier (e.g., abundance, growth rate, extinction rate, biological structure), IBMs can provide information in population-level ecological risk assessments about every organism-level attribute included in the model, such as the realized production of offspring per female. Because they require simulation, *i*-state configuration IBMs provide probabilistic information about the selected attributes, yielding both estimates of central tendencies and variability. Thus risk estimates (e.g., the probability that abundance falls below some threshold level) can be extracted easily from model output.

IBMs typically require large amounts of high-quality data concerning the biology and behavior of the species modeled. Because of this and their relative complexity and sophistication, use of models in this class to support risk management decisions may be limited primarily to higher tiers of ecological risk assessment. Further, use of *i*-state distribution models requires a great deal of mathematical sophistication, although *i*-state configuration models are less demanding in this regard because conceptual models are fairly easily and directly translated into set of rules in the

Modeling for Risk Management Scenarios: Agricultural Pesticide Registration

Appendix 2 describes a tiered application of population models to support registration decisions about a fictitious cereal crop insecticide and its ecological risks to populations of common skylark, *Alauda arvensis*. A screening tier assessment uses a relatively sensitive biologically structured model, whereas a higher tier assessment uses a spatially explicit individual-based model reflecting Danish agroecosystem landscapes. Both models employ standard toxicity data from premarket assessments and assumed to be provided by the pesticide manufacturer. Toxicological effects are modeled on both survivorship and reproduction of skylark. Although the screening tier model assumes only direct effects, the higher tier model allows indirect effects related to farming practices, changes in food availability, and so on.

This case study attempted to mimic the registration decision process. After interpretation of assessment results by risk assessors, risk managers provided commentary on the usefulness of the tiered modeling approach to support decision-making. The study illustrates how modeling approaches to population-level ecological risk assessment can be used in a tiered format to support risk management decisions.

programming language. As a tool for higher tiers of risk assessment, however, IBMs offer a great deal of flexibility (with respect to populations and attributes modeled) and ecological realism to the analysis, and offer much promise for supporting environmental decision-making.

Metapopulation Models

Degree of biological aggregation distinguishes unstructured, biologically structured and individual-based models (Figure 9.1). Yet, models in those 3 classes can assume spatial context to be unimportant (i.e., uniform) and the population panmictic. In contrast, metapopulation models take into account the disjunct spatial distribution of many natural populations, which can be segregated into subunits or subpopulations that are linked by some level of migration (see Chapter 6). Extinction and recolonization drive the dynamics of local subpopulations. Such models do not necessarily incorporate habitat quality and stressors explicitly; alternatively, the environment can be described mathematically as a series of more-or-less identical patches. Patch treatment varies from presence and absence modeling to estimating demographic rates for each patch in the system, and with the incorporation of stochasticity, density dependence, and other forms of ecological realism. In addition, metapopulation models may be applied on either an individual-based or population level. The defining feature of metapopulation models is the presence of movement of individuals between patches as either migration or dispersal. Metapopulation models can be used to explore the influence of 1) the size, spacing, and density of patches; 2) rates and forms of movement among patches; and 3) rates of extinction within patches on population persistence and spatial distribution. Levins (1969, 1970) offers early considerations of the metapopulation modeling approach. More recent and complete treatments are given in Hanksi and Gilpin (1997), Hanski (1999), and Akçakaya and Regan (2002). Metapopulation models tend toward enhanced ecological realism over simple spatially unstructured models by reflecting spatial arrangement and context of population subunits, and can be used to explore general relationships between spatial structure and population dynamics.

Metapopulation models tend toward enhanced ecological realism over simple spatially unstructured models by reflecting spatial arrangement and context of population subunits, and can be used to explore general relationships between spatial structure and population dynamics.

Examples of models in this class include the classic Levins model, lattice models, the incidence function model, and source-sink dynamics models. As with IBMs described earlier, the assumptions and specific formulations of metapopulation models vary depending on the specific problem and model used. For example, spatially implicit patch occupancy models assume that all individuals are equal in terms of demographic parameters and likelihood of dispersal, that populations are in equilibrium, and in some models that distance between patches is irrelevant (the mean field assumption, Hanski 1999). If dispersal distance is assumed to be important, then the effect of distance between patches on successful dispersal must be estimated, most often with a mathematical function based on existing data. Models incorporating biological structure in individual patches, or that are individually based, have the same assumptions as those two model classes in addition to assumptions regarding dispersal. The refinements represented by these different approaches aside, Levins' (1969, 1970) original model illustrates the basic approach:

$$dP/dt = cP\,(1 - P) - eP \qquad \text{Eq. 10}$$

where P is the proportion of habitat patches occupied, c and e are constant rates describing patch colonization and extinction, respectively, and t is continuous time. Levins' model assumes a large number of discrete patches in the environment connected by migration, and ignores differences (such as size, suitability, and interpatch distances) among patches. At equilibrium, the proportion of patches occupied is:

$$P = 1 - (e/c) \qquad \text{Eq. 11}$$

and the metapopulation (or "population of populations") is expected to persist as long as $e/c < 1$; that is, as long as the colonization rate is greater than the extinction rate. Several subsequent metapopulation models were developed as derivatives of this original formulation.

Early in their development, metapopulation models were used primarily to investigate general relationships between spatial configuration and population dynamics (e.g., Levins 1969; Hanski 1991, 1994). However, metapopulation approaches are becoming popular as tools supporting conservation efforts. For example, Lindenmayer and Possingham (1996b) compared extinction risk to Leadbeater's possums *(Gymnobelideus leadbeteri*) under various timber harvest scenarios, examining forest stand age, spatial configuration of remnant patches, and patch size in 2 different forests. They also conducted a sensitivity analysis to examine the effects of estimating the poorly understood dispersal behavior of the species on model output. As another example, Akçakaya and Atwood (1997) used a metapopulation approach to evaluate population persistence of California gnatcatchers as a function of habitat availability. A third example is described by Sjogren-Gulve and Ray (1996), who modeled pool frog (*Rana lessonae*) occurrence in a metapopulation of ponds as a factor of multiple environmental variables including water pH, presence of a predator, and occurrence of timber harvest near the pond. The model proved to be quite accurate in predicting both regional metapopulation dynamics and local pond turnover. Metapopulation models have also been used to investigate community dynamics and diversity. For example, Bellwood and Hughes (2001) found the species diversity of Pacific coral reefs to be influenced by the availability of recruits from other reefs within 600 km. Although the metapopulation concept is of potential importance to population-level ecological risk assessment (Sherratt and Jepson 1993; Akçakaya 2001; Johnson 2002), examples from ecotoxicology are few. Some exceptions to this are described in Chapter 6, which relates the results of modeling exercises that suggest that contaminant exposure in a single patch can lead to ecologically significant effects in neighboring patches because of the movement of organisms among sites (Maurer and Holt 1996; Spromberg et al. 1998). This finding has implications for assessments supporting several risk management decision contexts, notably those related to hazardous waste sites and chemical registration.

Metapopulation models can be used to evaluate risks to the population-level attributes of persistence (time to extinction analyses), and spatial patterns of occupancy, but also include expected abundances in various patches, variations in abundance, occupancy rates, and movement rates. Such models have contributed tremendously to our understanding of population dynamics in patchy environments, and have found widespread applications in the fields of ecology and conservation biology. The simplest models (2-population source-sink, Levins' model, and the spatially implicit structured metapopulation models) are important conceptual tools but are of limited applied value. Conversely, the spatially explicit metapopulation models have been used effectively in specific applications, but do not generally extrapolate beyond the study system for which they were developed and are generally used to explore the relative consequences of various management scenarios (e.g., Noon and McKelvey 1992; Lindenmayer and Possingham 1996a; Drechsler et al. 2003). Although many of the spatially explicit metapopulation models are extremely realistic and describe actual population dynamics fairly well, the major drawback is that these models require a large amount of empirical data to parameterize them. In particular, the estimation of successful movement among patches is of critical importance (Ruckelhaus et al. 1997), but this parameter is among the most difficult to estimate in free-living populations. Promising developments include stable isotope enrichment analysis (Caudill 2003) and genetic markers (Arnaud 2003; Tero et al. 2003). Mark-recapture methods have also been applied to this problem (Hestbeck et al. 1991; Hanski et al. 2000; Purse et al. 2003). Estimation of dispersal and migration of vertebrates has been greatly aided by the development of better radio-telemetry equipment and methodology (see Chapter 8).

Metapopulation models developed primarily to explore general relationships among patch configuration, dispersal characteristics and population persistence are most useful in early tiers of population-level ecological risk assessment. Such formulations have limited data requirements (e.g., those primarily exploring occupancy of patches may require only patch size, presence-absence data or abundance of individuals per patch, and interpatch distances as input information), and provide insights that potentially can be applied in a variety of situations and contexts. With increasing assumptions about intrapatch population dynamics and added biological detail about the population modeled (with concomitant increases in data requirements), metapopulation models might also be useful at higher tiers of an assessment. All models require information on dispersal. This can be modeled simply as a probability of movement, but further realism can be incorporated (and data needs increased) if dispersal likelihood is modeled as density-dependent, if successful dispersal is a function of distance traveled and landscape traversed, or if other ecologically relevant variables are also taken into account. Incorporation of inaccurate information on dispersal can lead to high rates of prediction error in model results (Ruckelshaus et al. 1997). Simulation analyses of invertebrate movement across agricultural landscapes exposed to pesticide applications also highlight the importance of accurate dispersal estimation (Sherratt and Jepson 1993).

Spatially Explicit Models

Representing the extreme in spatial disaggregation (Fig. 9-1), and as their name implies, spatially explicit models (SEMs) consider the characteristics of the landscape—habitat types, quantities, qualities, and arrangements, together with the spatial distributions of environmental stressors in that landscape—explicitly. For this reason, SEMs have also been called landscape models. Extinction and recolonization of discrete patches are not key features of SEMs (as they are in metapopulation models). Rather, organisms are assumed to move more or less continuously across the modeled landscape as dictated by the movement rules assigned to them. For plants and other sedentary organisms, movement might be accomplished through seed or larval dispersal. Spatially explicit models simulate population dynamics in the landscape using any of the approaches described in the first 3 model classes here, though typically they use either biologically structured or IBM constructs appropriate to the level of sophistication reflected in the representation of the landscape. Recent discussions of SEMs can be found in Dunning et al. (1995), Kareiva et al. (1996), and Tilman and Kareiva (1997). Because they reflect specific landscapes (either real or simulated as scenarios) and organism movement in that landscape as controlled by species-specific properties (although movement patterns can also be evaluated as generic classes), SEMs emphasize ecological realism and accuracy of prediction over generalizability of model outcomes.

Because they reflect specific landscapes and organism movement in that landscape is controlled by species-specific properties, spatially explicit modeling emphasizes ecological realism and accuracy of prediction over generalizability of model outcomes.

SEMs vary in how they handle 3 basic components: the description of the landscape, organism movement across that landscape, and the demographic dynamics that occur in the landscape. The landscape itself can be represented as a more or less continuously varying surface with "grain" determined by the resolution of GIS input (e.g., Topping 1999), or as a matrix of internally homogeneous cells (of various sizes) contiguously arrayed on the landscape (e.g., Akçakaya and Atwood 1997; Schumaker 1998; Fuentes and Kuperman 1999). In these latter "cellular automaton" models, cells are classified by their habitat characteristics and concentrations of environmental stressors into a number of states. The state of a single cell in the matrix depends on the history of that cell, and the state of the neighboring cells. Cellular automata differ from (or perhaps, represent a special case of) metapopulation models in that organism movement generally is assumed to occur among adjacent cells in the matrix, as opposed to between any 2 patches with each time step.

Spatially explicit models use rules to describe the movements of organisms across the landscape. Such rules can be highly specific to the organism modeled, or can be described more generically. Often, movement rules include consideration of the quality of the immediate and surrounding landscape, in terms of habitat type, prey availability, occupancy by conspecifics, and so on. Thus organisms can "sample" the landscape and use that information to influence movement patterns. Alternatively, movement can be modeled as a random process (Wu et al. 2000), with organisms moving across the landscape in an uninformed manner, and occupying areas of unknown quality at the end of each time step. Organisms can also disperse through diffusion across the continuous landscape, at rates determined by local densities of individuals in clusters on that landscape (so-called reaction-diffusion models, which handle both space and time in a continuous fashion, are reviewed by Okubo 1980). In all cases, these rules determine the distribution of organisms in space, and are typically applied to assess the probability of an organism passing through or reaching a location in a landscape. Turchin (1989) provides a useful introduction to movement models in population modeling by considering the influence of aggregative effects in foraging. Wu et al. (2000) provide a good mathematical example of a movement model and consider the effect of persistence in direction of movement.

As a generalization, population dynamics in the landscape can be described in SEMs using unstructured, biologically structured, or individual-based modeling approaches. Clearly, however, the sophistication and complexity of landscape and movement components of the SEM, together with the biology of the population, will inform the selection of population dynamics method. For example, SEMs modeling the landscape in a more-or-less continuous fashion, and using movement rules that reflect adjacent landscape quality and other important environmental characteristics, are ideally suited to *i*-state configuration IBM approaches that capture the behaviors and responses of individual organisms (for example, Topping 1999; Topping et al. 2003a). Biologically structured population modeling approaches might be more suited to cellular automaton constructs for landscape description and movement (e.g., Schumaker 1998), which handle time and space more discretely. That said, spatially explicit IBMs likely represent the extreme in population model sophistication, complexity, and perhaps ecological realism. Their implementation is limited in the resolution of biology and environmental condition only by data availability or hardware and software constraints, rather than by theoretical constraints. Spatially explicit IBMs are suitable for describing spatiotemporal interactions at a fine scale, and are capable of generating spatial negative or positive feedback, an important feature in situations where such phenomenon are likely to occur. These qualities combine to create extremely realistic modeling representations of the real world.

Modeling for Risk Management Scenarios: Hazardous Waste Sites

Spatially explicit modeling (SEM) is usually thought of in the context of terrestrial systems, and most such models have been developed for such applications. However, water bodies, such as lakes, reservoirs, and oceans, also have spatial structure. The idea that SEMs could be applied to "seascapes" suggests their use in population-level ecological risk assessment of aquatic hazardous waste sites. Recently, Linkov et al. (2002) described an SEM of flounder exposure to contaminants associated with in-place sediments at a dredged material disposal site. Their exposure model could be combined with biologically structured or individual-based models population dynamics formulations to describe the population-level effects of in-place sediments at aquatic hazardous waste sites, extant and relic disposal sites, or other aquatic situations in which the spatial distribution of contaminants is heterogeneous.

Development and application of SEMs is a relatively recent phenomenon. Spurred in part by the growing recognition of the importance of spatial context to population dynamics and risk (see Chapter 6), and by the increasing availability of high-end computing capability, such models are being used at an increasing rate. For example, Kondoh (2003) uses a cellular automaton approach, coupled with individual behavior, to address tradeoffs between reproductive and predation rates in the coexistence of competing prey species in a spatially heterogenous system. Also using a cellular automaton, Schumaker et al. (2004) evaluate the consequences of 4 grow-out scenarios, reflecting differences in human population growth and zoning objectives, for development over the next 50 years in the Willamette Valley in western Oregon. Schumaker et al. modeled several species of mammal and bird, using a projection matrix approach to elucidate their population dynamics and spatial distributions resulting under each scenario. These results are helping to inform (human) community-level decisions about future development. Westerberg and Wennergren (2003) use a matrix model to distribute populations in space in the same way that a biologically structured model distributes populations over stages, thus retaining some of the analytical properties of that modeling approach. Vacher et al. (2003) use a more traditional approach in modeling pest adaptation to Bt cotton using a grid-based environment but assuming panmictic reproduction. Mooij et al. (2002) describe an individual-based model of intermediate complexity, considering interactions between hydrology and snail kite populations in the Florida Everglades. Topping et al. (2003b) used a detailed landscape model in combination with detailed behavioral models to investigate impacts of changes in agricultural practices on vole populations in Denmark, and Topping and Odderskær (2004) use a similarly constructed model to evaluate pesticide use on the dynamics of common skylark populations (see text box and Appendix 2). In a nice example of spatially explicit IBM use in conservation ecology and resource management, Turner and colleagues (Turner et al. 1993, 1994a,b; Turner and Romme 1994) modeled the dynamics of fire damage to vegetation in Yellowstone National Park, Wyoming, and the resulting effects on elk and bison populations. This information provided insights for fire management in an ecosystem shaped in large part by a natural fire regime.

As modeling tools, SEMs can be used to describe almost any of the population-level attributes important to environmental decision-making. Generally, the specific attributes modeled are determined by the model formulation describing population dynamics, but with the nuance of a spatial component. Hence, SEMs can describe the spatial distribution of, for example, population abundance relative to specified local areas or habitat types. When simulations are individually based, the total population and the states of individuals are continuously available, and hence the number and types of attributes describable is limited only by the power of the model used.

The extreme ecological realism and developmental flexibility of SEMs come with a cost: the detailed landscape descriptions, movement rules (potentially) specific to the population being modeled, and biological information needs combine to result in the most data intensive of the model classes considered in this chapter. As a general rule (particularly true for spatially explicit IBMs), models in this class are not mathematically tractable, and therefore can not be easily analyzed to create generalizable rules. The high demand for data and the mechanistic approach employed in model development also result in substantial technical demands on the modeler. Thus SEMs likely are best reserved for higher tiers of population-level risk assessment, when enhanced specificity and realism is necessitated by the decisions the models are intended to support.

Spatially explicit modeling, especially spatially explicit individual-based models, likely are best reserved for higher tiers of population-level risk assessment, when enhanced specificity and realism is necessitated by the decisions the models are intended to support.

TABLE 9.2
Incorporating genetics in population dynamic models

Model formulation	Incorporation of genetic variation
Biologically structured modeling approaches	Issue: impacts of genetic variation within populations Approach: incorporate inbreeding depression as reductions in vital rates (e.g., VORTEX)
Metapopulation modeling approaches	Issue: impacts of genetic variation among population subunits Approach: model dispersal and migration among population subunits as gene flow
Individual-based modeling approaches	Issue: impacts of genetic variation among individuals within populations Approach: allow use of landscape to physically isolate groups of individuals and monitor resulting changes in genetic diversity (e.g., Topping et al. 2003b)

INCORPORATING GENETIC ATTRIBUTES INTO POPULATION DYNAMICS MODELS

As represented by the taxonomy in Fig. 9-1, population modeling approaches can in part be categorized by their ability to incorporate differing levels and types of variation among populations, metapopulations and individuals. Genetic variation among these groups can be (but seldom has been) captured in population dynamics models at these same levels. In fact, models and methods to project and test central hypotheses uniting population genetics and dynamics are lacking, and their development remains an important challenge (e.g., Burger and Lynch 1995; Ashley et al. 2003). As evidence of the difficulty in developing such a unified approach to modeling populations, a recent review by Groom and Pascual (1998) concluded that of 58 population viability analyses published in Conservation Biology, only 7% included genetics. Despite the difficulties, increasing rates of environmental change (e.g., Palumbi 2001) create the need to incorporate evolution into population dynamic models, which often are used to make projections or predictions that span the time or generations sufficient for contemporary evolutionary change. Examples of "surprises" (i.e., unexpected adverse consequences) that have occurred when evolution has not been taken into account include captivity and hatchery conditions that select for traits that are nonadaptive in the wild, unexpected effects of introduced species, and increased levels of endangerment related to increased genetic differentiation, smaller than predicted effective population sizes (N_e; see Chapter 5) and localized adaptation in response to habitat fragmentation (reviewed by Ashley et al. 2003).

Current needs and the potential for contributions suggest that the integration of population genetic information into population dynamic modeling provides an important opportunity to advance population-level ecological risk assessment, and some general approaches along these lines have been developed (Table 9.2). For example, demographic models incorporate variation in fitness attributes between groups within a population, or within the population as a whole. Often, these models are structured to accommodate variation in vital rates among demographic groups. This structure can accommodate genetic information when differences in vital rates among populations or groups have a genetic basis. Specifically, the impact of inbreeding depression has been incorporated into projections of population persistence via theoretical or measured reductions in early life stage fitness. This approach has been used in population viability analysis of several threatened and endangered species of small population size (reviewed in Brook et al. 2000). VORTEX (http://pw1.netcom.com/~rlacy/vortex.html) population viability analysis software produced by R. Lacy (Department of Conservation Biology, Chicago Zoological Society, Brookfield, Illinois) and distributed freely to nonprofit use organizations by Chicago Zoological Society, has been developed for this purpose. However, similar strategies can be used to incorporate inbreeding effects into other matrix modeling approaches (e.g., as noted in Brook et al. 2002).

Current needs and the potential for contributions suggest that the integration of population genetic information into population dynamic modeling provides an important opportunity to advance population-level ecological risk assessment.

Interestingly, although population projections or predictions often span many generations, this approach has rarely been used to predict or describe the evolution of genetically based differences in vital rates between populations. Specifically, temporal changes in vital rates and life history could be incorporated into modeling projections. For example, changes in specific model parameters could reflect adaptation based on calculated selection coefficients for specific stressors. Using this approach, evolved tolerance to environmental contaminants, including pesticides and other toxic chemicals, could be predicted and then tested. In fact, simulations using simple and then more complex models were used to simulate genetic responses associated with chemical stress (mercury exposure), including forces of selection, and drift (Newman and Jagoe 1998). In addition, the implications of such changes to population growth rate in adapted versus nonadapted populations are described (but not tested) in a case study on estuarine fish adapted to a class of persistent, bioaccumulative, and toxic chemicals (reviewed in Nacci et al. 2002b).

Classical metapopulation modeling approaches describe physically disparate subunits that may perform differently, sometimes as a result of varying environments or habitats (Levins 1969; Hanski and Gilpin 1997; Spromberg et al. 1998). Essential to the concept of metapopulations is dynamic turnover; that is, colonization and extinction, and exchange among patches (Levins 1970). Metapopulation dynamics can reflect both within-patch processes (reproduction, survival) and between-patch processes. Population spatial structure and dynamics, integral to the metapopulation construct, has long been recognized as having a major influence on the maintenance and loss of genetic variation (e.g., Hanski and Gilpin 1997). From a genetic perspective, spatial structure affects population genetic structure, heterozygosity, and N_e. Because colonization involves founding by few individuals, metapopulation dynamics include effects of small population size. Theoretically, metapopulation dynamics can either reduce or increase population genetic diversity more than predicted for similarly sized but spatially unstructured populations. More specifically, factors influencing N_e of metapopulations include the number and carrying capacities of population patches, rates of extinction and colonization, the number and sources of founders, and the rate of gene flow between patches (Hanski and Gilpin 1997). In sum, the metapopulation approach provides a strong theoretical basis for predicting how multiple processes affect genetic variation and the distribution of genotypes among populations and across landscapes.

Thus metapopulation models provide conceptual frameworks and application mechanisms that permit the incorporation of variation from environmental stochasticity, random catastrophic events, demographic stochasticity, and genetics. Overall, metapopulation approaches can be used to estimate population persistence and extinction probabilities that consider interacting colonization-extinction probabilities for multiple patches, taking into account evolutionary processes. A few, relatively intensive studies demonstrate this approach (as reviewed by Hanski and Gilpin 1997). Despite well-developed theory and these examples, the relevance in this context of population genetic information based on traditional genetic (neutral) markers has been criticized. Specifically, neutral genetic population structure and change may not be informative relative to population persistence. However, metapopulation modeling does provide a theoretical mechanism to acknowledge and account for genetic adaptation to heterogeneous environments, for example, via competing effects of selection (to varying agents associated with a heterogeneous environment) and gene flow. The inclusion of quantitative genetic approaches into metapopulation modeling may provide an opportunity to incorporate ecologically important genetic variation into population consequences. Empirical studies provide an opportunity to test the accuracy of these predictions, and advance understanding of implications for population persistence of habitat quality and structure.

Similarly, intrapopulation variation in behavior, experiences and performance among individuals can be accommodated using individual-based modeling approaches (e.g., DeAngelis and Gross 1992). Inherent in these formulations is the assumption that important properties of populations emerge from individual variation, and that intrapopulation variation is an adaptive property (Lomnicki 1988). Consequently, these approaches provide a means to include genetically based individual variation. For example, if the fitness functions of particular genes and their occurrence in the population are known, the consequences of specific environmental conditions on population abundance and genetic structure can be predicted. Lomnicki (1988) provides an example of population genetic structural changes in a population subject to temporally fixed stress and frequency-dependent hard and soft selection. Although this approach provides opportunities to address theoretically interesting and ecologically important questions, its application is limited by our knowledge of gene function. Furthermore, model projections which incorporate genetic variation are often untestable because of the unrealistic assumptions about environmental stressors as selective agents: constant in magnitude and long-term (i.e., multiple generations) in duration. Alternatively, IBM approaches have been used on a limited basis to simulate effects of temporally and spatially varying environments.

A rather unique integration of individual-based modeling and classic genetic theory, that is, a genetically explicit IBM, was used to predict genome-wide changes in population genetic structure in response to a spatially and temporally varying environment (Topping et al. 2003b). This approach was used to predict changes in population genetic structure and diversity in response to changes in migration associated with varying agricultural field usage. This application demonstrated a mechanism to predict population consequences based on the behavior of individuals inhabiting a complex environment. Specifically, genetics for each individual (and thus the population at census intervals) were predicted based on behavior and a system of 16 unlinked loci, each with 4 potential alleles. Genes were assumed to be neutral, with linkage equilibrium and no mutations. Population size, number of alleles, allele frequencies, and expected heterozygosity (from allele frequencies) were monitored in the simulation. Population outcomes of abundance and heterozygosity were interpreted in the context of responses to spatial and temporal environmental variation. In short, increased environmental variation (small fields with frequent crop rotations) produced reduced genetic variation. Although these results are intriguing, the value of genome-wide neutral genetic variation to population persistence is not well understood. Integrating a quantitative trait locus approach with individual-based techniques may provide the mechanism to evaluate the consequences to population persistence of changes in the abundance and distribution of ecologically meaningful traits. As with other approaches, empirical studies provide an opportunity to test the accuracy of these predictions and advance understanding implications of habitat quality and structure to population persistence.

The increasing accessibility of molecular technology and potential utility of ecologically important genetic information may permit further exploration of mechanisms to integrate population genetics and dynamics. Such advances, together with appreciation of when unified modeling approaches are needed, should enhance the value of population-level ecological risk assessment in decision-making.

Despite its conceptual importance (see Chapter 5) and the availability of approaches with which to evaluate that importance, the role of genetics in the persistence of populations remains controversial (e.g., Frankham and Ralls 1998; Mann and Plummer 1999). However, it is at least prudent to acknowledge that genetic effects may be critical for some populations, such as small and isolated populations subject to inbreeding. Populations theoretically most vulnerable to inbreeding effects associated with low effective population sizes include those with fluctuating population sizes, variation in reproductive success and unequal sex ratio (Frankham 1995). Empirical correlations with low N_e include large body size and long generation time, which itself is correlated with high variation in reproductive success (Allendorf and Ryman 2002). In describing 3 idealized mammals, Mills and Smouse (1994) also concluded that those with low growth rates theoretically were most vulnerable to inbreeding depression effects.

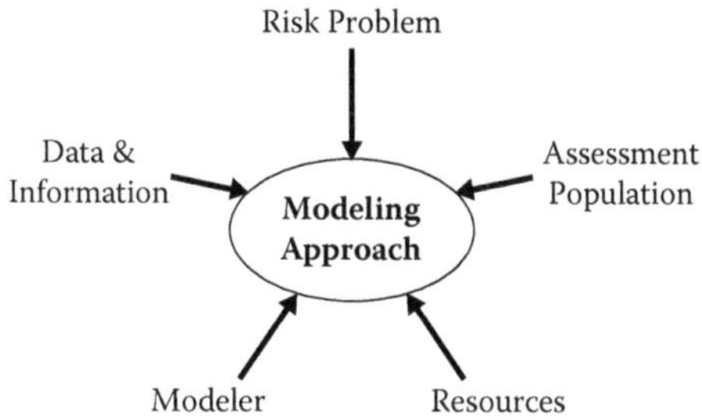

FIGURE 9.2 Factors affecting choice of models.

Historically, population genetics approaches have been founded on the use of neutral markers to describe features such as structure, dispersal, and N_e. However, these markers fail to account for quantitative characters on which selection can act, or "ecologically important" genetic variation (e.g., Reed and Frankham 2001). This recognized limitation may significantly undermine the relevance of predictions concerning the impact of genetic changes on population persistence. The increasing accessibility of molecular technology and potential utility of genetic information (as gene functions become known) may permit further exploration of mechanisms to integrate population genetics and dynamics. Such advances, together with appreciation of when unified modeling approaches are needed, should enhance the value of population-level ecological risk assessment in decision-making.

MODELING TO SUPPORT RISK MANAGEMENT AND DECISION-MAKING

How can the range of modeling approaches described in this chapter be best used to support decision-making? In the context of the ultimate use of model outputs in the risk decision (vis-a-vis diagnosis, projection, and prediction), at least 2 considerations are cogent to answering this question: population attribute selection and model selection. Whereas the attributes of populations have been discussed in the context of protection goals in Chapter 3, we focus here on model selection.

In a perfect world, the best modeling approaches would be applied in each risk assessment situation. However, that best model might not always be evident. Additionally, modeling is subjective, with different modelers making different decisions concerning assumptions, data uses, and so on. Thus model selection for a given risk assessment application depends on 5 basic considerations (Figure 9.2):

- The question or "risk problem" the model is intended to address in the "context of the decision" to be made: Is the problem generic or site-specific? What are the spatial and temporal characteristics of the problem? How will the results be used to support the decision?
- The "assessment population": What is the life history and demography of the population? Is density-dependence expected to be important? Are genetic changes expected? How does its spatial distribution compare to that of the stressors? What population attributes are important?

- The "data and information" available as model input: What are the nature, quantity, and quality of data available for the model?
- The "resources available" for modeling exercises: Are sufficient tools, time, and effort available to develop, evaluate, run, and interpret the model(s)?
- The experience and preferences of the "modeler": What experience does the individual performing the modeling analyses have relative to different approaches? Does he or she prefer certain approaches over others?

Forms of some of these questions are asked in the assessment framework developed in Chapter 10. They can be used to help focus model selection during problem formulation of an assessment, and their answers help to determine how model results should be used to support characterization and interpretation of risk.

Although it is now possible to build and execute extraordinarily complex models, these should not necessarily be the first tools a risk assessor considers. They require a great deal of data to parameterize, and creating and executing them is a resource-intensive process. Rather, model selection should first consider whether incorporating additional complexity to enhance model realism or accuracy is warranted by the problem and decision context. Put simply, appropriate modeling approaches to population-level ecological risk assessment might be best identified in a tiered assessment and decision-making format.

Modeling for Risk Management Scenarios: Hazardous Waste Sites

In support of a Remedial Investigation/Feasibility Study (RI/FS) at the Portsmouth Naval Shipyard (Maine), Gleason et al. (2000) modeled the population responses of the purple sea urchin, *Arbacia punctulata*, resulting from exposure to lead, a primary contaminant of concern in the estuarine waters surrounding the shipyard. They used standard bioassay data collected during site investigations to parameterize a relatively simple stage-classified model designed explicitly to reflect the life history stages and vital rates represented by the bioassays. The population growth rate exposure-response relationship resulting from this effort was used subsequently in a weight-of-evidence characterization of ecological risks at the site (Johnston et al. 2002).

Conceptually, a tiered application of population-level ecological risk assessment would consist of series of complete assessments arrayed from most general and broadly based (screening level) to most realistic, accurate and situation-dependent (definitive level). The types and complexity of modeling approaches used, the quality and quantity of data required, the population-level attributes evaluated, the nature of risk conclusions developed, the degree to which results can be extrapolated to other situations, and the degree of confidence in risk conclusions would vary across tiers. Screening-level tiers might model generic populations reflecting broad patterns of life history and demography and susceptibility to the stressors. They might use unstructured or biologically structured models with low data requirements, producing conclusions that are highly generalizable but that (perhaps) are accompanied with substantial uncertainty in relation to the risk problem they address. Conversely, definitive level assessments would use realistic models (such as certain biologically structured models, IBMs, and SEMs) that are accurate for particular situations (combinations of species and environmental settings), and therefore requiring potentially large amounts of high-quality data describing the life history and demography of the species and information about the environment. The risk estimates developed from definitive assessments would be very specific to the situation evaluated, and consequentially would have limited generalizability to other scenarios. Ideally, assessors could progress through increasingly more realistic and accurate assessment tiers until they reach a definitive assessment of risk, or at least reach a level of confidence in the risk conclusion that is acceptable to the decision-maker (relative to the costs of making a wrong decision). Practically, the availability of data and resources might prevent moving to a higher tier of analysis, at least until science addresses the particular knowledge gaps preventing more sophisticated modeling. Alternately, the process could be entered and completed at any single tier based on the requirements of the assessment and nature of available information. Throughout all tiers, the assessment population and its attributes would remain as the assessment endpoint. Although not described explicitly, the concept of tiering is reflected in the framework for population-level ecological risk assessment described in Chapter 10.

Conceptually, a tiered application of population-level ecological risk assessment would consist of series of complete assessments arrayed from most general and broadly based (screening level) to most realistic, accurate, and situation-dependent (definitive level). The types and complexity of modeling approaches used, the quality and quantity of data required, the population-level attributes evaluated, the nature of risk conclusions developed, the degree to which results can be extrapolated to other situations, and the degree of confidence in risk conclusions would vary across tiers.

An alternative tiered approach involves population-level assessment endpoints only within certain tiers of the assessment. Tiering might proceed from assessments of adverse effects on organism-level attributes at lower tiers, to population, community, or ecosystem attributes at higher tiers. This is the approach employed in the United States for pesticide registration, as described in Appendix 1. Still different are assessment approaches that do not employ tiers explicitly. These might involve multiple assessment endpoints representing different levels of biological organization including populations, with the conclusions about risk to each contributing to the decision process (perhaps in a weight-of-evidence analysis). In such cases, the population-level assessment might be performed only once, rendering the choice of modeling approach all the more critical.

Whether or not tiering is applied, some decision contexts may require a minimum of population modeling, with little exploration of population dynamics over time, and no consideration of demographic or spatial variability. Would a measure of population abundance or density suffice? In these situations, the resources required of data collection and complex modeling may not be warranted, and unstructured models may suffice. When sufficient data already exist for more complex modeling efforts, or when they can be obtained with a well-designed but not necessarily resource- or time-consuming sampling strategy, more complex modeling might be supported, but not without the concomitant costs of modeling and interpretation that they entail. If accuracy and realism are important elements to the problem, then more sophisticated models may be required. As reflected earlier in this chapter, dynamic properties including changes in abundances in age or stage classes, transient patterns of population growth rate, stochasticity in environmental and demographic rates, and spatial considerations all require greater complexity in the modeling approach.

Models should be more or less suited to the biology of the assessment population. For example, the dynamics of relatively sedentary organisms likely would not require the complexity inherent in metapopulation and spatially explicit formulations (although certain risk questions reflecting the effects of spatial variation in exposure might warrant incorporation of spatial considerations). In contrast, appropriate models for populations affected substantially by sink-source dynamics likely require some degree of spatial disaggregation. Thus, biological characteristics or processes known to influence population dynamics strongly in the context of the risk problem should be captured in the mechanisms reflected in the model. Even within classes of models, the degree to which the biology of the population is represented is an important consideration, having implications on model construction and parameterization, as is obvious with IBMs and demographically structured models.

The data and informational requirements of models deserve special consideration in selecting the appropriate analysis approach for assessment purposes. The degree to which these requirements can be satisfied determines, at least in part, which models may be successfully employed. In this regard, some of the models discussed earlier (e.g., unstructured) require only that the relationships between chemical exposure and effects on aggregate birth and death rate be known. However, such approaches do not easily permit incorporation of mechanistic information regarding toxicity or population control, and therefore are of limited use in examining scenarios involving short-term temporal variation in exposure, incremental effects of multiple stressors, or the ramifications of various remediation options. At the other end of the spectrum are formulations such as IBMs that require detailed knowledge of physiological and metabolic processes and how these influence the vital rates of fecundity and survivorship. The realism of these approaches is further enhanced through incorporation of density influences on or stochastic variation in these processes. Such data are obviously difficult to obtain (see Chapter 4), yet their inclusion into appropriate models permits the most detailed assessments.

Along these lines, consideration of the ease with which assessment population-specific information can be obtained is important (see Chapter 8). Does valuable information exist regarding temporal and spatial patterns of abundance? Perhaps the most cost-effective source of such data is historic data bases obtained for other purposes. Can the life history characteristics of critical life stages be determined from field collections or from the literature? Can sensitivity and the modes of toxicity be examined in the laboratory? Such determinations, if they can be made, are usually performed at increased cost. Thus the availability of information regarding the life history of the species and the dynamics of individual growth and development is crucial to determining model selection. Further, difficulties in obtaining information about populations affected by the risk management decision (e.g., at hazardous waste sites) may affect the ability to evaluate the veracity of estimates of risk.

The limiting nature of resources available for the analysis is an unfortunate reality. When time and money are no object (a situation never occurring in risk management and environmental decision-making), analysis of population-level effect might well involve the most complex, data-intensive modeling approaches available, presumably with enhanced confidence in assessment conclusions. Yet, tradeoffs must almost always be made among the resources expended obtaining necessary data, developing and analyzing models, and interpreting results, all against a backdrop of pressures for timely assessment conclusions to support the risk decision. The classes of models presented above vary in the level of resources needed for their development and implementation, independent of the population being assessed and the risk context itself. At the extremes are unstructured models, which are readily accessible in more or less standardized forms and packages, and individual-based models which require formulation and development on a case-by-case basis (with concomitant increases in resource requirements). Thus compromise may be required between desirable model qualities and abilities and the resources available for modeling. The appropriate balance point is almost certainly a function of the certainty required in assessment results and the cost of making a wrong management decision.

The limiting nature of resources available for the analysis is an unfortunate reality. Tradeoffs must almost always be made among the resources expended obtaining necessary data, developing and analyzing models, and interpreting results, all against a backdrop of pressures for timely assessment conclusions to support the risk decision.

Finally, the skills, expertise, and predilections of the modeler are critical to the selection of models to be used at any tier within the assessment. Most modelers have hands-on familiarity with only a subset of available formulations, which influences their proficiency with any given model. They vary in making and accepting assumptions about the biology of the assessment population and other aspects of the risk problem, a situation that can influence the analysis results and the uncertainties associated with them. This is in large part a quality assurance issue.

UNRESOLVED ISSUES AND RECOMMENDATIONS

Despite the relative maturity of population modeling in ecology, conservation biology, and resource management (see Chapter 7), several needs were identified during development of this chapter with respect to population approaches to population-level ecological risk assessment:

- Because a variety of modeling formulations exist, reflecting varying degrees of sophistication, complexity, data needs and analyst requirements, decision context–specific guidance may be needed to aid in their selection and use to support risk management decisions. Such guidance should consider selection of population models within a tiered assessment approach.
- Certain prospective applications of population-level ecological risk assessment, particularly those supporting regulatory decisions, might benefit from development of standardized packages or frameworks of population models arrayed in tiers of increasing sophistication and data requirements. Similar in concept to those developed by the FOCUS (Forum for the Co-ordination of Pesticide Fate Models and Their Use; FOCUS 1995) group for pesticide exposure models in the European Union, such frameworks could employ standard sets of assessment populations and spatial scenarios, together with built-in data sets describing the species' demography, behavior, and other aspects of their ecology. Early tiers would employ simple models and broad assumptions concerning the assessment populations, whereas more definitive tiers would employ spatially explicit IBMs and ecological realistic descriptions of populations and standardized landscapes. Standardization of this nature has the advantages of focusing initial research and data collection efforts, minimizing subsequent data collection and development needs, and normalizing the risk assessments undertaken by all parties in the decision process.
- Additional guidance is needed to help inform risk assessors, risk managers, and stakeholders about the importance of considering density dependence and population genetics in particular risk problems.
- Additional research is needed to help define when and how density dependence should be reflected in modeling approaches to risk assessment. Enhanced understanding of the ecological conditions in which density effects are important, and of the interactions of density dependent processes and stressor effects, should promote development of improved modeling formulations to account for density dependence.
- Additional research is needed toward developing unified models that reflect the interactions between population dynamics and population genetics.

10 A Framework for Population-Level Ecological Risk Assessment

Randy Wentsel, Nelson Beyer, Valery Forbes, Steve Maund, and Robert Pastorok

DEVELOPMENT OF ECOLOGICAL RISK ASSESSMENT FRAMEWORK

Since the early development of ecological risk assessment (ERA) methods, risk managers and assessors have worked within an assessment framework derived primarily from human health risk assessment. The general components of the framework consist of a problem formulation, exposure analysis, effects (e.g., toxicological) analysis, and risk characterization, with risk planning and communication steps as major points of interaction between risk managers and risk assessors. In 1992, the Risk Assessment Forum of the US Environmental Protection Agency (USEPA) released a *Framework for Ecological Risk Assessment*. This framework and subsequent guidance acknowledges population-level and higher (e.g., community and ecosystem) level risk assessments; however, the development of methods in ERA has focused primarily on organism-level endpoints (e.g., survival, growth and reproduction of organisms).

Other chapters of this book have discussed ecological protection goals for populations (Chapter 3) and guidance for population-level ERAs, including discussion of empirical approaches (Chapter 8), modeling methods (Chapter 9), and risk management and decision-making issues (Chapter 2). This chapter uses the information from those chapters and combines it with focused discussions on critical phases of the ERA framework to inform risk assessors on the application of population-level methods to ERA. Through the use of flow charts, problem formulation questions, and case study examples, this chapter will provide the assessor with guidance on how to consider and apply population ERA methods to specific assessments.

WHAT IS ERA?

ERA is a process for organizing and analyzing data, assumptions, and uncertainties to evaluate the likelihood of adverse ecological effects that may occur or are occurring as a result of exposure to 1 or more stressors. Stressors can be chemical, physical (e.g., habitat destruction), or biological (e.g., introduced species). ERA is important for environmental decision-making because it provides risk managers with an

TABLE 10.1
Comparison of terms used to describe the risk assessment frameworks employed in the United States and European Union

US Environmental Protection Agency (1998)	European Union (2000)
Problem formulation	Problem definition
Analysis	Risk analysis
Characterization of exposure	Exposure assessment
Characterization of ecological effects	Hazard identification
	Hazard characterization
Risk characterization	Risk characterization
Communicating results to the risk manager	Risk management
Risk management and communicating results to interested parties	Risk communication

approach to consider available scientific information along with other important factors to select a course of action. Recent guidance documents on frameworks for ERA include USEPA (1998) and European Union (EU) (2000). The EU guidance document discusses risk assessment in both terms of human health and environmental effects, so in some ways is more general than the USEPA document. However, the general principles of the ERA process discussed in these 2 documents are similar, despite differences in terminology. Comparisons of the 2 overarching approaches are included in Table 10.1. Recently, a framework for integrated human health and ecological risk assessment was developed by the World Health Organization (WHO 2001; Munns et al. 2003; Suter et al. 2003) that harmonizes the 2 approaches for more holistic assessments of risk. More details regarding the specific definitions of these approaches can be found in the source documents.

For the purposes of this document, the following approach (based on the USEPA definitions) will be used to describe the risk assessment process for the purposes of identifying how population-level ERA could be used to inform the ERA process.

ERA includes 3 primary phases — problem formulation, analysis, and risk characterization. Problem formulation is the initial phase of the process, which includes the development of assessment endpoints, conceptual models, and an analysis plan. Assessment endpoints are explicit expressions of the actual environmental values that are to be protected. They link the risk assessment to management concerns. Assessment endpoints include both a valued ecological entity and an attribute of that entity that is important to protect and is potentially at risk (e.g., nesting and feeding success of piping plovers or areal extent and patch size of eelgrass). Potential interactions between assessment endpoints and stressors are explored by developing conceptual models that link anthropogenic activities with stressors and evaluate interrelationships among exposure pathways, ecological effects, and ecological receptors. The analysis plan justifies what will be done and what will not be done in the assessment, describes the data and measures to be used in the risk assessment, and indicates how risks will be characterized, including how measures of effects will be related to assessment endpoints.

The analysis phase, which follows problem formulation, includes 2 principal activities — characterization of exposure and characterization of ecological effects.

The process is flexible, and interaction between the ecological effects and exposure evaluations is critical. Both activities include an evaluation of available data for scientific credibility and relevance to assessment endpoints and the conceptual model. In exposure characterization, data analyses describe the sources of stressors, the distribution of stressors in the environment, and the contact or cooccurrence of stressors with ecological receptors. In ecological effects characterization, data analyses may evaluate stressor-response relationships or evidence that exposure to a stressor causes an observed response. The products of analysis are summary profiles that describe exposure and the stressor-response relationships that relate measures of exposure and measures of effect.

During risk characterization, risks are estimated and interpreted and the strengths, limitations, assumptions, and major uncertainties are summarized. Risks are estimated by integrating exposure and stressor-response profiles using a wide range of techniques, such as comparisons of point estimates or distributions of exposure and effects data, process models, or empirical approaches such as field observational data.

Other Frameworks for Population-Level ERA

Despite that populations are named as targets of protection in a variety of legislative contexts (USEPA 1997; EU 1997; Chapter 2), and that methods for assessing the state of populations are fairly well developed, there have been very few efforts to formally incorporate population-level risk assessment into existing ERA frameworks. Examples of available population ERA frameworks include population viability analysis (PVA) (Shaffer 1990), a European Commission framework for risk assessment for animal populations (SSC 2003), and frameworks developed by researchers at Oak Ridge National Laboratory (O'Neill et al. 1982; Efroymson et al. 2004).

Population Viability Analysis

Population viability analysis (PVA) is a process of identifying the threats faced by a species population and evaluating the likelihood that the species population will persist for a given time into the future (Shaffer 1990). Its primary use is in the protection of threatened and endangered species and the stepwise process has steps that are applicable for population-level ERA. Ault et al. (1998) describe a framework, which used PVA, for population-level risk assessment developed to assess risks of hydraulic entrainment of commercial and noncommercial fish and invertebrates by dredging. Drechsler et al. (2003) used PVA in combination with a stochastic metapopulation model to predict effects of different management actions, taking into account uncertainty in parameter estimates.

Commercially available software for PVA has been developed that links spatial data from a geographic information system (GIS) with metapopulation models for estimating risk of species extinction, time to extinction, and other relevant population-level endpoints (RAMAS GIS; http://www.ramas.com/ramas.htm). Drechsler et al. (2003) used PVA in combination with a stochastic metapopulation model to predict effects of different management actions, taking into account uncertainty in parameter estimates.

Although PVA can provide a useful framework for population-level ERA in many situations (see Chapter 7), that it focuses on extinction risk to some extent limits its usefulness. Certain effects of chemicals and other environmental stressors on populations, such as reductions in carrying capacity and weaken density dependence (e.g., Forbes et al. 2003), may have important ecological implications but are not easily addressed by PVA.

Scientific Steering Committee of the European Commission's Health and Consumer Protection Directorate-General

The Scientific Steering Committee (SSC) of the European Commission's Health and Consumer Protection Directorate-General published a report on risk assessment for animal populations with emphasis on wildlife (SSC 2003). The main aim of the report was to compare the premises and methodologies employed for assessing animal health risks (i.e., largely for farm animals) and those employed for assessing population risks focusing on mammals and birds. The report considers approaches used for assessing infectious disease risk, immunological medicines, microorganisms used as active substances for plant protection products, industrial chemicals, pesticides, pharmaceuticals, and feed additives. The report describes the extent to which population risk assessment for wildlife is incorporated into the different risk assessment frameworks. It concludes that risk for animal populations "is assumed as an essential part of several risk assessment processes." With respect to industrial chemicals, it is noted that no specific protocols or guidance for assessing the risks to animal populations are defined in the Technical Guidance Document for industrial chemicals, and this aspect is not a key issue in chemical registration. With regard to pesticides, the report notes that the environmental risk assessment for pesticides focuses on the protection of nontarget species, which in most cases is concerned with effects at the population level. Whereas the risks of pesticides for wild populations are included as part of the ecotoxicologic assessment, the risks to domestic animal health are covered, simultaneously to human health impacts, in the toxicological and residue assessments. For the most part, effects of pesticides on populations of wildlife are estimated from toxicological information gathered from standard test species, extrapolated to wildlife species using fixed factors, or in some cases, statistical distributions of toxicity data. Exposure may be modeled from information on chemical persistence, bioaccumulation potential, and usage, usually starting with worst-case assumptions followed by more realistic refinements (e.g., that an animal does not spend 100% of its time feeding on treated fields) should the initial estimated risks fail accepted trigger values (Annex VI of Directive 91/414/EEC). Thus population-level risks are neither measured nor modeled, but are assumed on the basis of risks to organisms.

Oak Ridge National Laboratory Framework

Efroymson et al. (2004) proposed a preliminary ecological framework for population-level ERAs, which they applied to a case study evaluating the effects of habitat removal and fragmentation from petroleum exploration and production activities on population densities and extinction probabilities of vertebrates. The framework provides guidance on the selection of assessment endpoint for populations and on the analysis of the potential for exposure to contaminants. The premise of the proposed framework is that incorporating spatial considerations into exposure estimates should result in more ecologically realistic and less conservative exposure assessments.

A FRAMEWORK FOR POPULATION-LEVEL ERA

Here we propose a framework developed specifically to provide detailed guidance for population-level ERA. While keeping within the more generally accepted framework for ERA (e.g., USEPA 1998), our intention is to focus on the specific considerations, methodologic options, and decision criteria that are unique to population-level assessments. Providing more structure to risk assessors on the consideration of problem formulation, analysis, and risk characterization approaches and activities relative to population-level assessment endpoints will benefit performance of population-level ERAs. We aim thereby to guide the process, increasing both reproducibility and transparency, and to set expectations for reporting, interpretation and communication of results.

ERA has typically straddled the divide between adapting human health risk assessment methods and ecological assessment methods. As the field has matured, more complex methods (e.g., probabilistic analyses, food web models) have been used. Whereas typical ERAs may infer risks to populations using organism-level responses measured in toxicity tests, such as decreases in reproduction or individual growth, population-level attributes, such as abundance, are being considered as more appropriate for many assessments. Increasingly, support for more ecological relevance in ERA and the use of population and community analysis techniques have been put forward (Landis 2000; Bartell et al. 2003; Pastorok et al. 2003). This framework attempts to provide context on where population-level ERA may not be appropriate and, where, through a series of focusing questions the application of population-level ERA may be beneficial.

Overview of the Framework

The framework presented in Figure 10.1 follows that described in USEPA (1998). Although all the sections are addressed, the problem formulation and risk characterization sections were determined to be the most critical for consideration of populations in ERA and in the application of their results in management decisions. The problem formulation phase includes discussions on where population-level ERA may or may not be appropriate, where it can be considered at screening-level or higher tiers, and where it could be a primary or supporting line of evidence. Guidance on selecting tools to estimate risks to populations is provided through a series of questions.

As with any new tool, describing and communicating population-level risk requires the development of new approaches and methods for effectively characterizing the risk for managers and stakeholders. Methods for the translation of model outputs into estimates of risk are provided. The use of population-level ERA also brings different strengths and weakness, from more common ERA methods in the

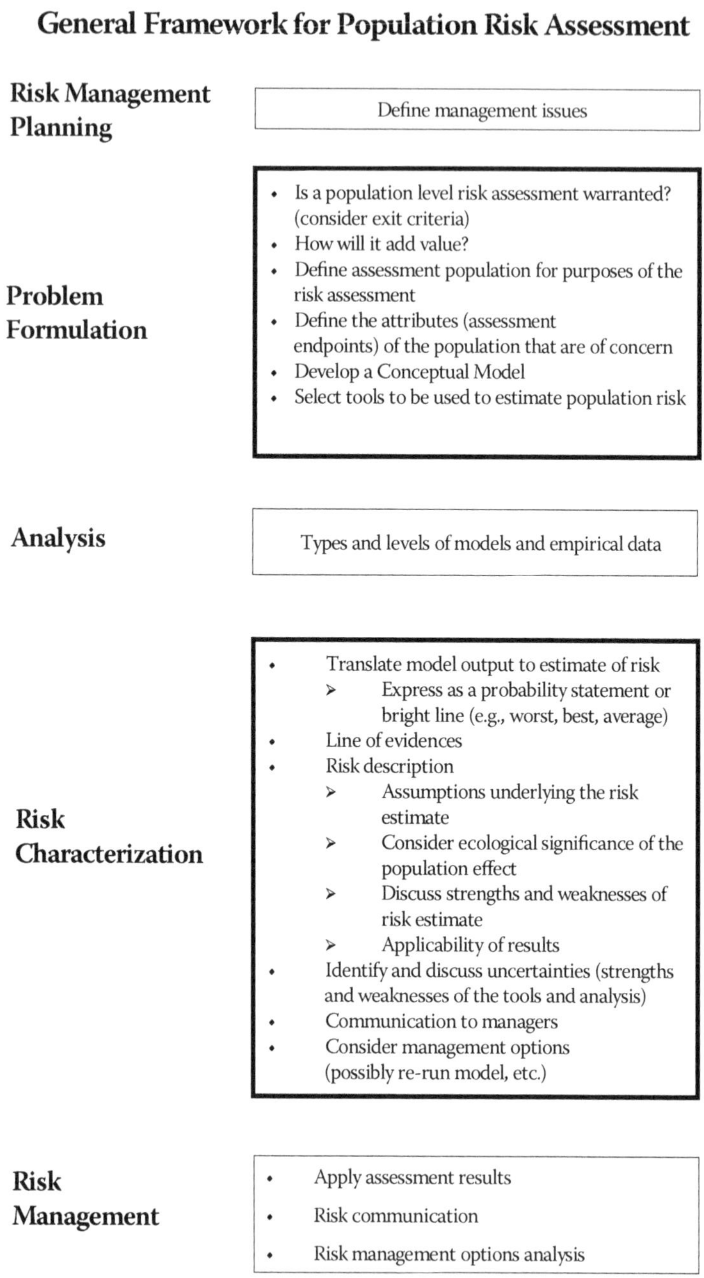

FIGURE 10.1 A general framework for population-level ecological risk assessment.

consideration of uncertainty and sensitivity analysis. These also need to be effectively communicated in the risk characterization phase.

Defining the Management Issues and Decision Criteria

Management goals are formulated as assessment objectives that provide structure and direction for the ERA. The development of specific assessment objectives can be used as a starting point for problem formulation. Assessment objectives assist stakeholders involved in the decision making of an a priori determination of what constitutes an unacceptable risk to a population or a de minimis risk.

Problem Formulation

Here we describe some of the components of problem formulation that are particularly relevant to population-level ERAs. We begin with a discussion of the role of conceptual models. We then elaborate on the concept of an "assessment population." Finally, we provide a discussion of the selection and phrasing of assessment endpoints for use in population-level ERAs and the selection of measures to evaluate those endpoints.

Conceptual models are commonly used to inform risk assessment activities because these can identify relationships between sources, exposure pathways, and the ecological receptors. Conceptual models can also be illustrated in pictorial diagrams, trophic diagrams, or stylized linear or branched process flow diagrams that link sources of stressors, transport processes, and receptors. These are intended to depict all potential exposure pathways. Dialogue among stakeholders is used to identify each complete pathway. Also, incomplete pathways are identified and eliminated from further consideration. The conceptual model is useful in the identification of potential assessment species and should be useful in constructing arguments for or against a suite of assessment and measurement endpoints.

Evaluation of the conceptual model is an important step in determining the appropriateness of a population-level ERA. "Thought experiments" can be used to anticipate interactions among components of a food web to anticipate direct and indirect effects a stressor might exert on populations of the different species of interest. Overlays of transport and fate models for chemicals or analogous descriptions of biological or physical stressors can also guide investigators (stakeholders) through a systematic review of likely exposure scenarios. Considerations of spatial and temporal scales of interactions should be included to identify assessment species and the specific measures that would be useful to obtain in the ERA given the constraints reflected in the management goals.

The conceptual models may be a useful way to capture important scale information for the problem. Clearly defining the scale of an assessment including the operational definition of populations is an important step in the problem formulation. Within a site or landscape, the scale may be refined further based on the percentage of the land within the scale of the analysis that provides habitat for a given species.

The selection of scales is important because it frames the problem and will influence how the problem is evaluated in later steps. The scientific working group assigned to evaluate methods for conducting viability assessments under the National Forest Management Act (NFMA) addressed the issue of temporal and spatial scale of risk assessments (Andelman et al. 2001). Their recommendations are worth noting here. Two factors that enter into the choice of species for viability assessments include 1) the spatial and temporal scale of proposed activities and 2) the potential for cumulative effects as a result of combinations of activities through time and across the landscape, including actions on lands adjacent to National Forest land. How do the species distributions match up with the spatial scales of proposed actions? It is rarely appropriate to conduct viability analyses at the spatial scale typical of project-level decisions. Assessments at broader scales will usually be both more meaningful biologically and more cost-effective. For site-specific ERAs, the selection of scales is important. For species whose distributions are much larger in scale than the proposed actions, it may be appropriate to refer to a regional scale viability assessment if one is available. If not, the assessor should consider undertaking a regional scale assessment. For species with distributions of about the same scale as the proposed actions, the assessor should plan for a viability assessment for this proposal, unless one is already available for a sufficiently similar situation. For species whose distributions are much smaller than the scale of the proposed activities, as may be the case for narrow endemics or species specializing on small, patchily distributed habitats, plan to do site-specific assessments for any management activities that will affect areas where the species occurs.

The site assessment, including habitat and land surveys and a review of surrounding areas to identify adjacent features, support the selection of a scale for the assessment. In some cases, the size of the site may be so small that the assessment might focus exclusively on endangered species within the site boundaries. In other cases, the site may be larger, contamination may migrate, and the site boundaries might be adjacent to other potential habitat or communities such as a riverway, lake, or forest. The scale may be selected based on a number of different factors. The nature and fate and transport potential of the contamination will likely be an important consideration. The location of the site within the landscape will also be an important factor. A site that is in the middle of a wildlife corridor would require a different scale of analysis than a site in the middle of a major metropolitan center. The type of contamination and impacted media may also influence the selection of the scale of analysis. Nonbioaccumulating contaminants in a more stable medium like subsurface soil would present different scaling considerations compared to a highly bioaccumulative compound in an aqueous medium. Finally, contaminants that have a specific effect on a specific, sensitive species may guide the selection of the analytical scale. With an understanding of the scale of the site, the stakeholders can develop the problem formulation further.

In some cases, there is guidance on how to define space and time scales for assessment populations. With regard to temporal and spatial scales, the United States Fish and Wildlife Service (1973) developed guidelines for classifying species as either endangered or threatened, as defined by the Endangered Species Act of 1973 (ESA) (USFWS 1973). The Endangered Species Act states that the following factors determine whether or not a species should be listed as endangered or threatened (Nicholopoulos 1999): the present or threatened destruction, modification, or curtailment of the species' habitat or range; overuse for commercial, recreational, scientific, or educational purposes; disease or predation; the inadequacy of existing

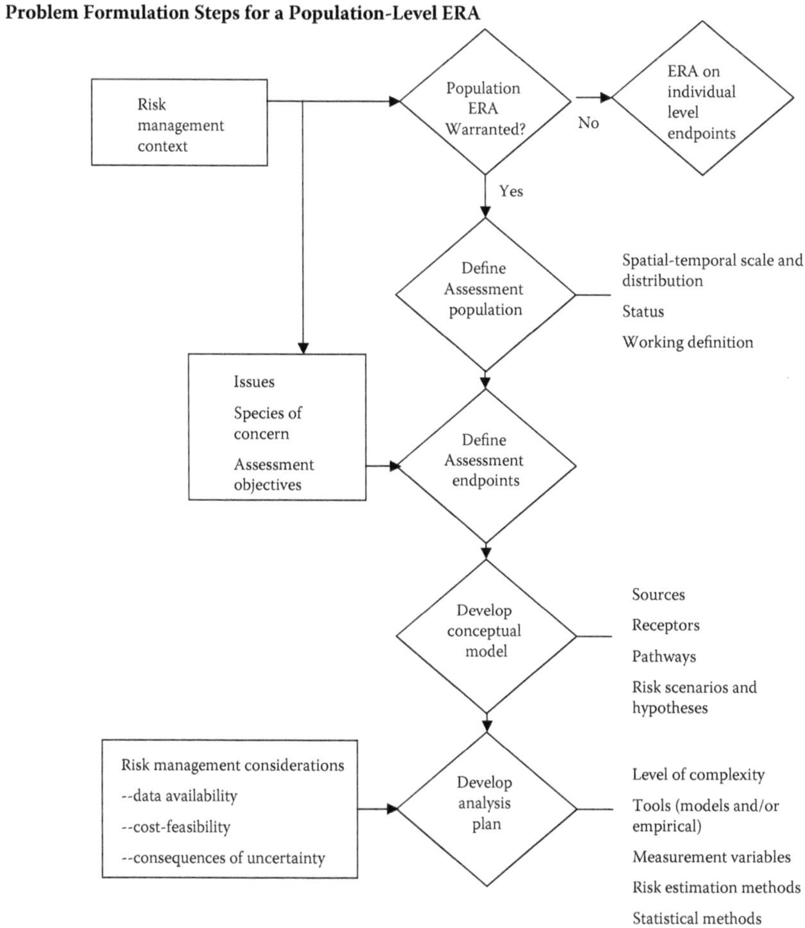

FIGURE 10.2 Problem formulation flow chart.

regulatory mechanisms; and other natural or manmade factors affecting the species' continued existence. This classification system is primarily subjective. These protocols take into account current and future threats and management activities that may cause population decline. It is important to note that there are no prescriptions for consideration of temporal or spatial scale.

In Chapter 3, the term "assessment population" was proposed for use in risk assessments that evaluate risks to populations. This concept is discussed further later in this chapteras it relates to a framework for population-level ERAs.

Figure 10.2 presents the flow of steps an assessor should take in the consideration and application of methods for population-level assessments. It highlights how risk management issues, assessment objectives, spatiotemporal scale and distribution, data availability, and other problem formulation factors frame what assessment endpoints are identified and what analysis plan is appropriate. The remaining text in this section provides further context on the boxes presented in the flow chart.

Is a Population-Level Risk Assessment Warranted?

Assessors may decide that population endpoints are inappropriate or unnecessary for particular ERAs.

1) Environmental laws and regulations may preclude a population endpoint in some ERAs.
2) The species evaluated might be endangered or otherwise highly valued and the individual organisms might be considered the appropriate assessment entity.
3) In some assessments, a population-level endpoint is not relevant. Examples would include those in which managers have identified tumors in fish or abnormalities in amphibians as primary concerns of the public. A resource such as air quality, as it relates to visibility in a park, may bear no relation to populations.
4) Population endpoints may be impractical as measurement endpoints using an empirical approach. The guidance for conducting risk assessments at Superfund sites (USEPA 1997) points out that even in the absence of the stressors examined in an assessment, populations of at least some kinds of organisms fluctuate so greatly that it is impractical to quantify the effect of a stressor. The authors mention populations of small mammals and fish as especially variable. Given the variability inherent in some populations, it might take several years of data from reference sites to establish reliable bounds of reference populations. So although maintenance of population size may be a relevant assessment endpoint, it may be necessary to use organism-level measures of performance (e.g., survival, reproductive output) as measures of effects.
5) The quality of the habitat may be considered by the assessor as the resource to protect. In that case, there may be no specific local populations identified for the assessment.
6) Cost is also a consideration, and a population assessment might be considered too expensive in particular circumstances.

How Will It Add Value?

Before embarking on a population-level ERA, the risk assessor should consider whether employing a population-level approach will add value to the risk assessment (i.e., improve the quality of the risk analysis and risk management process above and beyond what would be achieved without using population-level ERA). In so doing, the first consideration is to decide whether a population-level risk assessment is warranted (see Is a Population-Level Risk Assessment Warranted?).

Clearly, the problem formulation stage will give an indication of whether population-level approaches will add value. If the problem formulation identifies that individual organisms are the key focus of the assessment then population-level ERA may not add value (although, even if the risk assessment is focused on organisms, those individuals are normally part of a population, and population-level ERA could help to inform the decision-making for an assessment of impact on individuals (e.g., by looking at competition, considering longer term impact by investigating reproductive rate, etc.). Several examples of how population-level ERA could add value in a number of areas are described.

Population-level ERA can be used as an additional line of evidence for supporting the conclusions of risk assessment based on laboratory or field monitoring and experimental data. For example, effects observed on key organism-level endpoints in laboratory studies (e.g., mortality, growth and reproduction, which are relevant to population processes) could then be analyzed to assess the potential for effects on population growth rate (e.g., Ferson et al. 1996). In microcosm and mesocosm studies, experimental constraints are sometimes such that it is only possible to look at effects and recovery under a very limited set of environmental conditions. Use of population-level ERA tools may permit the application of such data to a wider set of environmental conditions. In field monitoring studies which are carried out under a limited set of environmental conditions (e.g., over a limited number of years), it may be possible to use population-level ERA to forecast responses over a longer time period.

Population-level ERA may also help to assist in identifying further data that need collection, for example by identifying key life stages or processes for which effects information is needed, providing information about critical exposure periods in an organism's life-history.

Population-level ERA approaches could also potentially help as a screening tool. For example, for a particular group of organisms of concern, it may be possible to identify specific life-history attributes that could be modeled to assess the potential impact of a toxicant. In so doing, such approaches may help to identify species or life stages for additional study and guide further experimental work (Forbes et al. 2001).

Define Assessment Population for Purposes of the Risk Assessment

Spatial and temporal scale and distribution need to be carefully considered. The assessment population may be a small component of the "true" biologic population. It even may be a small component of the relevant metapopulation. The operational definition for an assessment population may be those members of the species residing, foraging, or otherwise using the specific area of interest in the assessment. Generally, with very large assessment areas or at regional scales, the definitions of assessment population and biologic population should converge.

Define the Properties and Attributes (Assessment Endpoints) of the Population of Concern

Assessment endpoints are chosen to reflect environmental values that are protected by law, that provide critical resources, or that provide an ecological function that would be significantly impaired (or that society would perceive as having been impaired) if the resource were altered (USEPA 1998). From the set of ecological receptors identified at the site, specific receptors (habitats, local populations, individual threatened and endangered organisms, soil ecosystem processes or functions) are selected. A variety of criteria may be used to select assessment endpoints including, but not limited to the following:

- species vital to the structure and function of the food web (e.g., principal prey species or species that are major food items for principal prey species);
- rare, endangered, or threatened species or those protected under various legal statutes;
- species that exhibit a marked toxicological sensitivity to the stressor;
- economically important or societally valued species;
- species with unique life histories (those that fill unique ecological niches) and/or feeding habits or representatives of a particular guild; and
- species common on or near the site.

Some of these assessment endpoints may be framed in terms of populations and others might be framed in terms of other levels of biological organization. This is important, as any assessment may involve a mix of assessment endpoints.

Develop the Conceptual Model

A conceptual model is a description of the relationship between the ecological system, assessment endpoints, receptors, and their exposure to the stressors being assessed. It can be portrayed as a pictorial or written description of the relationship. Conceptual models include the risk hypotheses that describe the relationship between the stressor, exposure pathways, and response of the ecological receptor. Issues common to conceptual frameworks include source of the stressors, pathways, receptors (species, life stages), and potential effects. When considering population-level endpoints, the spatiotemporal scale of the assessment and distribution of the population are among the important considerations. The scenarios presented in Appendix 1 present various examples of population-level conceptual models and the issues that need to be included in their development.

Select Methods to Be Used to Estimate Population Risk

After development of the conceptual model that defines the boundaries of the population-level ERA to be conducted, the risk assessor will need to determine the types of methods that are most appropriate to achieve the assessment objective. The first step here is to determine whether an empirical approach or a modeling approach, or some combination of the two, is feasible on the basis of available data and features of the population and environment under consideration. The scenarios in Appendix 1 demonstrate, using realistic examples, how the answers to these questions can guide the approach to be used. In making this decision, the risk assessor should consider the following.

1) *Is the assessment retrospective or prospective?* By definition, prospective risk assessment involves predicting the likelihood of impacts before they

occur. Therefore, these kinds of risk assessment will naturally rely heavily on modeling. However, in the case of site-specific prospective risk assessments, it may be desirable to gather empirical information to determine the present state of the populations (i.e., before a potential impact). Retrospective assessments are, as a rule, site specific and therefore tend to be amenable to empirical approaches for assessing risk to the assessment populations.

2) *Is it possible to measure the selected assessment endpoints directly?* In Step 4 of the Framework, assessment endpoints for the population of concern were identified. Whereas some endpoints might be straightforward to measure in field populations (e.g., size and age structure in certain fish species), other endpoints may not be as amenable to direct measurement (e.g., juvenile survival) and may need to be modeled.
3) *Are there already available data that would be more amenable to be used in a modeling versus empirical assessment?* Before planning an extensive sampling programme or developing a unique model to be used in the assessment, the risk assessor should consider whether there are any existing data that could provide input to the assessment and the extent to which existing models can be adapted to the system under consideration. Both the quality and quantity of existing data need to be evaluated. Model considerations may include accessibility, level of expertise required, hardware requirements, degree of validation needed, cost of software and computing time, scientific credibility, and realism.
4) *Is the analysis to be used at an earlier or later phase of a tiered assessment?* Clearly, limitations of funding and time will place limits on the scope of any risk assessment. However, in the initial (screening) stages of a risk assessment, the levels of accuracy and precision needed may not be very great. It may be sufficient to demonstrate that a risk estimate based on simple, worst-case modeled scenarios does not exceed some threshold of acceptance. In contrast, for more refined risk assessments, greater levels of accuracy and situation specificity are often desirable, and it can in such situations be worthwhile to collect the necessary empirical data under realistic field conditions in such situations.
5) *Does body size and geographic distribution of the species make it prohibitively expensive to do a field survey or experiment? Is the population size large enough that it can be sampled appropriately (i.e., would statistical power be sufficient)?* For very large and/or very sparsely distributed species, it may not be feasible to sample the population appropriately using empirical methods. For such species a modeling approach is likely to be more practical.
6) *Is generation time short enough to allow empirical population assessment methods?* Depending on the time available for conducting the risk assessment in relation to the generation time of the population of interest, certain empirical methods for population-level assessment may be eliminated from consideration. In such cases it might be possible to combine limited empirical information (e.g., on age or size structure) with a modeling approach.

7) *Does the population have a particular status (e.g., endangered) that would constrain field sampling?* For species populations having a special protection status, it may not be practical, ethical, or permitted to sample them empirically (although some types of nondestructive sampling methods may be allowed). Risk assessments for these kinds of populations would tend to rely heavily on a modeling approach.
8) *How great is the need to generalize the results beyond the populations being assessed?* For retrospective, site-specific risk assessments (e.g., a hazardous waste site) it is often sufficient to estimate likelihood of impact for the populations at the site, under the set of conditions existing at the site. Empirical approaches, where feasible, may provide direct estimates of risk or impact with the highest level of confidence for such specific risk assessments. However, in other cases (many prospective risk assessments) it may be desirable to generalize the results of the assessment to other populations in other systems. In such cases models can provide a powerful tool for generalizing the assessment and can facilitate quantitative assessments of parameter uncertainties and sensitivities.

Empirical Methods

After a risk assessor has determined that the assessment shall partly or entirely rely on empirical methods, the following questions will need to be considered (see also Chapter 8 for detailed methodology).

1) *Is the species amenable to field surveys? Field experiments? Laboratory experiments?* As a rule, larger-bodied species are more amenable to direct observation in the field (e.g., by trapping, mark-recapture, visual counting), whereas smaller-bodied or more short-lived species are more practical to employ in laboratory or mesocosm experiments. Controlled field experiments can, in principle, be designed for many different kinds of species populations, but species characteristics (such as body size, habitat preference) will place constraints on the type of attributes (individual behavior, population density) that can be measured under controlled field conditions.
2) *Is it more appropriate to use a repeated measures design or a 1-time sampling of the population?* Depending on the characteristics of the species and the endpoints being assessed it may be more appropriate or feasible to sample the population once (cross-sectional sampling) or several times (repeated measures design). Issues such as independence of samples, confounding of variables and consideration of interactions among independent variables need to be considered before conducting any measurements. Guidance on such issues can be found in the biostatistics literature, for example, in Scheiner and Gurevitch (1993) and Underwood (1997).
3) *Is there an appropriate reference site that can be used for comparison?* A common type of sampling design involves comparing the populations of interest with an appropriate reference or control population that is

assumed or demonstrated not to be impacted. Whereas identification of appropriate reference conditions is fairly straightforward in controlled experiments, it can be more of a challenge in observational studies in which the relevant features of the sites being compared are not under the assessor's control and are likely to vary in some degree unrelated to the presumed stressors. It can be advantageous, when possible, to include more than 1 reference site and more than 1 affected site for comparison. In any case, application of various multivariate statistical methods may help to identify potentially confounding variables that could contribute to differences between populations unrelated to stressors of interest.

4) *What habitat features need to be considered (e.g., statistically blocked) to separate effects of stressors from natural factors?* In both field and laboratory experiments, the grouping of experimental units subjected to similar environmental conditions can be a powerful way to control for unwanted sources of variation and a useful way to accommodate environmental heterogeneity. There are various methods for blocking, but they can have costs in terms of loss of statistical power. A good overview of the options and their costs can be found in Scheiner and Gurevitch (1993).
5) *Which features of the organism's behavior should influence study design?* Issues such as habitat preference, diurnal or seasonal migration, or social interactions may influence which part of a population is being sampled and can add a confounding source of variability or bias to population samples. Sampling at the same time of day or in the same season may minimize this potential problem. Determining which aspects of an organism's behavior need to be accounted for requires at least some knowledge of the species' biology and ecology.
6) *Are there genetic issues that need to be considered in the sampling (e.g., heterogeneity)?* Genetic considerations may come into play in assessing risks for very small populations in which population bottlenecks may be a concern. If an aspect of a population being assessed (e.g., size at a given age) has a marked genetic component, and genotypes cannot be assumed to be randomly distributed within the population (e.g., if there is a form of social structure such that families live close together), then particular care may need to be taken to obtain a random sample of genotypes from the population. Genetic information may additionally provide insight into such issues as metapopulation structure, gene flow, and mechanisms of tolerance or susceptibility.

Modeling Approaches

Should a risk assessor decide that the assessment shall be partly or entirely based on a modeling approach, the following questions will need to be considered (see also Chapter 9 for detailed methodology).

1) *What is known about the key lifecycle traits of the species?* Knowledge about a species' life-history features clearly is necessary for selecting values for

model input parameters. It can also be important for deciding whether or not particular models may be used. For example a simple 2-stage model that divides the life cycle into an adult and a juvenile stage and that calculates population growth rate on the basis of 5 lifecycle variables (Calow and Sibly 1990) assumes that fecundity remains fairly constant over the length of the reproductive period. For species in which fecundity either increases or declines with age this model may not be the best choice, and a matrix approach (Caswell 2001) may be more appropriate. For many species information on key lifecycle traits may be lacking, particularly for field populations. In some cases it may be possible to calculate values for certain lifecycle traits by assuming a given rate of population growth. For example, if juvenile survival probability in the field is unknown, it is sometimes possible to calculate what it would be if population size was approximately stable.

2) *How much generality, realism, and precision are needed?* For certain types of risk assessment (e.g., premarketing notifications), estimates of likely adverse effects are intended to be general (i.e., for the aquatic environment), and requirements of precision and realism are relatively low. It is often sufficient to demonstrate that worst-case estimates of potential effects do not exceed predefined threshold criteria. For such assessments, very simple population models, that ignore many of the potential complexities operating on actual field populations, may be the most appropriate choice. In contrast, when the aim is to assess risk for a particular population at an actual site (e.g., risk to an endangered species population occupying a hazardous waste site), requirements of realism and precision are high, whereas the need to generalize is low. In such cases, it is both feasible and desirable to employ more sophisticated population models that incorporate various biotic and abiotic complexities specific to the population under consideration.
3) *Does the organism have discrete life stages (e.g., egg, larvae, pupa, adult) that could be modeled with an age- or stage-structured model?* Choice of an age-based or stage-based population model may be constrained by the ability to morphologically distinguish and sample different ages or stages. This will limit the structure of the population model and the number of classes that are included in it.
4) *Is the analysis to be used at an earlier or later phase of a tiered assessment?* Models employed at lower tiers of the risk assessment process typically require less accuracy and realism than models employed for higher tier, more refined risk assessments. The former generally have to answer the question, "Does a predicted effect, based on worst-case assumptions exceed predefined cutoff criteria?", whereas the latter may need to derive a statement of the likelihood of some adverse effect given more realistic assumptions about exposure and effects.
5) *Are the data of sufficient quality and quantity to provide useful input to the model?* Uncertainties in model input parameters can severely limit the usefulness of population models for risk assessment, and, as stated previously, parameter uncertainty has been a major challenge in the application

of PVA in conservation biology (see Chapter 7). Uncertainty and variability in model input parameters can be addressed using models that incorporate stochasticity (Maltby et al. 2001), and sensitivity analyses can be used to focus additional research efforts on reducing uncertainties in the most important variables (e.g., Dreshler et al. 2003).

6) *Are density-dependent or density-independent factors likely to be more important for limiting this population?* Incorporation of density dependence into population models adds an additional element of complexity and uncertainty, but may be critical if it is known that the population is under density-dependent limitation in the field (Chapter 4). Surprisingly little empirical information is available about the form of the relationship between density and population growth rate, and recognizing the operation of density dependence in real datasets has been the subject of much controversy (Sibly and Hone 2002). Whether density dependence is likely to ameliorate or exacerbate effects of chemical stressors on population dynamics remains a topic of active research (Forbes et al. 2001; Chapter 4). Modeling different (likely) forms of density dependence, a combination with chemical effects on population dynamics, may be one approach to determining whether or not ignoring density effects is likely to lead to under- or overestimation of population-level risks.
7) *Does the spatial structure of the population itself or the spatial characteristics of the stressors dictate that a spatially explicit approach be used?* Various population models exist that incorporate spatial structure (e.g., Topping and Odderskaer 2004; Chapter 9), and these are clearly most appropriate for site-specific risk assessments in which some detailed knowledge of the spatial distribution of the population with respect to relevant habitat features is available.
8) *Are there genetic issues that need to be considered in the model (e.g., bottlenecks)?* Reduction in genetic diversity can be an important indicator of a response to previous exposure to stressors and of susceptibility to future impacts. Genetic diversity can be incorporated into population models and is often included in PVA (Maltby et al. 2001; Chapter 9).
9) *Do decision-makers accept the model enough to use it as the basis for management?* As the level of complexity of population models increases, the difficulty of explaining its derivation, underlying assumptions, and resulting output also increases. Assuming that the output of the model is intended to guide management strategies, it is essential that the model can be made understandable to managers and other stakeholders involved in the decision-making stage to the extent that they are willing to include it as a basis for risk management.
10) *What kind of resources are available for the modeling exercise?* A variety of practical constraints including the amount of time available for the risk assessment, level of modelling expertise required, accessibility to hardware and software, and availability of model input data can all influence, to varying degrees, the choice of models used in any risk assessment. These constraints should be recognized and made transparent to decision makers.

ANALYSIS

Population-level ERA differs from other kinds of ERA in the choice of assessment endpoints. As a result, measures of effect and the approaches used to characterize effects in the analysis phase of the assessment may also differ. In earlier chapters, measures of population-level effects, such as population abundance or growth rate, were distinguished from organism-level attributes like organism survival, growth, and reproduction. In this section, we discuss analytical approaches used to characterize population-level effects in the context of the framework for ecological risk assessment. Detailed descriptions of the analytical approaches are presented in Chapters 8 and 9. Throughout this chapter, references to "population" refer to an assessment population as defined previously.

Components of the Analysis Phase of a Population-Level ERA

Approaches for the analysis phase of population-level risk assessment include empirical methods and population models. Empirical methods include toxicity tests, field surveys of organism distribution and abundance, and statistical models used to extrapolate between organism-level measurement endpoints and population-level assessment endpoints (such as the nonmechanistic statistical models discussed in the following sections). Population models as defined here refer to mathematical representations of population-level attributes of a single species, often yielding explicit descriptions of the dynamics of population-level endpoints in time and space. A population-level ERA may be based on empirical methods, modeling, or both. Because population models may be useful for summarizing available data and developing hypotheses for an assessment, assessments that integrate both kinds of approaches are generally the most informative for risk management decision-making. Figure 10.3 shows the relationship among components of the analysis phase of an integrated assessment.

Certain steps in the analysis phase for a population-level ERA are the same regardless of whether the primary emphasis is on an empirical approach or modeling (Figure 10.3). For example, any population-level assessment should include development of a conceptual model, which describes the distribution of the population in space and time, its relation to habitat types and other environmental factors, as well as the stressors potentially affecting the population. Analysis of the life history and diet of each receptor species is critical to any population-level assessment. Moreover, defining the relationship of organism-level endpoints to population-level assessment endpoints forms the basis for developing stressor-response functions for a population-level ERA, regardless of whether the approach is primarily empirical or mainly modeling.

Applying Empirical Approaches

Empirical methods for population risk assessment include direct measurement and empirical extrapolation. Direct measurement methods may be based on field surveys of the distribution and abundance of organisms within an assessment population, toxicity (or other stressor) testing with population-level endpoints, and experimental manipulations of portions of an assessment population in the field. Empirical

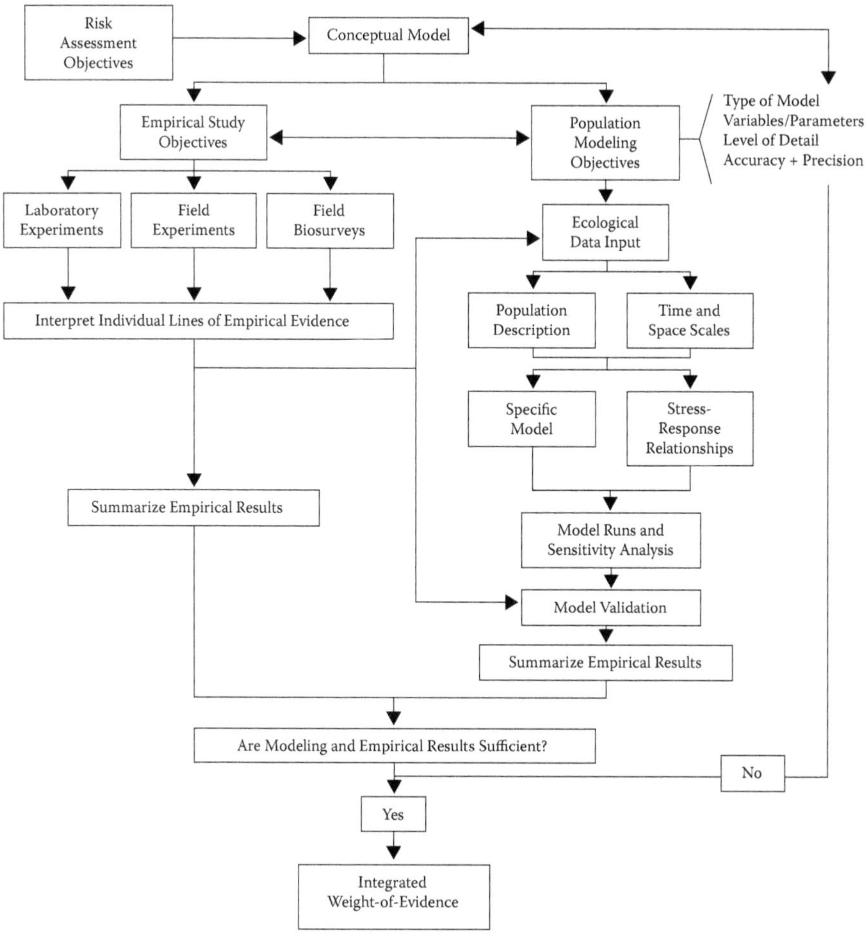

FIGURE 10.3 Overview of the analysis phase of a population-level ecological risk assessment.

extrapolation involves measurement of organism-level endpoints for a specific case and translation to population-level endpoints, which may be assessment endpoints. For example, one might measure survival, growth, and reproduction of an amphipod species in sediment toxicity tests at a contaminated site (i.e., site-specific measures of effect based on organism-level endpoints) and use regression relationships from the literature (i.e., not site-specific) to estimate effects on population densities of the amphipods in the field (i.e., a population-level assessment endpoint).

In a retrospective risk assessment for a contaminated site (similar to an impact assessment), such as the one described in Appendix 1, one obvious approach to characterize population-level effects is to compare a measure of population abundance at the site with conditions at a reference site using statistical hypothesis testing (*cf.* Green 1979; Legendre and Legendre 1998). Although a field survey of population abundance may be one of the most direct methods for population-level

ERA, several authors (e.g., Hurlbert 1984; Stewart-Oaten et al. 1986; Eberhardt and Thomas 1991) have pointed out the problems associated with an observational approach. An advantage of experimental field studies is that treatments can be replicated, increasing the confidence that observed differences are due to the treatment. Nevertheless, observational field studies provide an excellent "reality check" on the results of other approaches. In their ERA guidelines, USEPA (1998) states: "A major advantage of field surveys is that they provide a reality check on other risk estimates, since field surveys are usually more representative of both exposures and effects (including secondary effects) found in natural systems than are estimates generated from laboratory studies or theoretical models."

Multiple lines of evidence may be part of any single population risk assessment, depending on the receptors and habitats of interest, the level of the assessment (e.g., screening vs. detailed), and available funding. The risk assessment plan prepared during problem formulation should specify how data from multiple lines of evidence based on empirical approaches at the population level would be integrated, including how newly collected data for organism- and population-level endpoints relate to any inputs to population models (see Integrating the Results of Empiric and Modeling Approaches). Implementing the approaches specified in the assessment plan requires careful planning to ensure efficient collection of data for multiple species and endpoints that may be part of any one assessment.

A key part of any ERA, including those with population-level endpoints is characterization of habitat quality and distribution. For population-level assessments, spatial analysis of habitat metrics relative to the distribution of the assessment population and levels of physical and chemical stressors is critical. Spatial analysis techniques include (e.g., Akçakaya 1995, 2001; Longley and Batty 1996; With et al. 1997; Carlin et al. 2000; Ebert 2001; Bertazzon et al. 2003) the following:

- overlay of organism and habitat distributions using a GIS or other mapping tool;
- modeling spatial correlation between species or between organisms, habitats, and stressors;
- contouring (e.g., by use of spatial correlation methods such as krieging) of habitat characteristics, stressor levels, or population metrics; and
- evaluation of landscape indices to quantify habitat distributional patterns.

Further details on empirical approaches to population-level assessments can be found in the ecological assessment literature, such as Warren-Hicks et al. (1989), Sokal and Rohlf (1995), Wiens and Parker (1995), USEPA (1996), Southwood and Henderson (2000), and Scheiner and Gurevitch (2001).

Applying Population Models

Population modeling during the analysis phase of an ERA can be used to translate the results of toxicity tests on organism-level traits (e.g., fecundity and survivorship) to population assessment endpoints (e.g., population growth rate, age structure).

Although some metapopulation models yield predictions of the presence or absence of individuals in habitat patches, the most basic output of population modeling efforts is a time series of estimates of population abundance (density). Age- and size-structured models also yield information on the age and size distributions of populations, respectively, and individual-based models provide information on a variety of population characteristics based on the distribution, abundance, and traits of individual organisms. Abundance estimates (and other population metrics) can then be converted to risk estimates (see Risk Characterization). Population modeling efforts may be limited by available data, in which case sensitivity analyses and testing of alternative models may become a primary focus of the assessment. Burgman et al. (1993), Akçakaya (1995, 2001), Caswell (1996), and Pastorok et al. (2002) provide guidance on implementing population models in the context of ERA (also see Chapter 9).

Derivation of Stressor-Response Relationships

If an empirical assessment approach is taken and the scale of the population assessment is large enough relative to the size of the population being studied, it may be possible to use data on gradients in stress levels and effects to derive stress-response relationships. On the other hand, when population effects are not detected at a site or there is not enough patchiness in the stressor levels and or the population in time or space, the assessment may reduce to a simple determination of whether effects are present or not. In general, modeling exercises should derive stressor-response relationships for each assessment endpoint, given the flexibility of population modeling and modern computing power.

Methods for derivation of stressor-response relationships for population endpoints do not differ from those used in other assessments. For assessments of contaminated sites where the stress is heterogeneously distributed in the environment, most analyses should rely on evaluation of spatial gradients in contamination and response endpoints. If localized groups of organisms with differing exposure levels can be identified within the assessment population, endpoints, such as fecundity, age structure, and abundance, can then be measured to develop exposure-response relationships.

Methods for developing stressor-response relationships include regression analyses (response regressed against stress level), statistical correlation analyses, process-based models such as population models, and expert judgment (e.g., examination of scatter plots) (USEPA 1998). The USEPA ERA guidelines (USEPA 1998) discuss evaluating causality based on criteria for observational data (Fox 1991) and additional criteria for experimental evaluation of causality modified from Koch's postulates (e.g., see Woodman and Cowling 1987). Spatial gradients in stress and response metrics can support causality determinations when there is a correlation between stress and adverse population responses in space.

Integrating the Results of Empirical and Modeling Approaches

An assessment may include both empirical measurements of population endpoints and modeling estimates of those same endpoints. Empirical and modeling results may represent multiple lines of evidence, which may be integrated in a weight-of-evidence approach to develop the final risk characterization (USEPA 1998). In some cases, the stressor-response relationships from each of these approaches may be compared and used together to increase confidence in the final analysis of exposure and risk.

Population models are excellent tools for integrating empirical data for organism- and population-level endpoints and evaluating various scenarios in a sensitivity analysis. Results can be used to guide further data collection either in the laboratory or the field. Modeling results may stimulate formulation of hypotheses for testing in microcosms, mesocosms, and/or field experiments. Empirical results for measurement endpoints may be used to validate population models that are used to derive population-level risk estimates in time and space. For example, population models developed and validated during a risk assessment at a chemically contaminated site may be used to evaluate the population-level effects of various remedial alternatives.

Additional tools may be needed for integrating modeling results and data from any field or laboratory investigations. GIS models offer one practical solution for storing and analyzing large amounts of population data in a spatial context. Software such as RAMAS/GIS (Akçakaya 1995, 2001) and PATCH (Schumaker 1998), which have been used in population viability analysis, and multimodel systems, such as ATLSS (DeAngelis 1996), combine the capabilities of GIS with population modeling software.

RISK CHARACTERIZATION

Risk characterization (USEPA 1998) involves interpreting information on exposure and stressor-response relationships to estimate risk, evaluate uncertainties, and describe the significance of ecological risks. Suter (1996b) describes an approach for estimating ecological risks based on individual lines of evidence and then combining them through a weight-of-evidence process (also see Menzie et al. 1996; USEPA 1998). Risk characterization for population-level endpoints involves integrating empirical and modeling approaches. A population-level ERA may also include interpretation of measurement endpoints at the organism, population, community, and ecosystem levels in the context of potential effects on populations of interest.

RISK ESTIMATION

The first step in characterizing risks to a population is to estimate risks in terms of the probability of effects on the assessment endpoint. Whenever possible, a quantitative estimate of risk should be made for the population-level assessment endpoint directly from empirical measurements or modeling results. For example, "there is a 20% chance the population will decrease below the critical threshold of 15 organisms per acre" is an example of a statement of a risk estimate. Below, we explore other ways of quantitatively expressing risk from empirical data and modeling results.

In cases where the assessment endpoint cannot be measured or modeled directly, an extrapolation from measures of effects to the assessment endpoint must be made before a final risk estimate can be derived. This extrapolation may be qualitative or quantitative. Earlier, we gave an example of extrapolating organism-level endpoints, such as amphipod survival and reproduction in a sediment toxicity test, to a population-level assessment endpoint, such as amphipod population abundance (density) using a statistical regression model. Sometimes, a risk estimate may simply be a qualitative assessment of the probability of adverse effects on population endpoints.

Risk Estimates Derived from Empirical Approaches

Empirical assessments that focus on population assessment endpoints commonly rely on statistical tests of the differences in measurement means between treatments defined by various levels of stress and a control treatment (e.g., in laboratory or field experiments) or a reference population (e.g., in a field survey of population attributes). Although risk assessors commonly make qualitative statements about the presence or absence of effects and may quantify differences between treatment and control (or reference) populations, little attention has been given to developing quantitative risk estimates from field and laboratory assessments. Some possible approaches include the following:

- Comparing the distributions of values of the population assessment endpoint between treatment and reference (e.g., a control or benchmark from the literature) using a joint probability distribution (Barnthouse 1996).
- Estimating a population-level response and quantifying variability from stress-response relationships derived from toxicity tests or field gradient analyses (e.g., mean response and variability around a regression line relating a population assessment endpoint to stressor levels) (Suter 1993, 1996a).
- Inferring probabilities from Type I and Type II statistical errors in statistical hypothesis testing using toxicity test results or field survey data (Green 1979; but see Suter 1996a for cautions about a hypothesis testing approach).

Landis and Chapman (personal communication) are working on methods to interpret the results of the sediment quality Triad (chemistry, toxicity, and benthic community indicators) in a probabilistic framework.

Risk Estimates Derived from Population Models

Various authors (e.g., Spencer and Ferson 1997a,b; Bartell 1990) have presented population-level risk estimates calculated from the results of ecological modeling. Age or stage-based models can provide information on survival and reproduction effects in terms of population-level impacts (Barnthouse 1990, 1993). Life-stage extrapolation methods (Barnthouse 1990) can predict uncertainty and risk in terms of ranges of potential effects resulting from a given exposure or range of exposures. Pastorok et al. (2001) summarized ways of expressing risk estimates derived from multiple runs (Monte Carlo) of a population dynamics model based on the approach of Spencer and Ferson (1997) as follows (Figure 10.4):

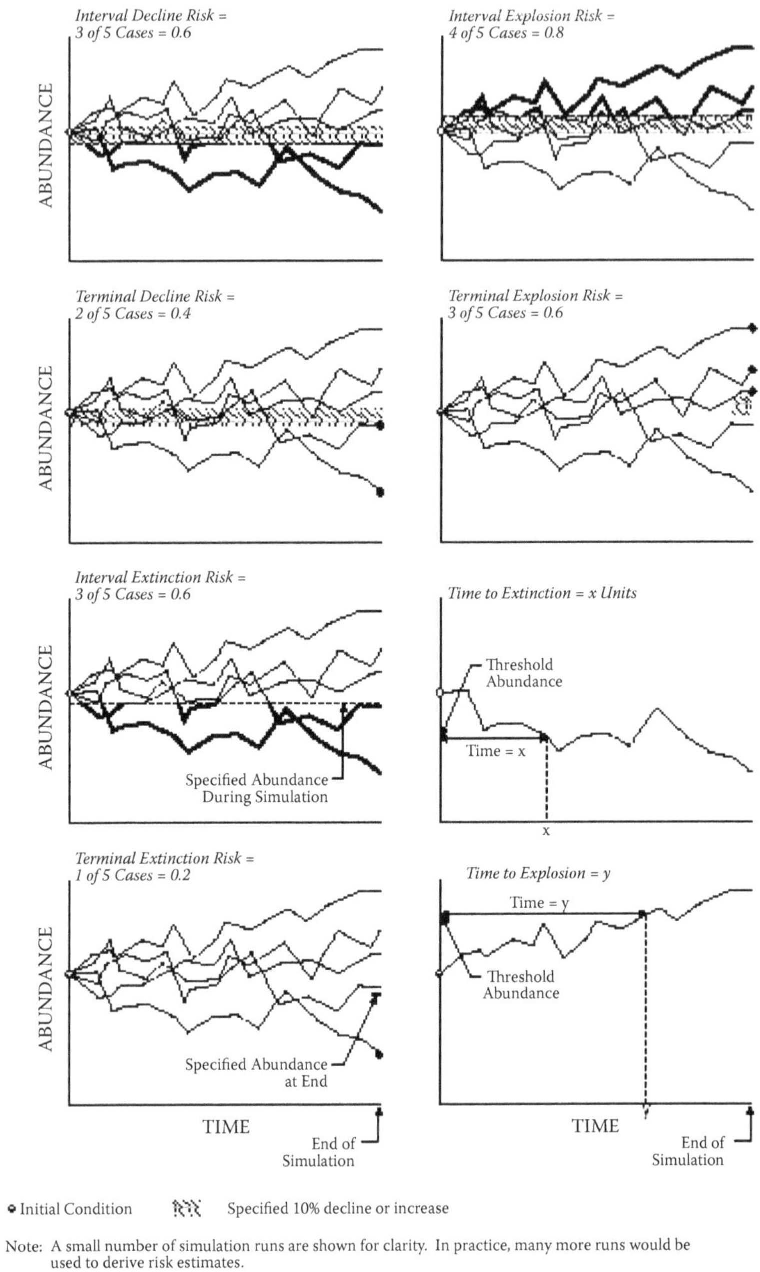

FIGURE 10.4 Examples of risk expressions for population-level ecological risk assessment.

- Interval decline risk: the probability of a population declining by as much as a given percentage of its initial value at any time during the period of prediction
- Interval extinction risk: the probability of a population falling as low as a given abundance at any time during the period of prediction
- Terminal decline risk: the probability of a population being as much as a given percentage lower than its initial value at the end of a simulation
- Terminal extinction risk: the probability of a population being as low as a given abundance at the end of a simulation
- Interval explosion risk: the probability of a population equaling or exceeding a given abundance at any time during the period of prediction
- Terminal explosion risk: the probability of a population being as great as or greater than a given abundance at the end of a simulation
- Time to extinction: the time required by a population to decrease to less than a given threshold abundance
- Time to explosion: the time required by a population to exceed a given threshold abundance

Risk Description

The risk description evaluates the lines of evidence for each estimate of risk and discusses the significance of the adverse effects on the assessment endpoints. The USEPA (1998) cites 3 factors that need to be considered — adequacy and quality of data, uncertainty associated with the data, how directly the risk estimate addresses questions in the risk assessment.

Akçakaya et al. (1999) presented an example of risk description where the rate of time to extinction was classified according to the level of concern (e.g., <20 years, endangered; 20 to 100 years, vulnerable; and >100 years, lower risk). This method establishes bright lines to communicate the population-level endpoint. These classifications were further characterized by statements of risk (e.g., endangered was a 20% probability of extinction within 20 years or 10 generations). Another descriptive method was to calculate the time to recovery of a population exposed to pesticides (ECOFRAM 1999). In Figure 10.4, the charts on internal and terminal decline risks present data on the risk of decline in population abundance. Simulated runs below a set range of abundance indicate increasing levels of concern as a higher percentage of simulations predict likely lower abundance of the population. To provide further information on the examples discussed in this paragraph, the assessor would need to discuss underlying assumptions of the risk estimate and strengths and limitations. What was the quality of the data that the model was based on, how well characterized were key parameters? Was a sensitivity analysis conducted and were steps taken to improve the quality of data supporting those parameters? What is the ecological significance of the risk estimate and how well does it relate to the assessment endpoints? Does the natural ecosystem variability impact the ability to detect stressor related effects?

Barnthouse et al. (1987, 1990) discussed risk description approaches based on age-structured models of fish populations. These models can provide initial tier

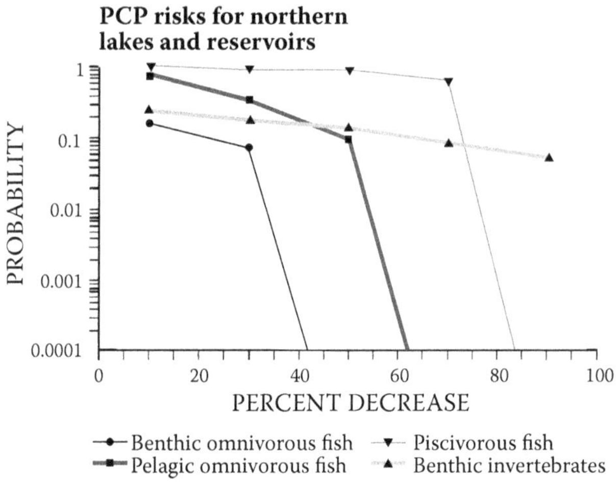

FIGURE 10.5 Graphic representation of risk estimates for fish exposed to pentachlorophenol (adapted from Bartell 2000).

information on future growth or decline in a population based on changes in survival and reproduction. The method is flexible and can incorporate density-dependence and environmental variation.

Evaluating adverse changes in a population also needs to include information on the nature and intensity of effects, spatial and temporal scale, and potential for recovery (USEPA 1998). Graphical presentations of risk estimates are especially useful for communicating the results of a risk assessment. Approaches used in the past for communicating organism-level risks can be easily applied to population risk estimates. One method of presenting risk estimates is to plot the probability of an effect against the magnitude. For example, Bartell (2000) presented risk estimates for fish populations in lakes potentially affected by pentachlorophenol in this way (Figure 10.5).

Bartell (1990) showed how uncertainty could be incorporated directly into graphical presentations of ERA expressed as the probability of a given decrease in fish production based on the framework of O'Neill et al. (1982). Risks to the fish populations were plotted on a 3-dimensional graph in relation to the exposure concentration and the coefficient of variation of the elements in the effects matrix. The "effects matrix" is a method for linking toxicity test data, with explicit estimates of parameter uncertainty, to a model of aquatic food web production.

The results of an ERA may be expressed as multiple lines of evidence for some assessment endpoints, which must be integrated somehow in the risk characterization phase. This will be true in cases of population risk assessment in which multiple assessment methods are used. In these cases, it may be necessary to integrate risk estimates from population models with those from empirical field surveys or experiments. Various weight-of-evidence schemes (e.g., Menzie et al. 1996) and multi-objective ranking systems (e.g., Bogardi and Duckstein 1992; Janssen 1992) are available that can be applied to the risk characterization. In fact, such approaches are needed for any risk assessment involving multiple lines of evidence regardless of whether the assessment is done on individual traits or population-level endpoints.

Identify and Discuss Variability and Uncertainty

The utilization of population-level methodologies in ERA will include different parameters that will have their own unique variability and uncertainty characteristics. Assessors will need to address these aspects when discussing the application of these tools (Pastorok et al. 2003). Types of uncertainty can include incomplete information, parameter uncertainty, form and structure of model expressions, use of point estimates, extrapolations, and species interactions. Environmental variability is more applicable in population-level assessments. The assessor will need to be aware of how variability will impact results of the assessment.

Communication to Managers

Although the measures discussed previously and others are commonly used in conservation biology (Burgman et al. 1993), they have received little attention in other programs, such as hazardous waste site assessment and pesticide evaluations. Risk managers in these latter programs may prefer simple estimates of the probability of given level of change in mean population abundance or stability. Graphical presentations can be useful tools for communication. These could include frequency of model runs leading to adverse effects at the population level, changes in mean population abundance and/or a measure of population variability, or extinction risk or time to recovery. These discussions can lead to the consideration by management of options or mitigation measures.

11 Issues and Recommendations

Wayne R. Munns, Jr., Lawrence W. Barnthouse, and Mary T. Sorensen

INTRODUCTION

The deliberations of the workshop made it clear that the intention, and often requirement, for protecting populations exist in the laws and regulatory processes of many jurisdictions. Further, the science supporting assessment of risks to population is rapidly developing in North America, Europe, and Asia and is already sufficient to support a diverse array of population-level assessments. Despite this, the relative paucity of population-level ecological risk assessments conducted in support of risk management decisions, and the lack of standard methodologies, suggest the need for continued improvement in the underlying science, its communication, and its acceptance. As important, there exists a clear need to make population-level risk an explicit consideration during formulation of the assessment activities intended to support environmental decisions.

In this final chapter, we summarize the principle issues remaining for full implementation and acceptance of population-level ecological risk assessment. These issues emerged during workshop deliberations and reflect similar discussions within the broader risk assessment community. In doing so, we also synthesize the recommendations of the work groups relative to these issues with the intention of advancing population-level ecological risk assessment and its use in environmental decision-making. Many of these issues can only be resolved in the context of specific regulatory and management uses; others are more generic and can be addressed through broad research, development, and communication efforts. Specific implications for the risk assessment research agenda are called out separately to highlight important scientific needs.

IMPLEMENTATION ISSUES

As explored in Chapter 2, protection of populations is the stated or implied goal of numerous policies and statutes. With a few notable exceptions (e.g., protection of threatened, endangered or particularly charismatic species), individual organisms generally have little legal significance. As a result, risk management decisions should focus on ensuring protection of populations within ecological communities as opposed to individual organisms. Yet, for a number of reasons, ecological risk assessment as a decision-support tool currently focuses primarily on adverse effects on organism-level attributes. This is particularly true for chemical risk management programs. Recognizing that populations possess characteristics and attributes uniquely different from organisms (Chapter 3), a focus on organism-level effects can misinform risk decisions. It can also miss spatial and temporal influences that may be important to meeting protection goals. The ultimate recommendations of the workshop are to continue development of population-level ecological risk assessment methods and to foster its current and future use in environmental decision-making to meet the intent of management goals. Although clearly warranted, the expanded focus to include population-level risk will represent a major philosophical and technical shift in implementation of ecological risk assessment as a decision-support tool.

We want to emphasize that not all environmental decisions require understanding of risks to populations. It is clear that various laws, regulations, and policies of several jurisdictions have prescribed protection of attributes of ecological entities other than populations, and we do not advocate that all decisions be based solely on consideration of population-level risk. However, appreciation of the ecological consequences of human activity and policy requires that society be cognizant of risks to populations (and by extension, communities and ecosystems). Such understanding can bring added value to the risk management process by suggesting, and informing selection among, a broader range of management options. Such understanding also is critical for sustainable resource use and environmental stewardship.

Decision Contexts Frame the Assessment

In all areas of environmental regulation, business, conservation, and resource management, the information needs of the risk management decision determine the nature of risk assessment activities. This is never more true than for situations within which management goals involve the protection of populations — there are no generic analysis approaches that will address all needs equally effectively. As illustrated throughout this book against the backdrop of the decision context scenarios of Appendix 1, assessment approaches necessarily will vary across programs and decisions contexts, taking differing forms and directions depending on the information needed. Framing the assessment appropriately is critical to its value in the decision process.

The US Environmental Protection Agency (USEPA) (1998) makes distinctions among 3 types of ecological risk assessment: 1) stressor-driven, in which concerns about environmental stressors and their sources elicit the assessment; 2) effects-driven, in which observed adverse effects (e.g., a sharp decline in the abundance of a wildlife population) trigger the assessment; and 3) values-driven, in which the goals for some ecological resource prompt the assessment. The objectives and approaches for population-level ecological risk assessment will vary across these situations. In emphasizing the relationships between protection goals and problem formulation, Chapter 3 recommends framing the assessment endpoints in a manner responsive to these situations. A goal of protecting the population as a valued resource (i.e., values-driven), or of minimizing the adverse effects of some environmental stressor (stressor-driven), may necessitate information about chemical, biological, and physical components required to support the assessment population (e.g., habitat quality and quantity). Chapter 3 recommends that these considerations be translated into assessment endpoints complementary to that specifically defined for the assessment population.

Defining Assessment Population, Spatial Scale, and Temporal Frame

Environmental decision-makers and stakeholders continue to struggle with how best to define populations relative to their management goals. The concept of assessment population, as proposed by USEPA (2004) and advocated in Chapter 3, offers a pragmatic solution to this problem, but still requires that the spatial boundaries of the assessment be defined operationally. This delineation it critical to the outcomes of assessment activities and the decisions they support. Although the spatial boundaries relevant to certain decisions may seem clear based on the scale of the proposed action (e.g., the boundaries of the site in hazardous waste remediation decisions), unless the range of the population coincides with (or is encompassed by) the geographic scope of the decision, the dynamics of the population may be influenced by processes operating elsewhere (as illustrated in Chapter 6). This suggests, for many instances, that the range and movement patterns characteristic of the population may be more important to defining assessment spatial scales than is the decision context itself (Chapter 3). Potential conflicts with the jurisdictional authority of the decision that this approach may create can be assuaged through assessment exercises that bound the severity of risk estimated. For example, modeling analyses might include one that assumes the entire population to reside within site boundaries and that spatial structure does not exist, and one that places site influences within the context of a broadly distributed metapopulation. The results of such bounding analyses can be used to inform the decision maker of the extent and severity of population-level risk associated with the site by itself as well as in the context of the surrounding landscape, and the effectiveness of various decision options. Of course, the preceding assumes that the decision is focused on the stressor (or stressors at the site in the example), that is, it is a stressor-driven assessment (USEPA 1998). When the management goal is oriented toward protection of populations of species (i.e., a value-driven assessment), definition of the assessment population becomes more straightforward — it is the entire population or goal-relevant portion thereof.

As explored in Chapters 3 and 6, heterogeneity in environmental quality experienced by the assessment population will influence its responses to stressors. Spatial variation in the concentrations of those stressors can have profound influences on population attributes. Although this may influence selection of empirical and modeling analysis activities more than it does selection of the spatial boundaries for the assessment population, the studies communicated in Chapter 6 demonstrate the importance of "actions-at-a-distance" in determining overall population risk. Thus we recommend that consideration be given to the spatial patterns of stressors and population subunits when defining the assessment population.

Questions also arise with respect to the temporal scales appropriate to consider in assessment activities. In Chapter 3, we recommend that time frames relevant to management goals and actions, in addition to ecological response and recovery times, be acknowledged explicitly when defining the assessment population as part of the assessment endpoint. Management goals involving long-term persistence, or sustainability, of populations require up-front agreement among risk managers, stakeholders, and other interested parties about the meaning of "long term." As described in Chapters 4 and 5, the existence of compensatory processes acting over various time scales can confound analysis activities and their interpretation. Early specification of the temporal scales relevant to decision-making will help to ensure that analysis activities provide information appropriate to the decision.

Balance of Empirical and Modeling Approaches

An issue for any assessment is the relative effort that should be allocated among empirical and modeling approaches for population-level ecological risk assessment. Certain risk problems may lend themselves primarily to empirical analysis (e.g., risk to abundance of small mammal populations residing on a hazardous waste site), while others are best addressed through modeling analysis and simulation (e.g., risk to long-term persistence of a valued wildlife population). Although it probably is appropriate to approach modeling results with a healthy degree of skepticism, they do offer a range of insights not attainable from purely empirical efforts. The appropriate balance of the 2 approaches depends on several factors (e.g., resources available, suitability of data collection methods for the assessment population, availability of models and modelers), and most certainly must be decided on a case-by-case basis. We recommend that the most informative risk assessments will involve both analysis approaches to greater or lesser degree, providing lines of evidence that collectively increase the confidence that decision makers can place in risk conclusions.

Relevant to this issue are the data required to perform either type of analysis. Clearly, use of empirical approaches to characterize effects assumes that sufficient data are available or obtainable. Yet, modeling analyses can be conducted with fewer constraints. The oft-repeated refrain of "we don't need no stinkin' data!" offered by at least 1 workshop participant is, of course, false bravado. Although theoretical exercises can be conducted using models in the absence of data, meaningful use of models in population-level ERA requires some degree of realism and accuracy with respect to the data used. Additionally, acceptance by decision makers and stakeholders of analysis results is facilitated when situation-dependent data and information are used. Both Chapters 8 and 9 make a point of emphasizing the interplay of data and models, and describe the mutually supporting roles the 2 approaches play.

Interpreting Significance of Population-Level Effects

Measuring or predicting population-level effects is only part of the challenge for the risk assessment or risk management process — the significance of changes in population attributes because of exposure to stressors must be interpreted to understand risks. Appreciation of this significance is crucial to developing, executing, and interpreting population-level ecological risk assessments. Often paraphrased as "so what?," answering this question is not straightforward and presents complex challenges that have yet to be overcome by science.

As noted by others (e.g., Maltby et al. 2001; Munns and Mitro 2004), and except in extreme situations (e.g., extinction, extirpation), ecological significance of population-level effects is difficult to establish for a number of reasons. First, populations display variation in their attributes: abundances fluctuate (perhaps around some long-term average abundance) because of environmental and demographic stochasticity, age structures change as reflections of year-class strengths, sex ratios vary randomly (or otherwise) around values of 1:1, and so on. A challenge is to distinguish changes caused by anthropogenic stressors against this background of this natural variability. Historically, inferential statistics has been used to make such distinctions. This has lead to designation of seemingly arbitrary "bright lines" that form the basis of regulatory decisions (Chapman et al. 2002). Yet, detection of statistical differences depends on the particulars of experimental designs, natural variation, and the magnitude of "treatment" differences, and do not necessarily equate with changes that are important ecologically. Establishing ecological significance is also hindered by the fact that ecological systems typically vary in near-continuous fashion, and in the absence of thresholds or obvious breakpoints, the importance of subtle changes is difficult to establish. Also confounding interpretation of ecological significance is the existence of compensatory mechanisms, such as homeostatic acclimation of individuals, genetic adaptation (Chapter 5), and density dependence in vital rates (Chapter 4) and migration (Chapter 6), that can ameliorate adverse effects over the short or long term. Finally, and perhaps most important, there are few (if any) "values" that ecological systems place on themselves. In the absence of such values, ecological change is often interpreted in the context of societal desires, preferences, needs, and policies.

Generic guidance for interpreting significance has been offered as it pertains to ecological risk assessment (Harwell et al. 1994; USEPA 1998). Paraphrased in the context of populations from Harwell et al. (1994), an ecologically significant change is one that is important to the structure or function of the population, exceeds natural variation, and is of sufficient type, intensity, extent, or duration to be important to society. In Chapter 2, we recommend that a priori determinations be made of what constitutes an acceptable risk to a population. Such determinations will necessarily rely heavily on the site- and context-specific legal and policy issues and somewhat less so on scientific underpinnings. We also recommend attempting to identify and capture commonalities within and across decision contexts into a basic set of operational narrative criteria of acceptable risk. These could be used directly as generically acceptable risk criteria, or to guide development of context-specific criteria. Such criteria could facilitate a common basis for policy decisions by multiple agencies or programs. Given our current state of understanding of population ecology, and in the absence of objective, quantifiable criteria against which to judge ecological significance, determination of acceptable risk will continue to rely on sound professional judgment reflecting the considerations offered by Harwell et al. (1994) and others. In large part, the answer to the "so what" question must come from society, because it is within this context that the adversity of effects ultimately will be judged. Considerable work remains to advance the theory and practice of population ecology to the point where the ecological significance question can be answered satisfactorily in all cases.

Guidance, Training, and Acceptance

One of the greatest outstanding challenges to acceptance of population-level ecological risk assessment as a decision-support tool is its understanding and acceptance by risk managers and stakeholders. Several work groups recommended actions to facilitate this understanding and acceptance. These ranged from developing standardized analysis tools and the logical constructs within which to use them (e.g., population models arrayed in tiers of increasing sophistication and data requirements), to developing procedural guidance for individual analysis approaches or the entire assessment, to engaging decision-makers and the public through education, training, and dialogue.

Particularly in a regulatory context, standardization and guidance are critical to provide a common understanding about procedures, interpretation and decision-making. We recommend that a series of guidance documents be developed for population-level ERA that collectively explain, with appropriate detail and rigor, the availability, use, and interpretation of analysis tools. Included would be guidance for selecting among population models, methods for estimating their parameters, and advice for model use and interpretation. Guidance in population data collection and analysis may be needed that addresses the unique aspects of aquatic and terrestrial populations in empirical studies. Additional guidance is needed to help inform risk assessors, risk managers, and stakeholders about the importance of considering density dependence and population genetics in particular risk contexts. Further, broad guidance is needed for conducting entire assessments within various legal and policy contexts and how their results should be communicated. The step-by-step approach offered in Chapter 10 goes a long way in this regard. With further development, and tailored specifically to meet a variety of regulatory and management needs, this framework could form the basis of general procedural guidance.

Training and education are especially important for use and acceptance of population-level ecological risk assessment. Risk assessment practitioners would benefit from enhanced opportunities for formalized training (Chapter 2 suggests a number of mechanisms) in population ecology theory, empirical field and laboratory methods, and modeling. Additionally, the rapid development of emerging technologies (e.g., Geographic Information Systems (GIS), spatial statistics, molecular methods) requires continued education in their use to ensure that practitioners and assessors are informed and sufficiently skilled. But, education of risk managers and decision-makers is equally as important. Such training should include overviews of the conceptual and technical aspects of a population-level ERA, but should also emphasize methods for effective interaction with risk assessors that facilitates obtainment of risk assessment results maximally useful for informed decisions. All parties should receive training that enhances the risk communication process. With regard to this last point, acceptance of population-level ecological risk assessment by stakeholders and the general public would be enhanced through development of effective risk communication strategies. In Chapter 2 we recommend including public and professional outreach in such strategies as a means to improve the clarity and transparency of the decision-making process. We also recommend fostering dialogue groups, with representatives from all sectors of environmental science and decision making, to initiate and facilitate such discussions.

IMPLICATIONS FOR RESEARCH AND DEVELOPMENT

As is evident from the descriptions in Chapters 8 and 9, a firm scientific foundation is in place from which to assess risks to populations. Yet, as with almost all decision-support tools and methodologies, the capabilities, acceptance and use of population-level ecological risk assessment will benefit from enhancements to the underlying science. In addition to the call for standardization of current methodologies by several of the work groups, advancements in selected areas of science were identified as being particularly important. Of these, perhaps the most critical reflect a need for enhanced realism in assessment methods with respect to compensatory processes operating within populations and the use of heterogeneous environments by populations.

Compensatory processes and mechanisms, including density-dependent feedbacks in demographic and migration rates, can have profound influences on how populations respond to anthropogenic stressors. Analyses that fail to recognize these influences can mislead the decision processes they support, particularly when those decisions reflect goals of population persistence and sustainability. Acknowledging uncertainty by the chemical risk assessment community about the relative importance of density dependence, in Chapter 4, we recommend increased experimental evaluation of density-toxicant interactions. Although density-toxicant interactions may be species and chemical specific in their details, evaluation of a broad set of such studies may lead to recognition of general patterns, perhaps based on mechanisms of toxicity and life history characteristics. Such patterns could inform problem formulation considerations about the importance of density dependent feedbacks for any given assessment endpoint and the forms they should take in analysis activities. The importance of density-dependence has long been recognized by fishery and wildlife biologists, and many of the methods used to account for density dependence in fishery and wildlife management models and conservation biology may be directly applicable to ecological risk assessments. At the very least, the experience gained from these related disciplines provides a valuable point of departure for other types of assessments.

Over the mid- to long-term, evolutionary responses of populations to changing selection pressures caused by human activity can also affect the accuracy and completeness of risk predictions. As recommended in Chapter 5, incorporation of modern advances in molecular biology in analysis activities should improve understanding of how populations respond to anthropogenic stressors. We recommend that this knowledge be incorporated into unified modeling platforms that integrate population dynamics and population genetics to enhance the predictive and diagnostic power of risk assessments.

Adding richness to the problem, populations typically exist within complex landscapes that themselves vary in space and time. Whether assessments are stressor-driven, effects-driven, or values-driven, the degree to which spatial heterogeneity is reflected in analysis activities influences the accuracy and generality of results produced. Although not all decisions necessarily require the support of detailed spatial analyses, questions pertaining to when such analyses are needed are largely unanswered. As concluded in Chapter 6, further development and refinement of spatial analysis methods will support application of risk assessment approaches to a wider array of environmental management questions (e.g., community development land use planning). To help meet these needs, we recommend that substantial research attention be given to development and standardization of spatial modeling and analysis methods, and that evaluations be made of the relative merits of incorporating spatial considerations in particular environmental decision contexts.

Taken collectively with the specific research and development needs outlined throughout this text, these 3 areas — density dependence, evolutionary change, and landscape complexity — define the topics of a research agenda for population-level ecological risk assessment. As emphasized in Chapter 2, this agenda has both basic and applied components. Basic research is necessary to increase our understanding of population biology and ecology in general, as well as of the population ecology of specific valued species. But because risk assessment is a practical tool for environmental decision making, the results of basic research must be translated into workable tools and methods that are understood by practitioners and decision makers alike. Thus there are roles to play by the academic, government, and business communities — and by scientific societies in addition to SETAC. Further, there exist rich knowledge bases, sets of tools, and experiences in the fields of conservation biology and resource management from which to learn and build (Chapter 7). Greater interaction and collaboration among all of these groups is a necessary ingredient to advancing the science of population-level ecological risk assessment and its use in environmental decision-making. The collective expertise of knowledgeable scientists from around the world is needed to advance the science and practice of population-level ecological risk assessment. This international workshop, which was attended by participants from 3 continents and 10 individual countries, shows the value of interactions between scientists who have different backgrounds but share common interests. We hope that this cooperation will continue long after the workshop itself has been forgotten.

References

Adams SM, DeAngelis DL. 1987. Indirect effects of early bass-shad interactions on predatory population structure and food web dynamics. In: Kerfoot WC, Sih A, editors. Predation: direct and indirect impacts on aquatic communities. Hanover (NH): University Press of New England. p. 103–117.

Akçakaya HR. 1995. RAMAS/GIS: linking landscape data with population viability analysis (ver. 2.0t). Setauket (NY): Applied Biomathematics. Available from: http://www.ramas.com/ramas.htm.

Akçakaya HR. 1998. RAMAS GIS: linking landscape data with population viability analysis (ver. 3.0). Setauket (NY): Applied Biomathematics.

Akçakaya HR. 2000a. Population viability analyses with demographically and spatially structured models. Ecol Bull 48:23–38.

Akçakaya HR. 2000b. Viability analyses with habitat-based metapopulation models. Popul Ecol 42:45–53.

Akçakaya HR. 2002. RAMAS/GIS: linking spatial data with population viability analysis, version 4.0. Setauket (NY): Applied Biomathematics.

Akçakaya HR. 2001. Linking population-level risk assessment with landscape and habitat models. Sci Total Environ. 274:283–291.

Akçakaya HR, Atwood JL. 1997. A habitat-based metapopulation model of the California gnatcatcher. Conserv Biol 11:422–434.

Akçakaya HR, Burgman MA, Ginzburg LR. 1999. Applied population ecology: principles and computer exercises. Sunderland (MA): Sinauer Associates, Inc.

Akçakaya HR, Burgman MA, Kindvall O, Sjogren-Gulve P, Hatfield J, McCarthy M, editors. 2004. Species conservation and management: case studies. New York (NY): Oxford University Press.

Akçakaya HR, Regan R. 2002. Population models—metapopulations. In: Pastorok A, Bartell SB, Ferson S, Ginzburg LR, editors. Ecological modeling in risk assessment. Boca Raton (FL): Lewis Publishers. p. 83–93.

Allan JD, Daniels RE. 1982. Life table evaluation of chronic exposure of *Eurytemora affinis* (Copepoda) to kepone. Mar Biol 66:176–184.

Allee WC. 1931. Animal aggregations. A study in general sociology. Chicago (IL): University of Chicago Press.

Allee WC. 1938. The social life of animals. London (UK): William Heinemann.

Allee WC, Emerson AE, Park O, Park T, Schmidt KP. 1949. Principles of animal ecology. Philadelphia (PA): W.B. Saunders.

Allendorf FW, Ryman N. 2002. The role of genetics in population viability analysis. In: Beissinger SR, McCoullough DR, editors. Population viability analysis. Chicago (IL): The University of Chicago Press. p. 50–85.

Allison PD. 1995. Survival analysis using the SAS system: a practical guide. Cary (NC): SAS Institute Inc.

Altwegg R. 2003. Multistage density dependence in an amphibian. Oecologia 136:46–50.

Andelman SJ, Beissinger S, Cochrane J, Gerber L, Gomez-Priego P, Groves C, Haufler J, Holthausen R, Lee D, Maguire LA, et al. 2001. Scientific standards for conducting viability assessments under the National Forest Management Act: report and recommendations of the NCEAS working group. Santa Barbara (CA): University of California, National Center for Ecological Analysis and Synthesis. November.

Andrewartha HG, Birch LC. 1954. The distribution and abundance of animals. Chicago (IL): University of Chicago Press.

Arnason AN, Schwarz CJ. 1999. Using POPAN-5 to analyze banding data. Bird Study Suppl 46:S157–S168.

Arnaud JF. 2003. Metapopulation genetic structure and migration pathways in the land snail *Helix aspersa*: influence of landscape heterogeneity. Landscape Ecol 18:333–346.

Ashley MV, Willson MF, Pergams RW, O'Dowd DJ, Gende SM, Brown JS. 2003. Evolutionarily enlightened management. Biol Conserv 111(2):115–123.

[ASTM] American Society of Testing and Materials. 2002. Risk-based corrective action (RBCA) standard guide for the protection of ecological resources. Available from: http://www.astm.org/cgi-bin/SoftCart.exe/DATABASE.CART/REDLINE_PAGES/E2205.htm?L+mystore+uvst1142.

Auld TD, Denham AJ. 1999. The role of ants and mammals in dispersal and post-dispersal seed predation of the shrubs *Grevillea* (Proteaceae). Plant Ecol 144:201–213.

Auld TD, Keith DA, Bradstock RA. 2000. Patterns in longevity of soil seedbanks in fire prone communities of southeastern Australia. Aust J Botany 48:539–548.

Ault JS, Lindeman KC, Clarke DG. 1998. FISHFATE: population dynamics models to assess risks of hydraulic entrainment by dredges. DOER Technical Notes Collection (TN DOER-E4), US Army Engineer Research and Development Center, Vicksburg, MS. Available from: http://www.wes.army.mil/el/dots/doer.

Avise JC. 2000. Phylogeography: the history and formation of species. Cambridge (MA): Harvard University Press.

Baird DJ, Burton GA Jr. 2001. Ecological variability: separating natural from anthropogenic causes of ecosystem impairment. Pensacola (FL): SETAC Press.

Baker RJ, Bickham AM, Bondarkov M, Gaschak SP, Matson CW, Rodgers BE, Wickliffe JK, Chesser RK. 2001. Consequences of polluted environments on population structure: the bank vole (*Clethrionomys glareolus*) at Chornobyl. Ecotoxicology 10:211–216.

Ballard JWO, Chernoff B, James AC. 2002. Divergence of mitochondrial DNA is not corroborated by nuclear DNA, morphology, or behavior in *Drosophila simulans*. Evolution 56:527–545.

Barata C, Baird DJ, Soares AMVM. 2002. Demographic responses of a tropical cladoceran to cadmium: effects of food supply and density. Ecol Appl 12:552–564.

Barker RR, Fletcher D, Scofield P. 2002. Measuring density dependence in survival from mark-recapture data. J Appl Stat 29:305–313.

Barnthouse LW, Suter II GW, Rosen AE. 1990. Risks of toxic contaminants to exploited fish populations: influence of life history, data uncertainty, and exploitation intensity. Env Tox and Chem 9:297–311.

Barnthouse LW. 1993. Population-level effects. In: Suter GW II, editor. Ecological risk assessment. Boca Raton (FL): Lewis Publishers. p. 247–274.

Barnthouse LW. 1996. Modeling ecological risks of pesticide application: a review of available approaches. Washington (DC): US Environmental Protection Agency, Office of Pesticide Programs.

Barnthouse LW. 2004. Quantifying population recovery rates for ecological risk assessment. Environ Toxicol Chem 23:500–508.

Barnthouse LW, Glaser D, Young J. 2003. Effects of historic PCB exposures on the reproductive success of the Hudson River striped bass population. Env Sci Technol 37:223–228.

Barnthouse LW, Sueter GW II, Rosen AE, Beauchamp JJ. 1987. Estimating responses of fish populations to toxic contaminants. Environ Toxicol Chem 6:811–824.

Barnthouse LW, Suter GW II, Rosen AE. 1990. Risks of toxic contaminants to exploited fish populations: influence of life history, data uncertainty and exploitation intensity. Environ Toxicol Chem 9:297–311.

Barnthouse LW, Sorensen MT, Thompson M. 2004. Committee effort to compile population-level ERAs. Society of Environmental Toxicology and Chemistry. Available from: http://www.setac.org/eraag/era_pop_index.htm.

Barnthouse LW, O'Neill RV, Bartell SM, Suter GW II. 1986a. Population and ecosystem theory in ecological risk assessment. In: Poston TM, Purdy R, editors. Aquatic toxicology and environmental fate. Philadelphia (PA): American Society for Testing and Materials. p. 82–96.

Barnthouse LW, Suter GW II, Bartell SM, Beauchamp JJ, Gardner RH, Linder E, O'Neill RV, Rosen AE. 1986b. User's manual for ecological risk assessment. Environmental Sciences Division Publication No. 2679. US Environmental Protection Agency, Office of Research and Development, Washington, DC, and Oak Ridge National Laboratory, Oak Ridge, TN.

Barnthouse LW, Suter GW II, Rosen AE. 1989. Inferring population-level significance from individual-level effects: an extrapolation from fisheries science to ecotoxicology. In: Suter GW II, Lewis MA, editors. Aquatic toxicology and environmental fate, 11th volume. ASTM STP 1007. p. 289–300.

Barnthouse LW, Biddinger GR, Cooper WE, Fava JA, Gillett JH, Holland M, Yosie TF. 1998. Sustainable environmental management. Jacksonville (FL): SETAC Press.

Bartell SM. 1990. Ecosystem context for estimating stress-induced reductions in fish populations. Am Fish Soc Symp 8:167–182.

Bartell SM. 2000. Review of aquatic ecosystem modeling approaches for risk assessment. Presentation to Workshop on Review of Ecological Models for Chemical Risk Assessment, May 2000, Fairmount Hot Springs, MT. Sponsored by American Chemistry Council, Washington, DC.

Bartell SM, Pastorok RA, Akcakaya HR, Regan H, Ferson S, Mackay C. 2003. Realism and relevance of ecological models used in chemical risk assessment. Hum Ecol Risk Assess 9:907–938.

Beacham TD, Schweigert JF, MacConnachie C, Labaree K, Miller KM. 2001. Population structure of herring (*Clupea pallasi*) in British Columbia: an analysis using microsatellite loci. Research Document 2001/128 Canadian Science Advisory Secretariart ISSN 1480-4883.

Begon M, Harper JL, Townsend CR. 1990. Ecology: individuals, populations and communities. Boston (MA): Blackwell Scientific Publications.

Beissinger SR. 2002. Population viability analysis: past, present and future. In: Beissinger SR, McCullough DR, editors. Population viability analysis. Chicago (IL): Chicago University Press. p. 5–17.

Beissinger SR, McCullough DR, editors. 2002. Population viability analysis. Chicago (IL): Chicago University Press.

Beissinger SR, Westphal MI. 1998. On the use of demographic models of population viability in endangered species management. J Wildl Manage 62:821–841.

Bellwood DR, Hughes TP. 2001. Regional-scale assembly rules and biodiversity of coral reefs. Science 292:1532–1535.

Benton MJ, Guttman SI. 1990. Effects of allozyme genotype on the survivorship of mayflies (*Stenonem femoratum*) exposed to copper. J North Am Benth Soc 9:271–276.

Berry D. 1996. Statistics: a Bayesian perspective. Belmont (CA): Duxbury Press.

Berryman A. 2004. Mature population dynamics? Trends Ecol Evol 19:66–67.

Berryman A, Turchin P. 1997. Detection of delayed density dependence: comment. Ecology 78:318–320.

Bertazzon S, Carlon C, Critto A, Marcomini A, Zanetto G. 2003. Integration of spatial analysis and ecological risk assessment in a GIS environment: the case study of the Venetian lagoon contaminated sediments. 4th International Conference on Integrating GIS and Environmental Modeling (GIS/EM4): problems, prospects and research needs. Banff, Alberta (Canada), September 2–8, 2000. Available from: http://www.colorado.edu/research/cires/banff/pubpapers/122/.

Beverton RJH, Holt SJ. 1957. On the dynamics of exploited fish populations. Min Agri Fish Food 19:1–533.

Bickerstaff K. 2004. Risk perception research: socio-cultural perspectives on the public experience of air pollution. Environ Int 30:827–840.

Bijlsma R, Bundgaard J, Boerema AC. 2000. Does inbreeding affect the extinction risk of small populations? Predictions from *Drosophila*. J Evolut Biol 13:502–514.

Bijlsma R, Bundgaard J, VanPutten WF. 1999. Environmental dependence of inbreeding depression and purging in *Drosophila melanogaster*. J Evolut Biol 12:1125–1137.

Bjornstad ON, Grenfell BT. 2001. Noisy clockwork: time series analysis of population fluctuations in animals. Science 293:638–643.

Bjornstad ON, Fromentin JM, Stenseth NC, Gjøsæter J. 1999a. Cycles and trends in cod populations. Proc Natl Acad Sci USA 96:5066–5071.

Bjornstad ON, Fromentin JM, Stenseth NC, Gjøsæter J. 1999b. A new test for density-dependent survival: the case of coastal codpopulations. Ecology 80:1278–1288.

Bjornstad ON, Begon M, Stenseth NC, Falck W, Sait SM, Thomson DJ. 1998. Population dynamics of the Indian meal moth: demographic stochasticity and delayed regulatory mechanisms. J Anim Ecol 67:110–126.

Bjornstad ON, Sait SM, Stenseth NC, Thompson DJ, Begon M. 2001. The impact of specialized enemies on the dimensionality of host dynamics. Nature 409:1001–1006.

Blockwell SJ, Taylor EJ, Jones I, Pascoe D. 1998. The influence of fresh water pollutants and interaction with Asellus aquaticus (L.) on the feeding activity of *Gammarus pulex*. Arch Environ Contam Toxicol 34:41–47.

Blockwell SJ, Maund SJ, Pascoe D. 1999. Effects of the organochloride insecticide lindane on the population responses of the freshwater amphipod *Hyalella azteca*. Environ Toxicol Chem 18:1264–1269.

Bogardi JJ, Duckstein L. 1992. Interactive multiobjective analysis embedding the decision maker's implicit preference function. Water Resour Bull 28:75–78.

Bonham CD. 1989. Measurements for terrestrial vegetation. New York (NY): John Wiley and Sons.

Boone MD, James SM. 2003. Interactions of an insecticide, herbicide, and natural stressors in amphibian community mesocosms. Ecol Appl 13:829–841.

Boone MD, Semlitsch RD. 2001. Interactions of an insecticide with larval density and predation in experimental amphibian communities. Conserv Biol 15:228–238.

Boone MD, Semlitsch RD. 2002. Interactions of an insecticide with competition and pond drying in amphibian communities. Ecol Appl 12:307–316.

Boonstra R, Bowman L. 2003. Demography of short-tailed shrew populations living on polychlorinated biphenyl-contaminated sites. Environ Toxicol Chem. 22:1394–1403.

Booy G, Hendriks RJJ, Smulders MJM, VanGroenedael JM, Vosman B. 2000. Genetic diversity and the survival of populations. Plant Biol 2:327–395.

Bormann BT, Martin JR, Wagner FH, Wood GW, Alegria J, Cunningham PG, Brookes MH, Friesma P, Berg J, Henshaw JR. 1999. Adaptive management. In: Johnson NC, Malk AJ, Sexton WT, Szaro R, editors. Ecological stewardship: a common reference for ecosystem management. Oxford (UK): Elsevier Science Ltd. p. 505–534.

Botkin DB. 1993. Forest dynamics: an ecological model. New York (NY): Oxford University Press.

Botkin DB, Janak JF, Wallis JR. 1972. Some ecological consequences of a computer model of forest growth. J Ecol 60:849–872.

Box GEP. 1976. Science and statistics. J Am Statistical Assoc 71:791–799.

Box GEP, Jenkins GM. 1970. Time series analysis, forecasting and control. San Francisco (CA): Holden-Day.

Breneman DH, Pontash KW. 1994. Stream microcosm toxicity tests: predicting the effects of fenvalerate on riffle insect communities. Environ Toxicol Chem 13:381–387.

Brook BW, O'Grady JJ, Chapman AP, Burgman MA, Akcayaka HR, Frankham R. 2000. Predictive accuracy of population viability analysis in conservation biology. Nature 6776:385–387.

Brook BW, Tonkyn DW, O'Grady JJ, Frankham R. 2002. Contribution of inbreeding to extinction risk in threatened species. Conserv Ecol 6:16.

Brooks SP. 1998. Markov chain Monte Carlo method and its application. Statistician 47:69–100.

Brooks SP, Catchpole EA, Morgan BJT, Harris MP. 2002. Bayesian methods for analyzing ringing data. J Appl Stat 29:187–205.

Brumfield RT, Beerli P, Nickerson DA, Edwards SV. 2003. The utility of single nucleotide polymorphisms in inferences of population history. Trends Ecol Evol 18:249–256.

Bruno JF, Stachowicz JJ, Bertness MD. 2003. Inclusion of facilitation into ecological theory. Trends Ecol Evol 18:119–125.

Buckland ST, Anderson DR, Burnham KP, Laake JL. 1993. Distance sampling: estimating abundance of biological populations. London (UK): Chapman and Hall.

Bulmer MG. 1975. The statistical analysis of density-dependence. Biometrics 31:901–911.

Bunnell FL. 1989. Alchemy and uncertainty: what good are models? Portland (OR): USDA Forest Service Pacific Northwest Research Station. General Technical Report PNW-GTR-232.

Burger R, Lynch M. 1995. Evolution and extinction in a changing environment: a quantitative genetic analysis. Evolution 49:151–163.

Burgman MA, Ferson S, Akcakaya HR. 1993. Risk assessment in conservation biology. London (UK): Chapman and Hall.

Burmaster DE, Anderson PE. 1994. Principles of good practice for the use of Monte Carlo techniques in human health and ecological risk assessments. Risk Anal 14:477–481.

Burnham KP, Anderson DR, White GC, Brownie C, Pollock KH. 1987. Design and analysis methods for fish survival experiments based on release-recapture. Bethesda (MD): American Fish Society Monograph Number 5.

Burnham KP, Anderson DR. 1998. Model selection and inference. New York (NY): Springer-Verlag.

Burnham KP, Anderson DR. 2002. Model selection and multi-model inference. 2nd ed. New York (NY): Springer-Verlag.

Burnham KP, White GC, Anderson DR. 1984. Estimating the effect of hunting on annual survival rates of adult mallards. J Wildl Manage 48:350–361.

Calow P, Sibly RM. 1990. A physiological basis of population processes: ecotoxicological implications. Funct Ecol 4:283–288.

Calow P, Sibley RM, Forbes V. 1997. Risk assessment on the basis of simplified life-history scenarios. Environ Toxicol Chem 16:1983–1989.

Campbell PJ, Aernold DJS, Brock TCM, Grandy NJ, Heger W, Heimbach F, Maund SJ, Streloke M. 1998. Guidance document on higher-tier aquatic risk assessment for pesticides (HARAP). Proceedings of the HARAP workshop, Lacanau Ocean, France 19–22 April 1998. SETAC-Europe publication.

Candolfi MP, Barrett KL, Campbell PJ, Forster R, Grandy N, Huet M-C, Lewis G, Oomen PA, Schmuck R, Vogt H, editors. 2001. Guidance document on regulatory testing and risk assessment procedures for plant protection products with non-target arthropods. Proceedings of the European Standard Characteristics of Beneficial Regulatory Testing (ESCORT 2) Workshop, Wageningen, The Netherlands, 21–23 March, 2000. SETAC Europe.

Candolfi M, Bigler F, Campbell P, Heimbach U, Schmuck R, Angeli G, Bakker F, Brown K, Carli G, Dinter A, et al. 2000. Principles for regulatory testing and interpretation of semi-field and field studies with non-target arthropods. J Pesticide Sci 73:141–147.

Cappuccino N, Price PW. 1995. Population dynamics: new approaches and synthesis. San Diego (CA): Academic Press.

Carlander KD. 1969a. Handbook of freshwater fishery biology. Vol. 1. Ames (IA): Iowa State University Press.

Carlander KD. 1969b. Handbook of freshwater fishery biology. Vol. 2. Ames (IA): Iowa State University Press.

Carlin BP, Louis TA. 1996. Bayes and empirical Bayes methods for data analysis. London (UK): Chapman and Hall.

Carlsen C, Critto A, Marcomini A. 2000. Risk based characterization of a contaminated industrial site using multivariate and geostatistical tools. Environ Pollut 111:417–427.

Carlsen TM, Coty JD, Kercher JR. 2004. The spatial extent of contaminants and the landscape scale: an analysis of the wildlife, conservation biology and population modeling literature. Environ Toxicol Chem 23:798–811.

Carr MR. 1996. Primer user manual: Plymouth routines in multivariate ecological research. Plymouth (UK): Plymouth Marine Laboratory.

Caswell H. 1982. Stable population structure and reproductive value for populations with complex life cycles. Ecology 63:1223–1231.

Caswell H. 1986. Life cycle models for plants. Lect Math Life Sci 18:171–233.

Caswell H. 1996. Demography meets ecotoxicology: untangling the population-level effects of toxic substances. In: Newman MC, Jagoe CH, editors. Ecotoxicology: a hierarchical approach. Boca Raton (FL): CRC Press/Lewis Publishers. p 255–292.

Caswell H. 2000. Life table response experiments in ecotoxicology. In: Kammenga J, Laskowski R, editors. Demography in ecotoxicology. New York (NY): John Wiley and Sons Ltd. p. 43–55.

Caswell H. 2001. Matrix population models: construction, analysis, and interpretation, 2nd edition. Sunderland (MA): Sinauer Associates, Inc.

Caswell H, John AM. 1992. From the individual to the population in demographic models. In: DeAngelis DL, Gross LJ, editors. Individual-based models and approaches in ecology: populations, communities and ecosystems. New York (NY): Chapman and Hall. p. 36–61.

Caswell H, Naiman RJ, Morin R. 1984. Evaluating the consequences of reproduction in complex salmonid life cycles. Aquaculture 43:123–134.

Catchpole EA. 1995. MATLAB: an environment for analyzing ring-recovery and recapture data. J Appl Stat 22:801–816.

Caudill CC. 2003. Measuring dispersal in a metapopulation using stable isotope enrichment: high rates of sex-biased dispersal between patches in a mayfly population. Oikos 101:624–630.

Chamberlin TC. 1890. The method of multiple working hypotheses. Reprinted 1965 Science 148:754–759.

Chan KS, Stenseth NC, Lekve K, Gjosæter J. 2003. Modeling pulse disturbance impact on cod population dynamics: the 1988 algal bloom of Skagerrak, Norway. Ecol Monogr 73:151–171.

Chandini T. 1989. Survival, growth and reproduction of *Daphnia carinata* (Crustacea: Cladocera) exposed to chronic cadmium stress at different food (*Chlorella*) levels. Environ Pollut 60:29–45.

Chapman PF, Ho K, Munns WR Jr, Solomon K, Weinstein MP. 2002. Issues in sediment toxicity and ecological risk assessment. Mar Pollut Bull 44:271–278.

Chapman PM, Wang F, Janssen CR, Goulet RR, Kamunde CN. 2003. Conducting ecological risk assessments of inorganic metals and metalloids: current status. Hum Ecol Risk Assess 9:641–697.

Chatfield C. 1996. The analysis of time series: an introduction, 5th ed. London (UK): Chapman & Hall.

Chevillon C, Bourguet D, Rousset F, Pasteur N, Raymond M. 1997. Pleiotropy of adaptive changes in populations: comparisons among insecticide resistance genes in *Culex pipiens*. Genet Res Cambridge 70:195–204.

Chevillon C, Raymond M, Guillemaud T, Lenormand T, Pasteur N. 1999. Population genetics of insecticide resistance in the mosquito *Culex pipiens*. Biol J Linnean Soc 68:147–157.

Clark T, Clark K, Paterson S, Mackay D, Norstrom RJ. 1988. Wildlife monitoring, modeling, and fugacity. Environ Sci Technol 22:120–127.

Clark ME, Rose KA, Levine DA, Hargrove WW. 2001. Predicting climate change effects on Appalachian trout: combining GIS and individual-based modeling. Ecol Appl 11:161–178.

Clarke KR, Warwick RM. 1994. Change in marine communities: an approach to statistical analysis and interpretation. Plymouth (UK): National Environmental Research Center.

Clarke KR, Warwick RM. 2001. Change in marine communities: an approach to statistical analysis and interpretation. 2nd ed. Plymouth (UK): PRIMER-E.

Clements WH, Newman MC. 2002. Community ecotoxicology. West Sussex (UK): John Wiley and Sons Ltd.

Cohen J. 1988. Statistical power analysis for the social sciences. Hillsdale (NJ): Lawrence Erlbaum Associates.

Cohen S. 2002. Strong positive selection and habitat-specific amino acid substitution patterns in Mhc from an estuarine fish under intense pollution stress. Mol Biol Evol 19:1870–1880.

Collins FS, Green ED, Guttmacher AE, Guyer MS. 2003. A vision for the future of genomics research. Nature 422:835–847.

Coltman DW, Pilkington JG, Smith JA, Pemberton JM. 1999. Parasite-mediated selection against inbred Soay sheep in a free-living, island population. Evolution 53:1259–1267.

Committee of Scientists. 1999. Sustaining the people's lands: recommendations for stewardship of the national forests and grasslands into the next century. Washington (DC): US Department of Agriculture.

Cooper AB, Hilborn R, Unsworth J. 2003. An approach for population assessment in the absence of abundance indices. Ecol Appl 13:814–828.

Costanza R, Fitz HC, Bartholomew JA, DeBellevue E. 1992. The everglades landscape model (ELM): summary of workshop II. Report to the Everglades Systems Research Division, South Florida Water Management District, West Palm Beach (FL).

Coulson T, Milner-Gulland EJ, Clutton-Brock T. 2000. The relative roles of density and climatic variation on population dynamics and fecundity rates in three contrasting ungulate species. Proc R Soc London Ser B 267:1771–1779.

Courchamp F, Clutton-Brock T, Grenfell T. 1999a. Inverse density dependence and the Allee effect. Trends Ecol Evol 14:405–410.

Courchamp F, Grenfell BT, Clutton-Brock TH. 1999b. Population dynamics of obligate cooperators. Proc R Soc London Ser B 266:557–564.

Coustau C, Chevillon C, French-Constant R. 2000. Resistance to xenobiotics and parasites: can we count the costs? Trends Ecol Evol 15:378–383.

Cox DR, Oakes D. 1984. Analysis of survival data. London (UK): Chapman & Hall.

Crnokrak P, Roff DA. 1999. Inbreeding in the wild. Heredity 83:260–267.

Crossman EJ. 1991. Introduced freshwater fishes: a review of the North American perspective with emphasis on Canada. Can J Fish Aquat Sci 48:46–57.

Crouse DT, Crowder LB, Caswell H. 1987. A stage-based population model for loggerhead sea turtles and implications for conservation. Ecology 68:1412–1423.

Crowder LB, Crouse DT, Heppell SS, Martin TH. 1994. Predicting the impact of turtle excluder devices on loggerhead sea turtle populations. Ecol Appl 4:437–445.

Cushing JM, Dennis B, Desharnais RA, Costantino RF. 1996. An interdisciplinary approach to understanding nonlinear ecological dynamics. Ecol Model 92:111–119.

Daniels S. 1994. Effects of cadmium toxicity on population dynamics of the blowfly *Lucilia sericata*. PhD thesis. UK: University of Reading.

Daniels RE, Allan JD. 1981. Life table evaluation of chronic exposure to a pesticide. Can J Fish Aquat Sci 38:485–494.

Darwin C. 1998. The origin of species by means of natural selection or the preservation of favored races in the struggle for life. New York (NY): The Modern Library. Originally printed in 1859.

Daskalov G. 1999. Relating fish recruitment to stock biomass and physical environment in the Black Sea using generalized additive models. Fish Res 41:1–23.

Day K, Kaushik NK. 1987. The adsorption of fenleverate to laboratory glassware and the alga *Chlamydomonas reinhardii*, and its effects on uptake of the pesticide by *Daphnia galeata mendotae*. Aquat Toxicol 10:131–142.

De Angelis DL, Gross LJ. 1992. Individual-based models and approaches in ecology: populations, communities and ecosystems. New York (NY): Chapman & Hall.

De Roos AM, McCauley E, Wilson WG. 1991. Mobility versus density limited predator-prey dynamics on different spatial scales. Proc Royal Acad Lond B 256:117–122.

De Roos AM, Persson L, McCauley E. 2003. The influence of size-dependent life-history traits on the structure and dynamics of populations and communities. Ecol Lett 6:473–487.

Demetrius L. 1975. Natural selection in age-structured populations. Genetics 79:535–544.

Demetrius L. 1975. Reproductive strategies and natural selection. Am Nat 109:243–249.

Dennis B, Munholland PL, Scott JM. 1991. Estimation of growth and extinction parameters for endangered species. Ecol Monogr 61:115–143.

Dennis B, Taper ML. 1994. Density dependence in time series observation of natural populations: estimation and testing. Ecol Monogr 64:205–224.

DeSante DF, Michel N., Kaschube .DR. Nott MP. 2005. The Monitoring Avian Productivity and Survivorship (MAPS) Program Annual Reports, 1989-2003. NBII/MAPS Avian Demographics Query Interface. Available at: http://www.birdpop.org/nbii/NBII-Home.asp (December 2006).

De Valpine P, Hastings A. 2002. Fitting population models incorporating process noise and observation error. Ecol Monogr 72:57–76.

Doak D, Kareiva P, Klepetka B. 1994. Modeling population viability for the desert tortoise in the western Mojave desert. Ecol Appl 4:446–460.

Dorazio RM, Johnson FA. 2003. Bayesian inference and decision theory—a framework for decision making in natural resource management. Ecol Appl 13:556–563.

Drechsler M, Frank K, Hanski I, O'Hara RB, Wissel C. 2003. Ranking metapopulation extinction risk: from patterns in data to conservation management decisions. Ecol Appl 13:990–998.

Dunning J. 1993. CRC handbook of avian body masses. Boca Raton (FL): CRC Press.

Dunning JB, Stewart DJ, Danielson BJ, Noon BR, Root TL, Lamberson RH, Stevens EE. 1995. Spatially explicit population models: current forms and future uses. Ecol Appl 5:3–11.

Easterling MR, Ellner SP, Dixon PM. 2000. Size-specific sensitivity: applying a new structured population model. Ecology 81:694–708.

Eberhardt LL, Thomas JM. 1991. Designing environmental field studies. Ecol Monogr 61:53–73.

Eberhardt LL. 2003.What should we do about hypothesis testing? J Wildl Manage 67:241–247.

Ebert DW. 2001. ATtILA—Analytical Tools Interface for Landscape Assessments. Landscape Ecology Branch, National Exposure Research Laboratory, US Environmental Protection Agency, Las Vegas, NV. Available from: http://www.epa.gov/nerlesd1/land-sci/pdf/011factsheet.pdf.

[ECOFRAMb] Ecological Committee on FIFRA Risk Assessment Methods. Aquatic Workgroup. 1999. ECOFRAM aquatic report. May 4, 1999. Available online at: www.epa.gov/oppefed1/ecorisk/index.htm.

[ECOFRAMa] Ecological Committee on FIFRA Risk Assessment Methods. Terrestrial Workgroup. 1999. ECOFRAM Terrestrial Draft Report. May 10, 1999. www.epa.gov/oppefed1/ecorisk/index.htm.

Edmands S, Harrison JS. 2003. Molecular and quantitative trait variation within and among populations of the intertidal copepod *Tigriopus californicus*. Evolution 57:2277–2285.

Efroymson RA, Carlsen TM, Jager HI, Kostova T, Carr EA, Hargrove WW, Kercher J, Ashwood TL. 2004. Toward a framework for assessing risk to vertebrate populations from brine and petroleum spills at exploration and production sites. In: Kapustka L, Galbraith H, Luxon M, Biddinger GR, editors. Landscape ecology and wildlife habitat evaluation: critical information for ecological risk assessment, land-use management activities, and biodiversity enhancement practices. ASTM International. West Conshohocken (PA): American Society for Testing and Materials (ASTM). STP 1458.

Ellison AM. 1996. An introduction to Bayesian inference for ecological research and environmental decision-making. Ecol Appl 6:1036–1046.

Ellner SP, Bailey BA, Bobashev GV, Gallant AR, Grenfell BT, Nychka DW. 1998. Noise and nonlinearity in measles epidemics: combining mechanistic and statistical approaches to population modeling. Am Nat 151:425–440.

Ellner SP, Fieberg J. 2003. Using PVA for management despite uncertainty: effects of habitat, hatcheries, and harvest on salmon. Ecology 84:1359–1369.

Emerson BC, Paradis E, Thebaud C. 2001. Revealing the demographic histories of species using DNA sequences. Trends Ecol Evol 16:707–716.

Emlen JM. 1989. Terrestrial population models for ecological risk assessment: a state-of-the-art review. Environ Toxicol Chem 8:831–842.

Emlen JM, Pikitch EK. 1989. Animal population dynamics: identification of critical components. Ecol Model 44:253–273.

[EC] Environment Canada. 1997. Environmental assessments of priority substances under the Canadian Environmental Protection Act. Guidance manual version 1.0. Chemicals Evaluation Division. PS/2/CC/3E.

Errington PL. 1946. Predation and vertebrate populations. Q Rev Biol 21:144–177.
European Chemicals Bureau 2003. Technical guidance document on risk assessment, Part II. European Commission Joint Research Centre. EUR20418. Ispra, Italy.
European Commission. 2001. White paper on the strategy for a future chemicals policy. 88. Brussels, Belgium.
[EWoFFT] European Workshop on Freshwater Field Tests. 1992. Summary and recommendations report. Potsdam, Germany.
[EU] European Union. 1997. Council Directive 97/57/EC of 21 September 1997; Establishing annex VI to Directive 91/414/EEC Concerning the placing of plant protection products on the market. Official Journal of the European Communities L265: 87–109.
EVS Environmental Consultants. 1999. Cherry Point screening level ecological risk assessment. Prepared for the Washington Department of Natural Resources. EVS Project No. 2/868-01.1. Seattle (WA): EVS Environmental Consultants.
Fahrig L. 1997. Relative effects of habitat loss and fragmentation on population extinction. J Wildl Manage 61:603–610.
Fahrig L. 2001. How much habitat is enough? Conserv Biol 100:65–74.
Falconer DS. 1989. Introduction to quantitative genetics, 3rd ed. Longman House, Burnt Mill (UK).
Fall A, Fall J. 2001. A domain-specific language for models of landscape dynamics. Ecol Model 141:1–18.
Feder ME, RA Krebs. 1997. Ecological and evolutionary physiology of heat-shock proteins and the stress response in *Drosophila*: complementary insight from genetic engineering and natural variation. In: Bijlsma R, Loeschcke V, editors. Environmental stress, adaptation and evolution. Basel (Switzerland): Birkhyuser Verlag. p. 155–173.
Ferry, L., R. Ferry, W. Isherwood, R. Woodward, T. Carlsen, Z. Demir, R. Qadir, and M. Dresen, 1999. *Final Site-Wide Feasibility Study for Lawrence Livermore National Laboratory Site 300*, Lawrence Livermore National Laboratory, Livermore, Calif. (UCRLAR-132609).
Ferson S, Ginzburg LR, Goldstein RA. 1996. Inferring ecological risk from toxicity bioassays. Water Air Soil Pollut 90:71–82.
Ferson S. 2002. Population models: scalar abundance. In: Pastorok RA, Bartell SM, Ferson S, Ginzburg LR, editors. Ecological modeling in risk assessment: chemical effects on populations, ecosystems, and landscapes. Boca Raton (FL): CRC Press. p. 37–54.
Ferson S, Burgman MA, editors. 2002. Quantitative methods in conservation biology. New York (NY): Springer.
Firko M, Hayes JL. 1990. Quantitative genetic tools for insecticide resistance risk assessment: estimating the heritability of resistance. J Econ Entomol 83:647–654.
Fleeger JW, Carman KR, Nisbet RM. 2003. Indirect effects of contaminants in aquatic ecosystems. Sci Total Environ 317:207–233.
Fletcher JS, Nellessen JE, Pfleeger TG. 1994. Literature review and evaluation of the EPA food-chain (Kenaga) nomogram: an instrument for estimating pesticide residues on plants. Environ Toxicol Chem 13:1383–1391.
Fliedner A, Remde A, Niemann R, Schafers C. 1997. Effects of the organotin pesticide azoclyclotin in aquatic microcosms. Chemosphere 35:209–222.
FOCUS. 1995. Leaching models and EU registration. EC Document 4952/VI/95. Brussels (Belgium): European Commission.
Foley P. 1994. Predicting extinction times from environmental stochasticity and carrying capacity. Conserv Biol 8:124–137.
Fontenot LW, Ayers TA, Day CH, Ellingston SB, Sager SL. 1998. Ecological risk assessment of endocrine-disrupting chemicals to amphibians. Atlanta, GA, April 20–22, ASTM 8th symposium on environmental toxicology and risk assessment: standardization of biomarkers for endocrine disruption and environmental assessment.

Fontenot LW, Noblet GP, Akins JM, Stephens MD, Cobb GP. 2000. Bioaccumulation of polychlorinated biphenyls in ranid frogs and northern water snakes from a hazardous waste site and a contaminated watershed. Chemosphere 40:803–809.

Forbes VE, Calow P. 1999. Is the per capita rate of increase a good measure of population-level effects in ecotoxicology? Environ Toxicol Chem 18:1544–1556.

Forbes VE, Calow P. 2002. Population growth rate as a basis for ecological risk assessment of toxic chemicals. Philos T Roy Soc B 357:1299–1306.

Forbes VE, Calow P, Sibly RM. 2001a. Are current species extrapolation models a good basis for ecological risk assessment? Environ Toxicol Chem 20:442–447.

Forbes VE, Calow P, Sibly RM. 2001b. Toxicant impacts on density-limited populations: a critical review of theory, practice, and results. Ecol Appl 11:1249–1257.

Forbes VE, Sibly RM, Linke-Gamenick I. 2003. Joint effects of population density and toxicant exposure on population dynamics of *Capitella* sp. I. Ecol Appl 13:1094–1103.

Foster PL. 2000. Adaptive mutation: implications for evolution. BioEssays 22:1067–1074.

Fowler CW, Baker JD. 1991. A review of animal population dynamics at extremely reduced population levels. 41. Report to the International Whaling Commission. Available at http://www.iwcoffice.org/publications/pubpre1999.htm.

Fox GA. 1991. Practical causal inference for ecoepidemiologists. J Toxicol Environ Health 33:359–373.

Fox GA, Gurevitch J. 2000. Population numbers count: tools for near-term demographic analyses. Am Nat 156:242–256.

Frankham R. 1995. Effective population size/adult population size ratios in wildlife: a review. Genet Res 66:95–107.

Frankham R, Ralls K. 1998. Conservation biology: inbreeding leads to extinction. Nature 392:441–442.

Frankham R, Balou JD, Briscoe DA. 2002. Introduction to conservation genetics. Cambridge (UK): Cambridge University Press.

Franklin AB, Anderson DR, Gutierrez RJ, Burnham KP. 2000. Climate, habitat quality, and fitness in northern spotted owl populations in northwestern California. Ecol Monogr 70:539–590.

Franklin AB, Shenk TM, Anderson DR, Burnham KP. 2001. Statistical model selection: an alternative to null hypothesis testing. In: Shenk TM, Franklin AB, editors. Modeling in natural resource management. Washington (DC): Island Press. p. 75–90.

Franklin IR, Frankham R. 1998. How large must populations be to retain evolutionary potential? Anim Conserv 1:69–73.

Franklin IR. 1980. Evolutionary change in small populations. In: Soule ME, Wilcox BA, editors. Conservation biology, an evolutionary-ecological perspective. Sunderland (MA): Sinauer Associates Inc. p. 135–149.

Freshman JS, Menzie CA. 1996. Two wildlife exposure models to assess impacts at the individual and population levels and the efficacy of remedial actions. Hum Ecol Risk Assess 2:481–498.

Fuentes MA, Kuperman MN. 1999. Cellular automata and epidemiological models with spatial dependence. Physica A 267:471–486.

Futuyma DJ. 1986. Evolutionary biology, 2nd ed. Sunderland (MA): Sinauer Associates, Inc.

Garcia-Villada L, Lopez-Rodas V, Banares-Espana E, Flores-Moya A, Agrelo M, Martin-Otero L, Costas E. 2002. Evolution of microalgae in highly stressing environments: an experimental model analyzing the rapid adaptation of *Dictyosphaerium chlorelloides* (Chlorophyceae) from sensitivity to resistance against 2,4,6-trinitrotoluene by rare preselective mutations. J Phycol 38:1074–1081.

[GRI] Gas Research Institute. 2002. Evaluating potential ecological impacts at exploration & production sites: a compilation of screening criteria. Report #02/2006. Gas Technology Institute. Chicago (IL).

Gelman A, Carlin JB, Stern HS, Rubin DB. 1995. Bayesian data analysis. New York (NY): Chapman and Hall.

Gentile JH, Gentile SM, Hairston NG Jr., Sullivan BK. 1982. The use of life-tables for evaluating the chronic toxicity of pollutants to *Mysidopsis bahia*. Hydrobiology 93:179–187.

Gentile JH, Gentile SM, Hoffman G, Heltshe JF, Hairston NG Jr. 1983. The effects of a chronic mercury exposure on survival, reproduction and population dynamics of *Mysidopsis bahia*. Environ Toxicol Chem 2:61–68.

Gervais JA, Anthony RG. 2003. Chronic organochlorine contaminants, environmental variability, and the demographics of a burrowing owl population. Ecol Appl. 13:1250–1262.

Gervais JA, Hunter CM, Anthony RG. 2005. Interactive effects of prey and p,p'DDE on burrowing owl population dynamics. Ecol Appl 16:666–677.

Getz WM. 1996. A hypothesis regarding the abruptness of density dependence and growth rate of populations. Ecology 77:2014–2026.

Gibbs PE, Bryan GW. 1986. Reproductive failure in populations of the dog-whelk, *Nucella lapillus*, caused by imposex induced by tributyltin from anti-fouling paints. J Mar Biol Assoc UK 66:767–777.

Giddings, JM, R. C. Biever, and K. D. Racke. 1997. Fate of chlorpyrifos in outdoor pond microcosms and effects on growth and survival of bluegill sunfish. Environ Toxicol Chem 16:2353–2362.

Giddings JM, Brock TCM, Heger W, Heimbach F, Maund SJ, Norman SM, Ratte HT, Schafers C, Streloke M. 1999. Community Level Aquatic System Studies—Interpretation Criteria (CLASSIC). Proceedings of the CLASSIC workshop held at the Fraunhofer Institute, Schmallenberg, Germany, 30 May–2 June 1999. Brussels (Belgium): SETAC-Europe publication.

Gillman MP, Hails R. 1997. An introduction to ecological modeling: putting practice into theory. Oxford (UK): Blackwell Science.

Gillman MP, Silvertown J. 1997. Population extinction and the uncertainty of measurement. In: Proceedings of the JNN/BES Symposium on the role of genetics in conserving small populations. Available at www.jncc.gov.uk/page-2556.

Gilpin M. 1996. Forty-eight parrots and the origins of population viability analysis. Conserv Biol 10:1491–1493.

Ginissel A, Augustin S, Courtin C, Pilate G, Lorme P, Bourguet D. 2003. Initial frequency of alleles conferring resistance to *Bacillus thuringiensis* poplar in a field population of *Chrysomela tremulae*. Proc R Soc London Ser B 270:791–797.

Ginzburg LR, Slobodkin LB, Johnson K, Bindman AG. 1982. Quasi-extinction probabilities as a measure of impact on population growth. Risk Anal 2:171–182.

Ginzburg LR, Ferson S, Akcakaya HR. 1990. Reconstructibility of density dependence and the conservative assessment of extinction risks. Conserv Biol 4:63–70.

Gleason TR, Munns WR Jr, Nacci DE. 1999. The influence of model time step on the relative sensitivity of population growth rate to survival, growth, and reproduction. Aspects Appl Biol 53:253–260.

Gleason TR, Munns WR Jr, Nacci DE. 2000. Projecting population-level response of purple sea urchins to lead contamination for an estuarine ecological risk assessment. J Aquat Ecosystem Stress Recov 7:177–185.

Gockel J, Kennington WJ, Hoffmann AA, Goldstein DB, Partridge L. 2001. Nonclinality of molecular variation implicates selection in maintaining a morphological cline of *Drosophila melanogaster*. Genetics 158:319–323.

Godfray HCJ, Chan MS. 1990. How insecticides trigger single-stage outbreaks in tropical pests. Funct Ecol 4:329–337.

Goel NS, Richter-Dyn N. 1974. Stochastic models in biology. New York (NY): Academic Press.

Goldingay R, Possingham H. 1995. Area requirements for viable populations of the gliding marsupial *Petaurus australis*. Conserv Biol 73:161–167.

Gomez A, Ceccine G, Snell TW. 1997. Effects of pentachlorophenol on the predator-prey interactions of two rotifers. Aquatic Toxicol 37:271–282.

Gotelli NJ. 1995. A primer of ecology. Sunderland (MA): Sinauer Associates.

Grant A. 1998. Population consequences of chronic toxicity: incorporating density dependence into the analysis of life table response experiments. Ecol Model 105:325–335.

Grant A. 2002. Pollution-tolerant species and communities: intriguing toys or invaluable monitoring tools? Hum Ecol Risk Assess 8:955–970.

Grant PR, Grant BR. 2002. Unpredictable evolution in a 30-year study of Darwin's Finches. Science 296:707–711.

Green RH. 1979. Sampling design and statistical methods for environmental biologists. New York (NY): John Wiley & Sons.

Grenfell BT, Wilson K, Finkenstaedt BF, Coulson TN, Murray S, Albon SD, Pemberton JM, Clutton-Brock TH, Crawley MJ. 1998. Noise and determinism in synchronized sheep dynamics. Nature 394:674–677.

Grimm V. 1999. Ten years of individual-based modelling in ecology: what have we learned and what could we learn in the future? Ecol Model 115:129–148.

Groenendijk D, Lucker SMG, Plans M, Kraak MHS, Admiraal W. 2002. Dynamics of metal adaptation in riverine chironomids. Environ Pollut 117:101–109.

Groom MJ, Pascual MA. 1998. The analysis of population persistence: an outlook on the practice of population viability. In: Fiedler PL, Kareiva PM, editors. Conservation biology for the coming decade. New York (NY): Chapman & Hall. p. 4–27.

Guthery FS, Lusk JJ, Peterson MJ. 2001. The fall of the null hypothesis: liabilities and opportunities. J Wildlife Manage 65:379–384.

Gutierrez AP. 1996. Applied population ecology: a supply-demand approach. New York (NY): John Wiley & Sons.

Hahn ME, Karchner SI, Franks DG, Merson RR. 2004. Aryl hydrocarbon receptor polymorphisms and dioxin resistance in Atlantic killifish (*Fundulus heteroclitus*). Pharmacogenetics 14:131–146.

Hahn ME. 1998. Mechanisms of innate and acquired resistance to dioxin-like compounds. Rev Toxicol 2:395–443.

Hakoyama H, Iwasa Y. 2000. Extinction risk of a density-dependent population estimated from a time series of population size. J Theor Biol 204:337–359.

Hakoyama H, Iwasa Y, Nakanishi J. 2000. Comparing risk factors for population extinction. J Theor Biol 204:327–336.

Hall, C., Neher, L. Wilder, L., Carlsen, T., Ashwood, T., Washington-Allen, R., and Efroymson, R., 2001. *Data Collection Protocol for Developing Geographical Information Systems to Investigate and Manage Ecological Impacts at Petroleum Exploration and Production Sites*, UCRL-ID-142371, Lawrence Livermore National Laboratory, January 2001.

Hallam TG, Lassiter RR, Li J, McKinney W. 1992. An approach for modelling populations with continuous structured models. In: DeAngelis DL, Gross LJ, editors. Individual-based approaches and models in ecology: populations, communities and ecosystems. New York (NY): Chapman & Hall. p. 312–337.

Hallam TG, Lassiter RR. 1994. Individual-based mathematical modeling approaches in ecotoxicology: a promising direction for aquatic population and community ecological risk assessment. In: Kendall RJ, Lacher TE, editors. Wildlife toxicology and population modeling. Boca Raton (FL): Lewis Publishers. p. 531–542.
Hallam TG. 1998. Applications of physiologically based population and community models to stress ecology. In: Waide JB, Gandy LM, compilers. Proceedings of the workshop on aquatic ecosystem modeling and assessment techniques for application within the US Army. Vicksburg (MS): US Army Engineer Watersways Experiment Station. Little Rock (AR): FTN Associates, Ltd. p. 109–118.
Hamers T, Krough PH. 1997. Predator-prey relationships in a two-species toxicity test system. Ecotox Environ Safe 37:202–212.
Hamilton WD. 1967. Extraordinary sex ratios. Science 156:477–488.
Hanotte O, Bradley DG, Ochieng JW, Verjee Y, Hill EW, Rege EO. 2002. African pastoralisms: genetic imprints of origins and migrations. Science 296:336–339.
Hansen FT, Forbes VE, Forbes, TL. 1999c. Effects of 4-n-nonylphenol on life-history traits and population dynamics of a polychaete. Ecol Appl 9:482–495.
Hansen F, Forbes VE, Forbes TL. 1999a. Using elasticity analysis of demographic models to link toxicant effects on individuals to the population level: an example. Funct Ecol 13:157–162.
Hansen TF, Stenseth NC, Henttonen H, Tast J. 1999b. Interspecific and intraspecific competition as causes of direct and delayed density dependence in a fluctuating vole population. Proc Natl Acad Sci USA. 96:986–991.
Hanski IA. 1991. Single-species metapopulation dynamics: concepts, models, and observations. Biol J Linn Soc 42:17–38.
Hanski IA. 1994. A practical model of metapopulation dynamics. J Anim Ecol 63:151–162.
Hanski IA. 1999. Metapopulation ecology. New York (NY): Oxford University Press.
Hanski IA, Gyllenburg M. 1993. Two general metapopulation models and the core-satellite species hypothesis. Am Nat 132:360–382.
Hanski IA, Gilpin ME. 1997. Metapopulation biology: ecology, genetics, evolution. San Diego (CA): Academic Press.
Hanski IA, Simberloff D. 1997. The metapopulation approach, its history, conceptual domain and application to conservation. In: Hanski IA, Gilpin ME, editors. Metapopulation biology: ecology, genetics and evolution. New York (NY): Academic Press. p. 5–26.
Hanski IA, Woiwood I, Perry J. 1993. Density dependence, population persistence, and largely futile arguments. Oecology 95:595–598.
Hanski IA, Alho J, Moilanen A. 2000. Estimating the parameters of survival and migration of individuals in metapopulations. Ecology 81:239–251.
Hare MP. 2001. Prospects for nuclear gene phylogeography. Trends Ecol Evol 16:700–706.
Harper FA, Smith FE, Macnair MR. 1997. Where is the cost in copper tolerance in *Mimulus guttatus*? Testing the trade-off hypothesis. Func Ecol 11:764–774.
Harrison S, Cappuchino N. 1995. Using density-manipulations experiments to study population regulation. In: Cappuchino N, Price PW, editors. Population dynamics new approaches and synthesis. San Diego (CA): Academic Press.
Hartl DL, Clark AG. 1997. Principles of population genetics. Sunderland (MA): Sinauer Associates, Inc.
Harwell M, Gentile J, Norton B, Cooper W. 1994. Issue paper on ecological significance. In: Ecological risk assessment issue papers, Risk Assessment Forum. Washington (DC): EPA/630/R-94/009.

Hassell MP, Latto J, May RM. 1989. Seeing the wood for the trees: detecting density dependence from existing life-table studies. J Anim Ecol 58:883–892.

Hastie TJ, Tibshirani RJ. 1990. Generalized additive models. London (UK): Chapman & Hall.

Hatch AC, Blaustein AR. 2000. Combined effects of UV-B, nitrate, and low pH reduce the survival and activity level of larval cascades frogs (*Rana cascadae*). Arch Environ Contam Toxicol 39:494–499.

Hatch SA. 2003. Statistical power for detecting trends with applications to seabird monitoring. Conserv Biol 111:317–329.

Heagler MG, Newman MC, Mulvey M, Dixon, PM. 1993. Allozyme genotype in mosquitofish, *Gambusia holbrooki*, during mercury exposure: temporal stability, concentration effects and field verification. Environ Toxicol Chem 12:385–395.

Hedrick PW, Kalinowski ST. 2000. Inbreeding depression in conservation biology. Annu Rev Ecol Syst 31:139–162.

Hedrick PW. 2001. Conservation genetics: where are we now? Trends Ecol Evol 16:629–636.

Hendriks AJ, Enserink EL. 1996. Modelling response of single-species populations to micro-contaminants as a function of species size with examples for waterfleas (*Daphnia magna*) and cormorants (*Phalacrocorax carbo*). Ecol Model 88:247–262.

Heppell SS, Caswell H, Crowder LB. 2000. Life histories and elasticity patterns: perturbation analyses for species with minimal demographic data. Ecology 81:654–665.

Hestbeck JB, Nichols JD, Malecki RA. 1991. Estimates of movement and site fidelity using mark-resight data of wintering Canada geese. Ecology 72:523–533.

Hilborn R, Walters CJ. 1992. Quantitative fisheries stock assessment: choice, dynamics, and uncertainty. London (UK): Chapman and Hall.

Hilborn R, Mangel M. 1997. The ecological detective: confronting models with data. Princeton (NJ): Princeton University Press.

Hoerger F, Kenaga EE. 1972. Pesticide residues on plants: correlation of representative data as basis for estimation of their magnitude in the environment. In: Coulston F, Korte F, editors. Global aspects of chemistry, toxicology and technology as applied to the environment. Vol. I. New York (NY): Thieme, Stuttgart, and Academic Press. p. 9–28.

Hoffman FO, Hammond JS. 1994. Propagation of uncertainty in risk assessments: The need to distinguish the difference between uncertainty due to lack of knowledge and uncertainty due to variability. Risk Anal 14:707–712.

Hoffmann AA, Hallas RJ, Dean JA, Schiffer DM. 2003. Low potential for climatic stress adaptation in a rainforest *Drosophila* species. Science 301:100–102.

Hoffmann AA, Merila J. 1999. Heritable variation and evolution under favorable and unfavorable conditions. Trends Ecol Evol. 14:96–101.

Holmes EE. 2001. Estimating risks in declining populations with poor data. Proc Natl Acad Sci USA. 98:5072–5077.

Holyoak M. 1994. Identifying delayed density dependence in time series data. Oikos 70:296–304.

Hope BK. 2000. Generating probabilistic spatially-explicit individual and population exposure estimates for ecological risk assessments. Risk Anal 20:573–589.

Hope BK. 2001. A case-study comparing static and spatially explicit ecological exposure analysis methods. Risk Anal 21:1001–1010.

Hsu JC. 1996. Multiple comparisons: theory and methods. London (UK): Chapman and Hall.

Hunter MD, Price PW. 1998. Cycles in insect populations: delayed density dependence or exogenous driving variables? Ecol Ent 23:216–222.

Hurlbert SH. 1984. Pseudoreplication and the design of ecological field experiments. Ecol Monogr 54:187–211.

Iwasa Y, Hakoyama H, Nakamaru M, Nakanishi J. 2000. Estimate of population extinction risk and its application to ecological risk management. Popul Ecol 42:73–80.

Iwasa Y, Hakoyama H. 1998. Extinction rate of a population with both demographic and environmental stochasticity. Theor Popul Biol 53:1–15.

Jackson RB, Linder CR, Lynch M, Purugganan M, Somerville S, Thayer SS. 2002. Linking molecular insight and ecological research. Trends Ecol Evol 17:409–414.

Janssen R. 1992. Multiobjective decision support for environmental management. Dordrecht (the Netherlands): Kluwer Academic Publishers.

Jaworska JS, Rose KA, Brenkert AL. 1997. Individual-based modeling of PCBs effects on young-of-the-year largemouth bass in southeastern USA reservoirs. Ecol Model 99:113–135.

Jensen A, Forbes VE, Parker ED. 2001. Variation in cadmium uptake, feeding rate, and life-history effects in the gastropod *Potamopyrgus antipodarum*: linking toxicant effects on individuals to the population level. Environ Toxicol Chem 20:2503–2513.

Joe M, Pollock KH. 2002. Separation of survival and movement rates in multi-state tag-return and capture-recapture models. J Appl Stat 29:373–384.

Johnson AR. 2002. Landscape ecotoxicology and assessment of risk at multiple scales. Hum Ecol Risk Assess 8:127–146.

Johnson DH. 1999. The insignificance of statistical significance testing. J Wildl Manage 63:763–772.

Johnson CJ, Gillingham MP. 2004. Mapping uncertainty: sensitivity of wildlife habitat ratings to expert opinion. J Appl Ecol 41:1032–1041.

Johnson BB, P Slovic. 1995. Presenting uncertainty in health risk assessment—initial studies of its effects on risk perception and trust. Risk Anal 15:485–494.

Johnston RK, Munns WR, Tyler PL, Marajh-Whittemore P, Finkelstein K, Munney K, Short FT, Melville A, Hahn S. 2002. Weighing the evidence of ecological risk of chemical contamination in the estuarine environment adjacent to the Portsmouth Naval Shipyard, Kittery, Maine, USA. Environ Toxic Chem 21:182–194.

Jongman RHG, ter Braak CJF, van Tongeren OFR. 1995. Data analysis in community and landscape ecology. Cambridge (UK): Cambridge University Press.

Kalmus H, Smith CAB. 1960. Evolutionary origin of sexual differentiation and the sex-ratio. Nature 186:1004–1006.

Kammenga JE, Van Gestel CAM, Hornung E. 2001. Switching life-history sensitivities to stress in soil invertebrates. Ecol Appl 11:226–238.

Kammenga JE, Busschers M, VanStraalen NM, Jepson PC, Bakker J. 1996. Stress induced fitness reduction is not determined by the most sensitive life-cycle trait. Funct Ecol 10:106–111.

Kangas A, Kurki S. 2000. Predicting the future of capercaillie (*Tetrao urogallus*) in Finland. Ecol Model 134:73–87.

Kanji GK. 2002. Statistical analysis of data from marked bird populations. J Appl Stat. 29:669.

Kapustka LA, Galbraith H, Luxon M, Yocum J, Adams B. 2004. Application of habitat suitability index values to modify exposure estimates in characterizing ecological risk. Landscape ecology and wildlife habitat evaluation. In: Kapustka H, Galbraith M, Luxon G, Biddinger R, editors. Landscape Ecology and Wildlife Habitat Evaluation: Critical information for ecological risk assessment, land-use management activities, and biodiversity enhancement practices. L.A. ASTM International. West Conshohocken (PA): American Society for Testing and Materials (ASTM). STP 1458.

Kaputska LA, Williams BA, Fairbrother A. 1996. Evaluating risk predictions at population and community levels in pesticide registration—hypotheses to be tested. Environ Toxicol Chem 15:427–431.

Kareiva P, Stark J, Wennergren U. 1996. Using demography theory, community ecology and spatial models to illuminate ecotoxicology. In: Baird DJ, Maltby L, Greig-Smith PW, Douben PET, editors. Ecotoxicology: ecological dimensions. London (UK): Chapman & Hall. p. 13–23.

Kaye TN, Pyke DA. 2003. The effect of stochastic technique on estimates of population viability from transition matrix models. Ecology 84:1464–1476.

Keller LF, Waller DM. 2002. Inbreeding effects in wild populations. Trends Ecol Evol. 17:230–241.

Kendall BE, Briggs CJ, Murdoch WW, Turchin P, Ellner SP, McCauley E, Nisbet RM, Wood SN. 1999. Why do populations cycle? A synthesis of statistical and mechanistic modeling approaches. Ecology 80:1789–1805.

Kendall RJ, Lacher TE. 1994. Wildlife toxicology and population modeling. Boca Raton (FL): Lewis Publishers.

Kendall WL, Pollock KH, Brownie C. 1995. A likelihood-based approach to capture-recapture estimation of demographic parameters under the robust design. Biometrics 51:293–308.

Kennedy JH, Ammann LP, Waller WT, Warren JE, Hosmer AJ, Cairns SH, Johnson PC, Graney RL. 1999. Using statistical power to optimize sensitivity of analysis of variance designs for microcosms and mesocosms. Environ Toxicol Chem 18:113–117.

Kersting, K. 1991. Microecosystem state and its response to the introduction of a pesticide. Verh Int Verein Limnol 23:1641–1646.

Kiesecker JM, Blaustein AR, LK Belden. 2001. Complex causes of amphibian population declines. Nature 410:681–684.

Kingsolver JG. 2001. The strength of phenotypic selection in natural populations. Am Nat 157:245–261.

Klerks PL. 2002. Adaptation, ecological impacts, and risk assessment: insights from research at Foundry Cove, Bayou Trepagnier, and Pass Fourchon. Hum Ecol Risk Assess 8:971–982.

Klerks PL, Moreau CJ. 2001. Heritability of resistance to individual contaminants and to contaminant mixtures in the sheepshead minnow (*Cyprinodon variegatus*). Environ Toxicol Chem 20:1746–1751.

Klerks PL, Weis J. 1987. Genetic adaptation to heavy metals in aquatic organism: a review. Environ Pollut 45:173–205.

Klok C, de Roos AM. 1996. Population level consequences of toxicological influences on individual growth and reproduction in *Lumbricus rubellus* (Lumbricidae, Oligochaeta). Ecotox Environ Safe 33:118–127.

Klüttgen B, Kuntz N, Ratte HT. 1996. Combined effects of 3,4-dichloroaniline and food concentration on life-table data of two related cladocerans, *Daphnia magna* and *Ceriodaphnia quadrangular*. Chemosphere 32:2015–2028.

Knight T. 2003. Floral density, pollen limitation, and reproductive success in *Trillium grandiflorum*. Oecology 137:557–563.

Koller G. 1999. Risk assessment and decision-making in business and industry: a practical guide. Boca Raton (FL): CRC Press.

Kolluru RV, Bartell SM, Pitblado RM, Stricoff RS. 1996. Risk assessment and management handbook for environmental, health, and safety professionals. New York (NY): McGraw-Hill.

Kondoh M. 2003. High reproductive rates result in high predation risks: a mechanism promoting the coexistence of competing prey in spatially structured populations. Am Nat 161:299–309.

Kooijman B, Bedaux J. 2000. Dynamic effects of compounds on animal energetics and their population consequences. In: Kammenga J, Laskowski R, editors. Demography in ecotoxicology. Chichester (UK): Wiley.

Kooijman SALM, Metz JAJ. 1984. On the dynamics of chemically stressed populations: the deduction of population consequences from effects on individuals. Ecotox Environ Safe 8:254–274.

Kopp RI, Guttman SI, Wissing TE. 1992. Genetic indicators of environmental stress in central mudminnow (*Umbra limi*) populations exposed to acid deposition in the Adirondack Mountains. Environ Toxicol Chem. 11:665–676.

Kovatch CE, Schizas NV, Chandle GT, Coull BC, Quattro JM. 2000. Tolerance and genetic relatedness of three meiobenthic copepod populations exposed to sediment-associated contaminant mixtures: role of environmental history. Environ Toxicol Chem 19:912–919.

Koztowski J. 1996. Optimal allocation of resources explains interspecific life-history patterns in animals with indeterminate growth. Proc Royal Soc London B 263:559–566.

Krebs RA, Feder ME, Lee J. 1998. Heritability of expression of the 70KD heat-shock protein in *Drosophila melanogaster* and its relevance to the evolution of thermotolerance. Evolution 52:841–847.

Krebs CJ. 2000. Ecological methodology. Longman House, Burnt Mill (UK).

Kriz C, Delorme P, Hodge V, Mulye H, Sebastien R, Hart C, Takacs P, François D, Kaminski D, MacQuarrie T. 2003. Assessment endpoints for ecological risk assessments of pesticides: a scientific perspective. In: Hedley K, Roe S, Niimi AJ, editors. Proceedings of the 30th Aquatic Toxicity Workshop, Ottawa, Ontario. 28 September–1 Oct 2003. Can Tech Rep Fish Aquat Sci 2510.

Kuhn A, Munns WR Jr, Poucher S, Champlin D, Lussier S. 2000. Prediction of population-level response from mysid toxicity test data using population modeling techniques. Environ Toxicol Chem 19:2364–2371.

Kuhn A, Munns WR Jr, Serbst JR, Edwards P, Cantwell MG, Gleason T, Pelletier M, Berry W. 2002. Evaluating the ecological significance of laboratory response data to predict population-level effects for the estuarine amphipod *Ampelisca abdita*. Environ Toxicol Chem 21:865–874.

Kuhn A, Munns WR Jr, Champlin D, McKinney R, Tagliabue M, Serbst J, Gleason T. 2001. Evaluation of the efficacy of extrapolation population modeling to predict the dynamics of *Americamysis bahia* populations in the laboratory. Environ Toxicol Chem 20:213–221.

Lacy RC. 1993. VORTEX: a computer simulation model for population viability analysis. Wildlife Res 20:45–65.

Lande R. 1988. Genetics and demography in biological conservation. Science 241:1455–1460.

Lande R. 1993. Risks of population extinction from demographic and environmental stochasticity and random catastrophes. Am Nat 142:911–927.

Lande R. 1994. Risk of population extinction from new deleterious mutations. Evolution 48:1460–1469.

Lande R. 1995. Mutation and conservation. Conserv Biol 9:782–791.

Lande R. 2002. Incorporating stochasticity in population viability analysis. In: Beissinger SR, McCullough DR, editors. Population viability analysis. Chicago (IL): The University of Chicago Press. p. 18–40.

Lande, R, Orzack SH. 1988. Extinction dynamics of age-structured populations in a fluctuating environment. Proc Nat Acad Sci USA 85:7418–7421.

Lande R, Engen S, Sæther B-E. 1998. Extinction times in finite metapopulation models with stochastic local dynamics. Oikos 83:383–389.

Landis WG. 2002. Uncertainty in the extrapolation from individual effects to impacts upon landscapes. Hum Ecol Risk Assess 8:193–204.

Landis WG. 2000. The pressing need for population-level risk assessment. SETAC Globe 1:44–45.
Landis WG, Lenart LA, Spromberg JA. 2000. Patch dynamics of horizontal gene transfer with application to the ecological risk assessment of genetically engineered organisms. Hum Ecol Risk Assess 6:875–899.
Landis WG, McLaughlin JF. 2000. Design criteria and derivation of indicators for ecological position, direction and risk. Environ Toxicol Chem 19:1059–1065.
Landis WG, Duncan PB, Hart Hayes E, Markiewicz AJ. 2004. A regional retrospective assessment of the potential stressors causing the decline of the Cherry Point Pacific herring run and alternative management endpoints for the Cherry Point Reserve. Hum Ecol Risk Assess 10:271–297.
Landis WG, Hart Hayes E, Markiewicz AM. 2003. Weight of evidence and path analysis applied to the identification of causes of the cherry point Pacific Herring Decline. In: Droscher T, Fraser DA, editors. Georgia Basin/Puget Sound Research Conference, 31 March–3 April 2003. Vancouver, British Columbia, Proceedings December 2003. Available from: http://www.psat.wa.gov/Publications/03_proceedings/PAPERS/ORAL/10f_landis.pdf. Accessed January 12, 2007.
Langton SD, Aebischer NJ, Robertson PA. 2002. The estimation of density dependence using census data from several sites. Oecology 133:466–473.
Laskowski R. 2000. Stochastic and density-dependent models in ecotoxicology. In Kammenga J, Laskowski R, editors. Demography in ecotoxicology. New York (NY): John Wiley and Sons. p. 57–71.
Lavie B, Nevo E. 1982. Heavy metal selection of phosphoglucose isomerase allozymes in marine gastropods. Mar Biol 71:17–22.
Lebreton JD, Burnham KP, Clobert J, Anderson DR. 1992. Modeling survival and testing biological hypotheses using marked animals: a unified approach with case studies. Ecol Monogr 62:67–118.
Lebreton JD, Colbert J. 1986. Users manual for program SURGE Version 2.0. CEFE, CNRS. Montpelier, France.
Lebreton JD, Hines JE, Pradel R, Nichols JD, Spendelow JA. 2003. Estimation by capture-recapture of recruitment and dispersal over several sites. Oikos 101:253–264.
Leffler JW. 1981. Aquatic microcosms and stress criteria for assessing environmental impact of organic chemicals. Subcontract No. T6411 (7197) 025. Report. Washington (DC): US Environmental Protection Agency, Office of Pesticides and Toxic Substances.
Lefkovitch LP. 1965. The study of population growth in organisms grouped by stages. Biometrics 21:1–18.
Legendre P, Legendre L. 1998. Numerical ecology. Amsterdam (The Netherlands): Elsevier Science BV.
Leslie PH. 1945. On the use of matrices in certain population mathematics. Biometrika 33(3):183–212.
Leslie PH. 1948. Some further notes on the use of matrices in population dynamics. Biometrika 35:213–245.
Levin L, Caswell H, Bridges T, DiBacco C, Cabrera D, Plaia G. 1996. Demographic responses of estuarine polychaetes to pollutants: life table response experiments. Ecol Appl 6:1295–1313.
Levins R. 1966. The strategy of model building in population biology. Am Sci 54:421–431.
Levins R. 1969. Some demographic and genetic consequences of environmental heterogeneity for biological control. Bull Entomol Soc Am 15:237–240.
Levins R. 1970. Extinction. In: Gerstenhaber M, editor. Some mathematical models in biology. Providence (RI): American Mathematical Society. p. 75–107.

Levinton JS, Klerks P, Martinez DE, Montero C, Sturmbauer C, Suatoni W, Wallace W. 1999. Running the gauntlet: pollution, evolution and reclamation of an estuarine bay and its significance in understanding population biology of toxicology and food web transfer. In: Whitfield M, editor. Aquatic life cycles strategies. Plymouth (UK): Plymouth Marine Institute of Biology.

Lewis EG. 1942. On the generation and growth of a population. Sankhya 6:93–96.

Liess M. 2002. Population response to toxicants is altered by intraspecific interaction. Environ Toxicol Chem 21:138–142.

Lindenmeyer DB. 1999. Future directions for biodiversity conservation in managed forests: indicator species, impact studies and monitoring programs. Forest Ecol Manage 115:277–287.

Lindenmayer DB, Burgman MA, Akçakaya HR, Lacy RC, Possingham HP. 1995. A review of the generic computer programs ALEX, RAMAS/space, and VORTEX for modeling the viability of wildlife populations. Ecol Model 82:161–174.

Lindenmayer DB, Possingham HP. 1996a. Modelling the inter-relationships between habitat patchiness, dispersal capability and metapopulation persistence of the endangered species, Leadbeater's possum, in south-eastern Australia. Landscape Ecol 11:79–105.

Lindenmayer DB, Possignham HP. 1996b. Ranking conservation and timber management options for Leadbeater's possum in southeastern Australia using population viability analysis. Conserv Biol 10:235–251.

Lingjærde OC, Stenseth NC, Kristoffersen AB, Smith RH, Moe SJ, Read JM, Daniels S, et al. 2001. Exploring the density-dependent structure of blowfly populations by nonparametric additive modeling. Ecology 82:2645–2658.

Link WA, Cam E, Nichols JD, Cooch EG. 2002. Of bugs and birds: Markov chain Monte Carlo for hierarchical modeling in wildlife research. J Wildl Manage 66:277–291.

Linke-Gamenick I, Forbes VE, Sibly RM. 1999. Density-dependent effects of a toxicant on life-history traits and population dynamics of a capitellid polychaete. Mar Ecol Prog Ser 184:139–148.

Linkov I, Burmistrov D, Cura J, Bridges TS. 2002. Risk-based management of contaminated sediments: consideration of spatial and temporal patterns in exposure modeling. Environ Sci Technol 36:238–246.

Lomnicki A. 1988. Population ecology of individuals. Princeton (NJ): Princeton University Press.

Long ER, MacDonald DD, Smith SL, Calder FD. 1995. Incidence of adverse biological effects within ranges of chemical concentrations in marine and estuarine sediments. Environ Manage 19:81–97.

Longley P, Batty M, editors. 1996. Spatial analysis: modeling in a GIS environment. New York (NY): John Wiley & Sons.

Lotka AJ. 1925. Elements of physical biology. Baltimore (MD): Williams and Wilkins Co.

Lubick N. 2003. New count of old whales adds up to big debate. Science 301:451.

Luck GW, Daily GC, Ehrlich PR. 2003. Population diversity and ecosystem services. Trends Ecol Evol 18:331–336.

Lynch M, Conery J, Burger, R. 1995. Mutation accumulation and the extinction of small populations. Am Nat 146:489–518.

MacArthur RH, Wilson EO. 1967. The theory of island biogeography. Princeton (NJ): Princeton University Press.

Macnair MR. 1991. Why evolution of resistance to anthropogenic toxins normally involves major gene changes. Genetica 84:213–219.

Macovsky LM. 1998. The effects of toxicant related mortality upon metapopulation dynamics: a laboratory model. Masters of science thesis. Bellingham (WA): Western Washington University.

Madenjian CP, Carpenter SR. 1991. Individual-based model for growth of young-of-the-year walleye: a piece of the recruitment puzzle. Ecol Appl 1:268–279.

Maelzer DA. 1970. The regression of log N_{n+1} on log N_n as a test of density dependence: an exercise with computer-constructed density-independent populations. Ecology 51:810–820.

Maltby L, Kedwards TJ, Forbes VE, Grasman K, Kammenga JE, Munns WR, Ringwood AH, Weis JS, Wood SN. 2001. Linking individual-level responses and population-level consequences. In: Baird DJ, Burton GA, editors. Ecological variability: separating natural from anthropogenic causes of ecosystem impairment. Pensacola (FL): SETAC. p. 27–82.

Malthus TR. 1798. An essay on the principle of population as it affects the future improvement of society. London (UK): Johnson.

Manel S, Schwartz MK, Luikart G, Taberlet P. 2003. Landscape genetics: combining landscape ecology and population genetics. Trends Ecol Evol 18:189–197.

Manly BFJ. 1997a. A method for the estimation of parameters for natural stage-structured populations. Res Popul Ecol 39:101–111.

Manly BFJ. 1997b. Randomization, bootstrap and Monte Carlo methods in biology, 2nd ed. London (UK): Chapman and Hall.

Mann CC, Plummer ML. 1999. A species fate, by the numbers. Science 284:36–37.

Marshall EA, Crowder LB. 1996. Assessing population responses to multiple anthropogenic effects: a case study with brook trout. Ecol Appl 6:152–167.

Marshall JS. 1962. The effects of continuous gamma radiation on the intrinsic rate of natural increase of *Daphnia pulex*. Ecology 43:598–607.

Marshall JS. 1978. Population dynamics of *Daphnia galeata mendotae* as modified by chronic cadmium stress. J Fish Res Board Can 35:461–469.

Martin TG, Kuhnert PM, Mengersen K, Possingham HP. 2005. The power of expert opinion in ecological models using Bayesian methods: impact of grazing on birds. Ecol Appl 15:266–280.

Massachusetts Department of Environmental Protection [MADER]. 1996. Guidance for disposal site risk characterization in support of the Massachusetts contingency plan. Chapter 9. Method 3 environmental risk characterization. Boston (MA): MADEP, Bureau of Waste Site Cleanup and Office of Research and Standards. BWSC/ORS-95-141.

Maund SJ, Taylor EJ, Pascoe D. 1992. Population responses of the freshwater amphipod crustacean *Gammarus pulex* (L.) to copper. Freshwater Biol 28:29–36.

Maurer BA, Holt RD. 1996. Effects of chronic pesticide stress on wildlife populations in complex landscapes: processes at multiple scales. Environ Toxicol Chem 15:420–426.

May RM. 1974. Stability and complexity in model ecosystems. Princeton (NJ): Princeton University Press.

May RM. 1975. Deterministic models with chaotic dynamics. Nature 256:165–166.

Mayfield H. 1961. Nesting success calculated from exposure. Wilson Bull 73:255–26.

Mayr E. 1970. Populations, species and evolution, an abridgement of animal species and evolution. Cambridge (MA): Harvard University Press.

McCallum H. 2000. Population parameters: estimation for ecological models. Oxford (UK): Blackwell Science Ltd.

McCarty LS, Mackay D. 1993. Enhancing ecotoxicological modeling and assessment. Environ Sci Technol 27:1719–1728.

McCauley E, Murdoch WW, Nisbet RM, Gurney WSC. 1990. The physiological ecology of *Daphnia*: development of a model of growth and reproduction. Ecology 71:703–715.

McCauley E, Nisbet RM, DeRoos AM, Murdoch WW, Gurney WSC. 1996. Structured population models of herbivorous zooplankton. Ecol Monogr 66:479–501.

McClung G, Sayre PG. 1994. Ecological risk assessment case study: risk assessment for the release of recombinant *Rhizobia* at a small-scale agricultural field site. In: A review of ecological assessment case studies from a risk assessment perspective. Washington (DC): US Environmental Protection Agency. Volume II. EPA-630-R-94-003.

McCullagh P, Nelder JA. 1989. Generalized linear models, 2nd ed. London (UK): Chapman & Hall.

McDonald TL, McDonald LL. 2002. A new ecological risk assessment procedure using resource selection models and geographic information systems. Wildlife Soc B 30:1015–1021.

McElhany P, Ruckelshaus MH, Ford MJ, Wainwright TC, Bjorkstedt EP. 2000. Viable salmonid populations and the recovery of evolutionarily significant units. US Department of Commerce, NOAA technical memo. Washington (DC): NMFS-NWFSC-42, 156.

McGee BL, Fisher DJ, Yonkos LT, Ziegler GP, Turley S. 1999. Assessment of sediment contamination, acute toxicity and population viability of the estuarine amphipod *Leptocheirus Plumulosus* in Baltimore Harbor, Maryland, USA. Environ Toxicol Chem 18:2151–60.

McKenzie JA, Batterham P. 1998. Predicting insecticide resistance: mutagenesis, selection and response. Philos Trans R Soc London Ser B 353:1729–1734.

McLaughlin JF, Hamilton JJ, Boggs CL, Ehrlich PR. 2002. Climate change hastens population extinctions. PNAS 99:6070–6074.

McLaughlin JF, Landis WG. 2000. Effects of environmental contaminants in spatially structured environments. Environmental contaminants in terrestrial vertebrates: effects on populations, communities, and ecosystems. In: Albers PH, editor. Pensacola (FL): SETAC.

Menzie CA, Henning MH, Cura J, Finkelstein K, Gentile J, Maughan J, Mitchell D, Petron S, Potocki B, Svirsky S, Tyler P. 1996. Special report of the Massachusetts weight-of-evidence workgroup: a weight of evidence approach for evaluating ecological risks. Hum Ecol Risk Assess 2:277–304.

Menzie CA, Wickwire WT. 2001. Defining populations: a key step in identifying spatial and temporal scales. Toxicol Industrial Health 17:223–229.

Merilä J, Crnokrak P. 2001. Comparison of genetic differentiation at marker loci and quantitative traits. J Evolution Biol 14:892–903.

Mertz DB. 1972. The *Tribolium* model and the mathematics of population growth. Ann Rev Ecol Syst 3:51–78.

Metz JAJ, Diekmann O. 1986. Dynamics of physiologically structured populations. New York (NY): Springer-Verlag.

Meyer R, Millar RB. 1999. Bayesian stock assessment using a state-space implementation of the delay difference model. Can J Fish Aquat Sci 56:37–52.

Meyer JN, DiGiuliuo RT. 2003. Heritable adaptation and fitness costs in killifish (*Fundulus beteroclitus*) inhabiting a polluted estuary. Ecol Appl 13:490–503.

Michod RE, Anderson WW. 1980. On calculating demographic parameters from age frequency data. Ecology 61:265–269.

Miller RR. 1972. Threatened freshwater fishes of the United States. Trans Amer Fish Soc 1972:239–252.

Mills LS, Doak DF, Wisdom MJ. 1999. Reliability of conservation actions based on elasticity analysis of matrix models. Conserv Biol 13:815–829.

Mills LS, Hayes SG, Baldwin C, Wisdom MJ, Citta J, Mattson DJ, Murphy K. 1996. Factors leading to different viability predictions for a grizzly bear data set. Conserv Biol 10:863–873.

Mills LS, Smouse PE. 1994. Demographic consequences of inbreeding in remnant populations. Am Nat 144:412–431.

Mills NJ. 2001. Factors influencing top-down control of insect pest populations in biological control systems. Basic Appl Ecol 2:323–332.

Mitton JB. 1997. Selection in natural populations. Oxford (UK): University Press.

Mladenoff DJ, Host E, Boeder J, Crow TR. 1996. LANDIS: a spatial simulation model of forest landscape disturbance, management and succession. In: Second international conference on integrating GIS and environmental modeling. Santa Barbara (CA): NCGIA. p. 175–179.

Moe SJ, Stenseth NC, Smith RH. 2001. Effects of a toxicant on population growth rates: sublethal and delayed responses in blowfly populations. Funct Ecol 15:712–721.

Moe SJ, Stenseth NC, Smith RH. 2002a. Density dependence in blowfly populations: experimental evaluation of non-parametric time-series modeling. Oikos 98:523–533.

Moe SJ, Stenseth NC, Smith RH. 2002b. Density-dependent compensations in blowfly populations give indirectly positive effects of a toxicant. Ecology 83:1597–1603.

Moe SJ, Kristoffersen AB, Smith RH, Stenseth NC. 2005. From patterns to processes and back: analysing density-dependent responses to an abiotic stressor by statistical and mechanistic modelling. Proc R Soc London Series B 272: 2133–2142.

Mooij WM, Bennetts RE, Kitchens WM, DeAngelis DL. 2002. Exploring the effect of drought extent and interval on the Florida snail kite: interplay between spatial and temporal scales. Ecol Model 149:25–39.

Moore CG, Stevenson JM. 1991. The occurrence of intersexuality in harpacticoid copepods and its relationship to pollution. Mar Pollut Bull 22:72–74.

Morgan MG, Henrion M. 1990. Uncertainty: a guide to dealing with uncertainty in quantitative risk and policy analysis. Cambridge (NY): Cambridge University Press.

Morris RF. 1959. Single-factor analysis in population dynamics. Ecology 40:580–588.

Morris WF, Doak DF. 2002. Quantitative conservation biology: theory and practice of population viability analysis. Sunderland, (MA): Sinauer Associates, Inc.

Morrow JE. 1980. The freshwater fishes of Alaska. Anchorage (AK): Alaska Northwest Publishing Co.

Mullet E, Ciutad N, Riviere-Shafighi S. 2004. Cognitive processes involved in the assessment of health hazards' severity. Health Risk Soc 6:277–288.

Munkittrick KR and LS McCarty. 1995. An integrated approach to ecosystem health management: top-down, bottom-up or middle-out? J. Aquatic Ecosystem Health 4:77–90.

Munkittrick KR and ME McMaster. 2000. Assessment of multiple stressors in aquatic ecosystems by directed assessment of cumulative effects using fish populations. p. 27–65 In Ferenc, SA and JA Foran (Eds.) *Multiple Stressors in Ecological Risk and Impact Assessment: Approach to Risk Estimation.* SETAC Press, Pensacola, FL.

Munns WR Jr. 2002. Axes of extrapolation in risk assessment. Hum Ecol Risk Assess 8:19–29

Munns WR Jr, Walker HA, Paul JF, Gentile JH. 1996. Ocean disposal at the 106-mile deepwater dumpsite as an ecological risk assessment case study. J Mar Environ Eng 3:279–297.

Munns WR Jr, Black DE, Gleason TR, Salomon K, Bengtson D, Gutjahr-Gobell R. 1997. Evaluation of the effects of dioxin and PCBs on *Fundulus heteroclitus* populations using a modeling approach. Environ Toxicol Chem 16:1074–1081.

Munns WR, Kroes F, Veith G, Suter II GW, Damstra T, Waters MD. 2003. Approaches for integrated risk assessment. Hum Ecol Risk Assess 9:267–272.

Munns WR Jr, Mitro MG. 2004. Assessing risks to populations at Superfund sites—characterizing effects on populations. Cincinnati (OH): US Environmental Protection Agency, Office of Research and Development.

Murdoch WW. 1994. Population regulation in theory and practice. Ecology 75:271–287.

Murray BG. 1999. Can the population regulation controversy be buried and forgotten? Oikos 84:148–152.

Nacci D, Coiro L, Champlin D, Jayaraman S, McKinney R, Gleason TR, Munns WR Jr, Specker JL, Cooper KR. 1999. Adaptations of wild populations of the estuarine fish *Fundulus heteroclitus* to persistent environmental contaminants. Mar Biol 134:9–17.

Nacci D, Coiro L, Champlin D, Jayaraman S, McKinney R. 2002a. Predicting the occurrence of genetic adaptation to dioxin-like compounds in populations of the estuarine fish *Fundulus heteroclitus*. Environ Toxicol Chem 21:1525–1532.

Nacci D, Gleason T, Gutjahr-Gobell R, Huber M, Munns WR Jr. 2002b. Effects of environmental stressors on wildlife populations: a modeling approach and case study. In: Newman MC, Roberts MH, Hale RC, editors. Coastal and estuarine risk assessment: risk on the edge. New York (NY): CRC Press Lewis Publishers. p. 247–272.

Nachman G. 2001. Predator-prey interactions in a nonequilibrium context: the metapopulation approach to modeling "hide-and-seek" dynamics in a spatially explicit tri-trophic system. Oikos 94:72–88.

Nakamaru M, Iwasa Y, Nakanishi J. 2002. Extinction risk of DDT to herring gull populations from DDT exposure. Environ Toxicol Chem 21:195–202.

Nakamaru M, Iwasa Y, Nakanishi J. 2003. Extinction risk to bird populations caused by DDT exposure. Chemosphere 53:377–387.

National Research Council. 1996. Understanding risk: informing decisions in a democratic society. Committee on Risk Characterization. Washington (DC): National Academy Press.

Nevo E, Noy R, Lavie B, Bieles A, Muchtar S. 1986. Genetic diversity and resistance to marine pollution. Biol J Linn Soc 29:139–144.

Newman D, Pilson D. 1997. Increased probablility of extinction due to decreased genetic population size: experimental populations of *Clarkia pulchella*. Evolution 51: 354–362.

Newman MC. 1998. Fundamentals of ecotoxicology. Chelsea (MI): Ann Arbor Press.

Newman MC, Jagoe R. 1998. Allozymes reflect the population-level effect of mercury: simulations of the mosquitofish (*Gambusia holbrooki* Girard) GPI-2 response. Ecotoxicology 7:141–150.

Nicholopoulos J. 1999. The endangered species-listing program. Endangered Species Bulletin 14:6–9.

Nichols JD. 1992. Capture recapture models: using marked animals to study population dynamics. Bioscience 42:94–102.

Nichols JD, Hines JE. 2002. Approaches for the direct estimation of lambda, and demographic contributions to lambda, using capture-recapture data. J Appl Stat 29:539–568.

Nicholson AJ. 1950. Population oscillations caused by competition for food. Nature 165:476–477.

Nicholson AJ. 1954. Compensatory reactions of populations to stresses, and their evolutionary significance. Aust J Zool 2:1–8.

Nisbet RM, Gurney WSC. 1982. Modelling fluctuating populations. Chichester (NY): Wiley Interscience.

Nisbet RM, Gurney WSC, Murdoch WW, McCauley E. 1989. Structured population models: a tool for linking effects at individual and population level. Biol J Linn Soc 37:79–99.

Noon BR. 2003. Conceptual issues in monitoring ecological resources. In: Busch DE, Trexler JC, editors. Monitoring ecosystems: interdisciplinary approaches for evaluating ecoregional initiatives. Washington (DC): Island Press. p. 27–72.

Noon BR, McKelvey KS. 1992. Stability properties of the spotted owl metapopulation in southern California. In: Verner J, McKelvey KS, Noon BR, Gutierrez RJ, Gould GI, Beck TW, editors. The California spotted owl: a technical assessment of its current status. Albany (CA): Pacific Southwest Research Station, USDA Forest Service General Technical Report PSW-GTR-133.

O'Brien SJ. 2000. Adaptive cycles: parasites selectively reduce inbreeding in Soay sheep. Trends Ecol Evol 15:7–9.

Odderskær P, Prang A, Poulsen JG, Elmegaard N, Andersen PN. 1997. Skylark reproduction in pesticide treated and untreated fields. Pesticide Research No 32. Denmark: Ministry of Environment and Energy.

[OECD] Organisation for Economic Co-operation and Development. 2006. Guidance document on simulated freshwater lentic field tests (outdoor microcosms and mesocosms). Report, Environment Directorate Test Guidelines Program, Paris, France.

Okubo A. 1980. Diffusion and ecological problems: mathematical models. Berlin (Germany): Springer-Verlag.

O'Neill RV, Gardner RH, Barnthouse LW, Suter GW, Hildebrand SG, Gehrs CW. 1982. Ecosystem risk analysis: a new methodology. Environ Toxicol Chem 1:167–177.

O'Neill RV, Rosen AE. 1986. User's manual for ecological risk assessment. Washington (DC): U.S. Environmental Protection Agency, Office of Research and Development; and Oak Ridge National Laboratory, Oak Ridge TN, Environmental Sciences Division Publication No. 2679.

[ORDEQ] Oregon Department of Environmental Quality. 2001. Guidance for ecological risk assessment: Levels I, II, III, IV. Portland (OR): ODEQ, Waste Management and Cleanup Division, Cleanup Policy and Program Development Section.

Orr R, McClung G, Peoples R, Williams JD, Meyer MA. 1999. Nonidigenous species. Ecological risk assessment in the federal government. Committee on Environmental and Natural Resources, Washington (DC): CENR/5–99/001.

Palumbi S. 2001. Humans as the world's greatest evolutionary force. Science 293:1786–1790.

Park T. 1937. Experimental studies of insect populations. Am Nat 71:21–33.

Parmar MKB, Machin D. Survival analysis: a practical approach. Chichester (UK): Wiley.

Pascual MA, Kareiva P, Hilborn R. 1997. The influence of model structure on conclusions about the viability and harvesting of Serengeti wildebeest. Conserv Biol 11:966–976.

Pastorok RA, Akcakaya HR, Regan H, Ferson SM, Bartell SM. 2003. Role of ecological modeling in risk assessment. Hum Ecol Risk Assess 9:939–972.

Pastorok RA, Bartell SM, Ferson S, Ginzburg LR. 2002. Ecological modeling in risk assessment: chemical effects on populations, ecosystems, and landscapes. Boca Raton (FL): CRC Press.

Paulsen CM, Fisher TR. 2003. Detecting juvenile survival effects of habitat actions: power analysis applied to endangered Snake River spring-summer chinook (*Oncorhynchus tshawytscha*). Can J Fish Aquat Sci 60:1122–1132.

Pearl R, Reed LJ. 1920. On the rate of growth of the population of the United States since 1790 and its mathematical representation. Proc Natl Acad Sci USA 6:275–288.

Pedersen HC, Steen H, Kastdalen L, Broseth H, Ims RA, Svendsen W, Yoccoz NG. 2003. Weak compensation of harvest despite strong density-dependent growth in willow ptarmigan. Proc R Soc London Ser B 271:381–385.

Pennsylvania Department of Environmental Protection. 1998. Ecological screening process. [PADEP]. Available from: www.state.pa.us/dep/subject/advcoun/cleanup/attachmentve3.doc. Accessed 10 November 2006.

Peterman RM. 1978. Testing for density-dependent marine survival in Pacific salmonids. J Fish Res Board Can 35:1434–1450.

Peters RH. 1983. The Ecological Implications of Body Size. Cambridge University Press, Cambridge.

Petersen RC, Petersen LBM. 1988. Compensatory mortality in aquatic populations—its importance for interpretation of toxicant effects. Ambio 17:381–386.

Peterson CH, Black R. 1988. Density-dependent mortality caused by physical stress interacting with biotic history. Am Nat 131:257–270.

Pidgeon N. 1998. Risk assessment, risk values and the social science programme: why we do need risk perception research. Reliab Eng Syst Safe 59:5–15.

Pielou EC. 1977. Mathematical ecology. New York (NY): John Wiley and Sons.

Platt JR. 1964. Strong inference. Science 146:1–7.

Plotnick RE, Gardner RH. 2002. A general model for simulating the effects of landscape heterogeneity and disturbance on community patterns. Ecol Model 147:171–197.

Pollard E, Lakhani KH, Rothery P. 1987. The detection of density-dependence form a series of annual censuses. Ecology 68:2046–2055.

Pollock KH, Nichols JD, Brownie C, Hines JE. 1990. Statistical inference for capture-recapture experiments. Wildlife Monogr 107:1–97.

Poole RW. 1974. An introduction to quantitative ecology. New York (NY): McGraw-Hill.

Possingham HP, Davies I. 1995. ALEX: a model for the viability analysis of spatially structured populations. Biol Conserv 73:143–150.

Possingham HP, Lindermayer DB, McCarthy MA. 2001. Population viability analysis. Encyclopedia Biodivers 4:831–843.

Posthuma L, Hogervorst RF, Joosse ENG, Van Straalen NM. 1993. Genetic variation and covariation for characteristics associated with cadmium tolerance in natural populations of the springtail *Orchesella cincta* (L.). Evolution 47:619–631.

Postma JF, Buckert-de Jong MC, Staats N, Davids C. 1994. Chronic toxicity of cadmium to *Chironomus riparius* (Diptera: Chironomidae) at different food levels. Arch Environ Contam Toxicol 26:143–148.

Postma JF, Van Kleunen A, Admiraal W. 1995. Alterations in life-history traits of *Cironomus riparius* (Diptera) obtained from metal contaminated rivers. Arch Environ Contam Toxicol 29:469–475.

Pradel R and Lebreton JD. 1991. Users manual for program SURGE Version 4.1. CEFE, CNRS. Montpelier, France.

Pradel R. 1996. Utilization of capture-mark-recapture for the study of recruitment and population growth rate. Biometrics 52:703–709.

Primack RB. 1998. Essentials of conservation biology, 2nd edition. Sunderland (MA): Sinauer Associates, Inc.

Punt AE, Hilborn R. 1997. Fisheries stock assessment and decision analysis: the Bayesian approach. Rev Fish Biol Fisher 7:35–63.

Purse BV, Hopkins GW, Day KJ, Thompson DJ. 2003. Dispersal characteristics and management of a rare damselfly. J Appl Ecol 40:716–728.

Purves WK, GH Orians. 1987. Life: the science of biology. Sunderland (MA): Sinauer Associates, Inc.

Quinn TJ II, Deriso RB. 1999. Quantitative fish dynamics. Biological Resource Management Series. New York (NY): Oxford University Press.

Ralph CJ, Sauer JR, Droege S. 1995. Monitoring bird populations by point counts. Albany (CA): USDA Forest Service Pacific Southwest Research Station. General Technical Report PSW-GTR-149.

Ray C, Hastings A. 1996. Density dependence: are we searching at the wrong spatial scale? J Anim Ecol 65:556–566.

Reed DH, Frankham R. 2001. How closely correlated are molecular and quantitative measures of genetic variation? A meta-analysis. Evolution 55:1095–1103.

Reed DH, Frankham R. 2003. Correlation between fitness and genetic diversity. Conserv Biol 17:230–237.

Regan HM, Akçakaya HR, Ferson S, Root KV, Carroll S, Ginzburg LR. 2003a. Treatments of uncertainty and variability in ecological risk assessment of single-species populations. Hum Ecol Risk Assess 9:889–906.

Regan HM, Auld TD, Keith DA, Burgman MA. 2003b. The effects of fire and predation on the long-term persistence of an endangered shrub, *Grevillea caleyi*. Biol Conserv 109:73–83.

Regan HM, Colyvan M, Burgman MA. 2002b. A taxonomy and treatment of uncertainty for ecology and conservation biology. Ecol Appl 12:618–628.

Regan HM, Hope BK, Ferson S. 2002a. Analysis and portrayal of uncertainty in a food web exposure model. Hum Ecol Risk Assess 8:1757–1777.

Regan TJ, Regan HM, Bomham K, Taylor RJ, Burgman MA. 2001. Modeling the impact of timber harvesting on a rare carnivorous land snail (*Tasmaphena lamproides*) in northwest Tasmania, Australia. Ecol Model 139:253–264.

Reiley MC, Stubblefield WA, Adams WJ, Di Toro DM, Hodson PV, Erickson RJ, Keating FJ, editors. 2003. Reevaluation of the state of the science for water-quality criteria development. Pensacola (FL): SETAC Press.

Rexstad EA, Burnham KP. 1991. Users guide for interactive program CAPTURE. Abundance estimation of closed animal populations. Fort Collins (CO): Colorado State University.

Ricker WE. 1954. Effects of compensatory mortality upon population balance. J Wildl Manage 18:45–51.

Ricker WE. 1954a. Stock and recruitment. Can J Fish Aquat Sci 11:559–623.

Ricker WE. 1975. Computation and interpretation of biological statistics of fish populations. Ottawa (Canada): Bulletin of the Fisheries Research Board of Canada, Bulletin 91:1–38.

Riechert SE, Hall RF. 2000. Local population success in heterogeneous habitats: reciprocal transplant experiments completed on a desert spider. J Evolut Biol 13:541–550.

Rivot E, Prevost E, Parent E. 2001. How robust are Bayesian posterior inferences based on a Ricker model with regards to measurement errors and prior assumptions about parameters? Can J Fish Aquat Sci 58:2284–2297.

Roberts SM. 1999. Practical issues in the use of probabilistic risk assessment. Hum Ecol Risk Assess 5:729–736.

Roff DA. 1984. The evolution of life history parameters in teleosts. Can J Fish Aquat Sci 41:989–1000.

Roff DA. 1997. Evolutionary quantitative genetics. London (UK): Chapman and Hall.

Roman J, Palumbi S. 2003. Whales before whaling in the North Atlantic. Science 301:508–510.

Rose KA, Cowan JH. 1993. Individual-based model of young-of-the-year striped bass population dynamics. I. Model description and baseline simulations. Trans Amer Fish Soc 122:415–438.

Rose KA, Cowan JH. 2003. Data, models, and decisions in US marine fisheries management: lessons for ecologists. Ann Rev Ecol Evol Syst 34:127–151.

Rose KA, Murphy CA, Diamond SL, Fuiman LA, Thomas P. 2003. Using nested models and laboratory data for predicting population effects of contaminants on fish: a step toward a bottom-up approach for establishing causality in field studies. Hum Ecol Risk Assess 9:231–257.

Rosenberg DK, McKelvey KS. 1999. Estimation of habitat selection for central-place foraging animals. J Wildl Manage 63:1028–1038.

Roush RT, McKenzie JA. 1987. Ecological genetics of insecticide and acaricide resistance. Ann Rev Entomol 32:361–380.

Rowe G, Beebee TJC. 2003. Population on the verge of a mutational meltdown? Fitness costs of genetic load for an amphibian in the wild. Evolution 57:177–181.

Ruckelshaus M, Hartway C, Kareiva P. 1997. Assessing the data requirements of spatially explicit dispersal models. Conserv Biol 11:1298–1306.

Sabo JL, Holmes EE, Kareiva P. 2004. Efficacy of simple viability models in ecological risk assessment: does density dependence matter? Ecology 85:328–341.

Sacchieri I, Kuussaari M, Kanakre M, Vikman P, Fortelius W, Hanski I. 1998. Inbreeding and extinction in a butterfly metapopulation. Nature 392:491–494.

Salice CJ, Miller TJ. 2003. Population-level responses to long-term cadmium exposure in two strains of the freshwater gastropod *Biomphalaria glabrata*: results from a life-table response experiment. Environ Toxicol Chem 22:678–688.

Samaniego FJ, Renau DM. 1994. Toward a reconciliation of the Bayesian and frequentist approaches to point estimation. J Am Stat Assoc 89:947–957.

Sample BE, Opresko DM, Suter II GW. Toxicological benchmarks for wildlife: 1996 revisions. Prepared by the Risk Assessment Program, Health Sciences Research Division, Oak Ridge National Laboratory for the US Department of Energy. ES/ER/TM-86/R3.

Sample BE, Arenal CA. 1999. Allometric models for interspecies extrapolation of wildlife toxicity data. B Environ Contam Tox 62:653–663.

Sample BE, Rose KA, Suter GW II. 1999. Estimation of population-level effects on wildlife based on individual-level exposures: influence of life history strategies. In: Albers P, Heinz G, Ohlendorf H, editors. Proceedings of symposium on environmental contaminants and terrestrial vertebrates: effects on populations, communities, and ecosystems. Pensacola (FL): SETAC Press.

Samson FB, Perez-Trejo F, Salwasser H, Ruggiero LF, Shaffer ML. 1985. On determining and managing minimum population size. Wildlife Soc B 14:425–433.

Sayyed AH, Wright DJ. 2001. Fitness costs and stability of resistance to *Bacillus thuringiensis* in a field population of the diamondback moth *Putella xylostella* L. Ecol Entomol 26:502–508.

Schaefer MB. 1954. Some aspects of the dynamics of populations important to the management of the commercial marine fisheries. Bull Inter-Am Trop Tuna Comm 1:25–56.

Schaefer, MB 1957. A study of the dynamics of the fishery for yellowfin tuna in the eastern tropical Pacific Ocean. Bull Inter-Am Trop Tuna Comm 2:245–268.

Scheiner SM, Gurevitch J, editors. 1993. Design and analysis of ecological experiments. New York (NY): Chapman & Hall.

Scheiner SM, Guretitch J, editors. 2001. Design and analysis of ecological experiments. London (UK): Chapman & Hall.

Schlueter MA, Guttman SI, Duan Y, Oris JT, Huang X, Burton GA. 2000. Effects of acute exposure to fluoranthene-contaminated sediment on the survival and genetic variability of fathead minnows (*Pimephales promelas*). Environ Toxicol Chem 19:1011–1018.

Schmidt K, Jensen K. 2000. Genetic structure and AFLP variation of remnant populations in the rare plant *Pedicularis palustris* and its relation to population size and reproductive components. Am J Biol 87:678–689.

Schmidt PS, Rand DM. 1999. Intertidal microhabitat and selection at *Mpi*: interlocus contrasts in the northern acorn barnacle, *Semibalanus balanoides*. Evolution 53:135–146.

Schoen DJ, Brownstein JS, White PA. 2002. Mutation load in natural populations of the sensitive fern *Onoclea sensibilis* exposed to soil mutagens. Ecol Appl 12:124–137.

Schumaker NH. 1998. A users guide to the PATCH model. Corvallis (OR): US Environmental Protection Agency. Environmental Research Laboratory. 600/R-98/135. Available online at: http://www.epa.gov/wed/pages/models.htm.

Schumaker N, Ernst T, White D, Baker J, Haggerty P. 2004. Projecting wildlife responses to alternative future landscapes in Oregon's Willamette Basin. Ecol Appl 14:381–401.

Schwarz CJ, Arnason AN. 1996. A general methodology for the analysis of capture-recapture experiments in open populations. Biometrics 52:860–873.

Schwarz CJ, Scweighert JF, Arnasion AN. 1993. Estimating migration rates using tag-recovery data. Biometrics 49:177–193.

[SSC] Scientific Steering Committee. 2003. Final report on the risk assessment for animal populations with emphasis on wildlife. Brussels (Belgium): European Commission Task Force.

Seber GAF 1982. The estimation of animal abundance and related parameters. New York (NY): MacMillan.

Seber GAF. 2001. Some new directions in estimating animal population parameters. J Agric Biol Environ Stat 6:140–151.

[SETAC] Society of Environmental Toxicology and Chemistry Europe. 1991. Guidance document on testing procedures for pesticides in freswhwater static mesocosms. Report. Society of Environmental Toxicology and Chemistry-Europe, Workshop at Monks Wood Experimental Station, Huntingdon, United Kingdom.

SETAC-RESOLVE. 1992. Proceedings of a Workshop on Aquatic Microcosms for Ecological Assessment of Pesticides (Wintergreen [VA], October 1991). SETAC Foundation for Environmental Education and the RESOLVE Program of the World Wildlife Fund.

Shaffer ML. 1987. Minimum viable populations: coping with uncertainty. In: Soule ME, editor. Viable populations for conservation. Cambridge, England (UK): Cambridge University Press. p. 69–86.

Shaffer, M. L. 1990. Population viability analysis. Conserv Biol 4: 39–40.

Shaffer ML, Samson FB. 1985. Population size and extinction: a note on determining critical population sizes. Am Nat 125:144–152.

Shaw AJ. 1999. The evolution of heavy metal tolerance in plants: adaptations, limits, and costs. In: Forbes VE, editor. Genetics and ecotoxicology, Philadelphia (PA): Taylor and Francis.

Shaw FH, Geyer CJ, Shaw RG. 2002. A comprehensive model of mutations affecting fitness and inferences for *Arabidopsis thaliana*. Evolution 56:453–463.

Shenk TM, Franklin AB, editors. 2001. Modeling in natural resource management: development, interpretation, and application. Washington (DC): Island Press.

Sherratt TN, Jepson PC. 1993. A metapopulation approach to modeling the long-term impact of pesticides on invertebrates. J Appl Ecol 30:696–705.

Sherratt TN, Roberts G, Williams P, Whitfield M, Briggs J, Shillabeer N, Maund SJ. 1999. A life-history approach to predicting the recovery of aquatic invertebrate populations after exposure to xenobiotic chemicals. Environ Toxicol Chem 18:2512–2518.

Shugart HH, Smith TM, Post WM. 1992. The potential for application of individual-based simulation models for assessing the effects of global change. Ann Rev Ecol Syst 23:15–38.

Sibly RM, Calow P. 1989. A life-cycle theory of responses to stress. Bio J Linn Soc 37:101–116.

Sibly RM. 1996. Effects of pollutants on individual life histories and population growth rates. In: Newman MC, Jagoe CH, editors. Ecotoxicology. A hierarchical treatment. Boca Raton (FL): Lewis Publishers. p. 197–223.

Sibly RM. 1999. Efficient experimental designs for studying stress and population density in animal populations. Ecol Appl 9:496–503.

Sibly RM, Hone J. 2002. Population growth rate and its determinants: an overview. Philos Trans R Soc Ser B 357:1153–1170.

Sibly RM, Smith RH. 1998. Identifying key factors using-contribution analysis. J Anim Ecol 67:17–24.

Sibly RM, Newton I, Walker CH. 2000a. Effects of dieldrin on population growth rates of sparrowhawks 1963–1986. J Appl Ecol 37:540–546.

Sibly RM, Williams TD, Jones MB. 2000b. How environmental stress affects density dependence and carrying capacity in a marine copepod. J Appl Ecol 37:388–397.

Sigler WF, Sigler JW. 1987. Fishes of the Great Basin. Reno (NV): University of Nevada Press.

Sih A, Jonasson BG, Luikart G. 2000. Habitat loss: ecological, evolutionary and genetic consequences. Trends Ecol Evol 15:132–134.

Sinclair ARE. 1996. Mammalian populations: fluctuation, regulation, life history theory and their implications for conservation. In: Floyd RB, Sheppard AW, De Barro PJ, editors. Frontiers of population ecology. Sydney (Australia): Commonwealth Scientific and Industrial Research Organization. p. 127–154.

Sjogren-Gulve P, Ray C. 1996. Using logistic regression to model metapopulation dynamics: large-scale forestry extirpates the pool frog. In: McCullough DR, editor. Metapopulations and wildlife conservation. Washington (DC): Island Press. p. 111–138.

Slovic P, Malmfors T, Mertz CK, Neil N, Purchase IFH. 1997. Evaluating chemical risks: results of a survey of the British Toxicology Society. Human Exp Toxicol 16:289–304.

Smith AT, Gilpin M. 1997. Spatially correlated dynamics in a pPika metapopulation. In: Hanski IA, Gilpin ME, editors. Metapopulation biology: ecology, genetics and evolution. New York (NY): Academic Press. p. 407–428.

Smith RH. 1973. The analysis of intra-generation change in animal populations. J Anim Ecol 42:611–622.

Smith RH, Daniels S, Simkiss K, Bell ED, Ellner S, Forrest B. 2000. Blowflies as a case study in non-linear population dynamics. In: Perry JN, Smith RH, Woiwod IP, Morse D, editors. Chaos in real data: the analysis of non-linear dynamics in short ecological time series. Dordrecht (The Netherlands): Kluwer Academic Publishers.

Smith RL. 1990. Ecology and field biology. 4th ed. New York (NY): Harper Collins Publishers, Inc.

Snell TW, Serra M. 2000. Using probability of extinction to evaluate the ecological significance of toxicant effects. Environ Toxicol Chem 19:2357–2363.

Spencer M, Ferson S. 1997. RAMAS ecotoxicology: ecological risk assessment for food chains and webs. Setauket (NY): Applied Biomathematics.

Sokal RR, Rohlf FJ. 1995. Biometry, 3rd ed. San Francisco (CA): Freeman.

Solow AR. 2001. Observation error and the detection of delayed density dependence. Ecology 82:3263–3264.

Sorensen M. 1994. Annotated reference compilation: conducting qualitative and quantitative ecological risk assessments at hazardous waste sites. Hum Ecol Risk Assess 1:497–513.

Sorensen M, Margolin J. 1998. Ecological risk assessment guidance and procedural documents: an annotated compilation and evaluation of reference materials. Hum Ecol Risk Assess 4:1085–1101.

Sorensen MT, Gala WR, Margolin JA. 2004. Chemical-specific risk characterization—selecting the right tool for the job. Hum Ecol Risk Assess 10:245–269.

Soule ME. 1987. Viable populations for conservation. Cambridge (UK): Cambridge University Press.

Southern HN. 1970. The natural control of a population of tawny owls (*Strix aluco*). J Zool 162:197–285.

Southwood TRE. 1987. Ecological methods. London (UK): Chapman and Hall.

Southwood TRE, Henderson PA. 2000. Ecological methods. Plymouth (UK): Blackwell Scientific Publishing.

Spencer M, Ferson S. 1997. RAMAS ecotoxicology: ecological risk assessment for structured populations, Setauket (NY): Applied Biomathematics.[Windows software and 81-page manual].

Spromberg JA, John BM, Landis WG. 1998. Metapopulation dynamics: indirect effects and multiple distinct outcomes in ecological risk assessment. Environ Toxicol Chem 17:1640–1649.

Spromberg JA, Birge WJ. 2005. Modeling the effects of chronic toxicity on fish populations: the influence of life-history strategies. Environ Toxicol Chem 24:1532–1540.

Spromberg JA, Meador JP. 2005. Relating results of chronic toxicity responses to population-level effects: modeling effects on wild Chinook salmon populations. Integr Environ Manage Assessment 1:9–21.

Stacey PB, Johnson VA, Taper MI. 1997. Migration within metapopulations: the impact upon local population dynamics. In: Hanski I, Gilipin ME, editors. Metapopulation biology: ecology, genetics, and evolution. San Diego (CA): Academic Press. p. 267–291.

Stahl RM. 2001. Risk management: ecological risk-based decision-making. Pensacola (FL): SETAC Press.

Starfield AM, Bleloch 1991. Building models for conservation and wildlife management. Edina (MN): Burgess International Group, Inc.

Starfield AM. 1997. A pragmatic approach to modeling for wildlife management. J Wildlife Manage 61:261–270.

Stark JD, Banken JAO. 1999. Importance of population structure at the time of toxicant exposure. Ecotox Environ Safe 42:282–287.

Stark JD, Tanigoshi L, Bounfour M, Antonelli A. 1997. Reproductive potential: its influence on the susceptibility of a species to pesticides. Ecotox Environ Safe 37:273–279.

Steidl RJ, Hayes JP, Schauber E. 1997. Statistical power analysis in wildlife research. J Wildlife Manage 61:270–279.

Stenseth NC, Falck W, Chan KS, Bjornstad ON, O'Donoghue M, Tong H, Boonstra R, Boutin S, Krebs CJ, Yoccoz NG. 1998. From patterns to processes: phase and density dependencies in the Canadian lynx cycle. Proc Natl Acad Sci USA 95:15430–15435.

Stenseth NC. 1999. Population cycles in voles and lemmings: density dependence and phase dependence in a stochastic world. Oikos 87:427–461.

Stenseth NC, Saitoh T, Yoccoz NG. 1998. Frontiers in population ecology of microtine rodents: a pluralistic approach to the study of population ecology. Res Popul Ecol 40:5–20.

Stephens PA, Sutherland WJ. 1999. Consequences of the effect for behavior, ecology and conservation. Trends Ecol Evol 14:401–405.

Stewart-Oaten A, Murdoch WW, Parker KR. 1986. Environmental impact assessment: "pseudoreplication" in time? Ecology 67:929–940.

Stockwell CA, Hendry AP, Kinnison MT. 2003. Contemporary evolution meets conservation biology. Trends Ecol Evolut 18:94–101.

Stokes TK, Gurney WSC, Nisbet RM, Blythe SP. 1988. Parameter evolution in a laboratory insect population. Theor Popul Biology 34:248–265.

Strand AE. 2002. METASIM 1.0: an individual-based environment for simulating population genetics of complex population dynamics. Mol Ecol Notes 2:373–376.

Sunnucks P. 2000. Efficient genetic markers for population biology. Trends Ecol Evol 15:199–203.

Suter GW II. 1993. Ecological risk assessment. Chelsea (MI): Lewis Publishers.

Suter GW II. 2007. Ecological risk assessment, 2nd ed. Boca Raton (FL): Lewis Publishers.

Suter GW II. 1996a. Abuse of hypothesis testing statistics in ecological risk assessment. Hum Ecol Risk Assess 2:331–349.

Suter GW II. 1996b. Risk characterization for ecological risk assessment of contaminated sites. ES/ER/TM-200. Oak Ridge (TN): Oak Ridge National Laboratory.

Suter GW II, Efroymson RA, Sample BE, Jones DS. 2000. Ecological risk assessment of contaminated sites. Boca Raton (FL): Lewis Publishers.

Suter GW II, Vermeire T, Munns WR Jr, Sekizawa J. 2003. Framework for the integration of health and ecological risk assessment. Hum Ecol Risk Assess 9:281–301.

Sutherland WJ. 2001. Sustainable exploitation: a review of principles and methods. Wildl Biol 7:131–140.

Sutherland WJ, Reynolds JD. 1998. Sustainable and unsustainable exploitation. In: Sutherland WJ, editor. Conservation science and action. Oxford (UK): Blackwell Science p. 90–115.

Sutton TM, Rose KA, Ney JJ. 2000. A model analysis of strategies for enhancing stocking success of landlocked striped bass populations. N Amer J Fish Manage 20:841–859.

Swindoll M, Stahl R, Ells S, editors. 2000. Natural remediation of environmental contaminants: its role in ecological risk assessment and management. Pensacola (FL): SETAC Press.

Tabor HK, Risch NJ, and Myers RM. 2002. Candidate-gene approaches for studying complex genetic traits: practical considerations. Nat Rev Genet 3:391–396.

Tanaka Y. 1998. Extinction probability and the ecological risk assessment. Proceedings of the first international workshop of risk evaluation and management of chemicals. Yokohama (Japan): Japan Science and Technology Corporation.

Tanaka Y, Nakanishi J. 2000. Mean extinction time of populations under toxicant stress and ecological risk assessment. Environ Toxicol Chem 19:2856–2862.

Tanhuanpaa M, Ruohomaki K, Uusipaikka E. 2001. High larval predation rate in non-outbreaking populations of a geometrid moth. Ecology 82:281–289.

Tannenbaum L. 2002. Terrestrial ERA: are we missing the forest for the trees? SETAC Globe 3:38–39.

Tannenbaum L. 2003. Can ecological receptors really be at risk? Hum Ecol Risk Assess 9:5–13.

Tannenbaum L, Johnson M, Bazar M. 2003. Application of the hazard quotient in remedial decisions: a comparison of human and ecological risk assessments. Hum Ecol Risk Assess 9:387–401.

Tatara CP, Mulvey M, Newman MC. 1999. Genetic and demographic responses of mosquitofish (*Gambusia holbrooki*) populations exposed to mercury for multiple generations. Environ Toxicol Chem 18:2840–2845.

Taub FB. 1969. A biological model of a freshwater community: a gnotobiotic ecosystem. Limnol Oceanog 14:136–142.

Taub FB. 1974. Closed ecological systems. Annu Rev Ecol Sys 5:139–160.

Taylor BL. 1995. The reliability of using population viability analysis for risk classification of species. Conserv Biol 9:551–558.

Taylor M, Feyereisen R. 1996. Molecular biology and evolution of resistance to toxicants. Mol Biol Evol 13:719–734.

Taylor EJ, Blockwell SJ, Maund SJ, Pascoe D. 1993. Effects of Lindane on the life-cycle of a freshwater macroinvertebrate *Chironomous riparius* Meighen (Insecta: Diptera). Arch Environ Contamination Toxicol 24:145–150.

Taylor RJ, Regan TJ, Regan HM, Burgman MA, Bonham KB. 2003. Impacts of plantation development, harvesting schedules and rotation lengths on the rare snail *Tasmophena lamproides* in northwest Tasmania: a population viability analysis. Forest Ecol Manage 175:455–466.

Ter Braak CFJ. 1995. Ordination. In: Jongman RHG, ter Braak CFJ, van Tongeren OFR, editors. Data analysis in community and landscape ecology. Cambridge (UK): Cambridge University Press.

Ter Braak CFJ. 1996. Unimodal models to relate species to environment. Wageningen (The Netherlands): DLO Agricultural Mathematics Group.

Ter Braak CJF, Smilauer P. 2002. CANOCO reference manual and CanoDraw for Windows user's guide: software for Canonical Community Ordination (version 4.5). Ithaca (NY): Microcomputer Power.

Tero N, Aspi J, Siikamaki P, Jakalaniemi A, Tuomi J. 2003. Genetic structure and gene flow in a metapopulation of an endangered plant species, *Silene tatarica*. Mol Ecol 12:2073–2085.

[TCEQ] Texas Commission for Environmental Quality. 2001. Guidance for conducting ecological risk assessments at remediation sites in Texas: RG-263 revised. Toxicology and risk assessment section. Formerly TNRCC (Texas Natural Resources Conservation Commission).

Thacker JRM, Jepson PC. 1993. Pesticide risk assessment and non-target invertebrates: integrating population depletion, population recovery and experimental design. Bull Environ Contam Toxicol 51:523–531.

Theodorakis CW. 2003. Establishing causality between population genetic alterations and environmental contamination in aquatic organisms. Hum Ecol Risk Assess 9:37–58.

Thomas CD, Hanski I. 1997. Butterfly metapopulations. In: Hanski IA, Gilpin ME, editors. Metapopulation biology: ecology, genetics and evolution. New York (NY): Academic Press. p. 359–386.

Thomas L, Laake JL, Strindberg S, Marques FFC, Buckland ST, Borchers DL, Anderson DR, Burnham KP, Hedley SL, Pollard JH. 2002. Distance 4.0. University of St. Andrews (UK): Research Unit for Wildlife Population Assessment. Available from: http://www.ruwpa.st-and.ac.uk/distance/. Accessed January 12, 2007.

Thomlinson R. 1965. Population dynamics: causes and consequences of world demographic change. New York (NY): Random House.

Thorrold SR, Latkoczy C, Swart PK, Jones CM. 2001. Natal homing in a marine population. Science 291:297–299.

Tilman D, Kareiva P, editors. 1997. Spatial ecology—the role of space in population dynamics and interspecific interactions. Princeton (NJ): Princeton University Press.

Topping CJ. 1999. An individual-based model for dispersive spiders in agroecosystems: simulations of the effects of landscape structure. J Arach 27:378–386.

Topping CJ, Odderskær P. 2004. Modeling the influence of temporal and spatial factors on the assessment of impacts of pesticides on skylarks. Environ Toxicol Chem 23:509–520.

Topping CJ, Ostergaard S, Pertoldi C, Bach LA. 2003b. Modeling the loss of genetic diversity in vole populations in a spatially and temporally varying environment. Ann Zool Fennici 40:255–267.

Topping CJ, Sibly RM, Açakaya HR, Smith GC, Cocker DR. 2005. Risk assessment of UK skylark populations using life-history and individual-based landscape models. Ecotoxicology 14(8):925–936.

Trautman MB. 1957. The fishes of Ohio. Columbus (OH): Ohio State University Press.

Tuljapurkar S, Caswell H, editors. 1997. Structured-population models in marine, terrestrial, and freshwater systems. New York (NY): Chapman and Hall.

Turchin P. 1989. Population consequences of aggregative movement. J Anim Ecol 58:75–100.

Turchin P. 1995. Population regulation: old arguments and a new synthesis. In: Cappuccino N, Price PW, editors. Population dynamics: new approaches and synthesis. San Diego (CA): Academic Press.

Turchin P, Berryman AA. 2000. Detecting cycles and delayed density dependence: a comment on Hunter & Price (1998). Ecol Ent 25:119–121.

Turchin P, Oksanen L, Ekerholm P, Oksanen T, Henttonen H. 2000. Are lemmings prey or predators? Nature 405:562–565.

Turner MG, Wu Y, Romme WH, Wallace LL. 1993. A landscape simulation model of winter foraging by large ungulates. Ecol Model 69:163–184.

Turner MG, Hargrove WH, Gardner RH, Romme WH. 1994a. Effects of fire on landscape heterogeneity in Yellowstone National Park, Wyoming. J Veget Sci 5:731–742.

Turner MG, Wu Y, Romme WH, Wallace LL, Brenkert A. 1994b. Simulating winter interactions between ungulates, vegetation and fire in northern Yellowstone Park. Ecol Appl 4:472–496.

Turner MG, Romme WH. 1994. Landscape dynamics in crown fire ecosystems. Landscape Ecol 9:59–77.

Twigg LE, Lowe TJ, Martin GR, Wheeler AG, Gray GS, Griffin SL, O'Reilly CM, Robinson DJ, Hubach PH. 2000. Effects of surgically imposed sterility on free-ranging rabbit populations. J Appl Ecol 37:16–39.

Uchmanski J. 2000. Resource partitioning among competing individuals and population persistence: an individual-based model. Ecol Model 131:21–32.

Underwood AJ. 1989. The analysis of stress in natural populations. Biol J Linnean Soc 37:51–78.

Underwood AJ. 1997. Experiments in ecology. Cambridge (UK): Cambridge University Press.

[USEPA] US Environmental Protection Agency. 1992. Framework for ecological risk assessment. Risk Assessment Forum. Washington (DC): USEPA/630/R-92/001.

[USEPA] US Environmental Protection Agency. 1993a. Chlorinated paraffins ecological risk characterization. Administrative Docket Center. Washington (DC): USEPA-102-B-063-66.

[USEPA] US Environmental Protection Agency. 1993b. Wildlife exposure factors handbook. Office of Research and Development. Washington (DC): USEPA-600-R-93-187a.

[USEPA] US Environmental Protection Agency. 1994. Managing ecological risks at EPA: issues and recommendations for progress. Washington (DC): USEPA-600-R-94-183.

[USEPA] US Environmental Protection Agency. 1996. Biological criteria: technical guidance for streams and small rivers (revised edition). Washington (DC): Available online at: http://www.epa.gov/bioindicators/html/bioltech.html. Accessed January 12, 2007.

[USEPA] US Environmental Protection Agency. 1997a. Mercury study report to congress, Vol. V: an ecological assessment of anthropogenic mercury emissions in the United States. Washington (DC): Office of Air Quality Planning and Standards and Office of Research and Development. USEPA-452-R-97–008.

[USEPA] US Environmental Protection Agency. 1997b. Ecological risk assessment guidance for superfund: process for designing and conducting ecological risk assessments. Interim final. Office of Solid Waste and Emergency Response Directive. Washington (DC): 9285.7-25 PB97-963211.

[USEPA] US Environmental Protection Agency. 1997c. National ambient air quality standards for ozone: final rule. July 18. Washington (DC): 62 FR 38856–38896.

[USEPA] US Environmental Protection Agency. 1998. Guidelines for ecological risk assessment. Washington (DC): USEPA-630-R-95-002F.

[USEPA] US Environmental Protection Agency. 1999. Issuance of final guidance: ecological risk assessment and risk management principles for superfund sites. Office of Solid Waste and Emergency Response Directive. Washington (DC): USEPA-9285-7-28-P.

[USEPA] US Environmental Protection Agency. 2000. Amended guidance on ecological risk assessment at military bases: process considerations, timing of activities, and inclusion of stakeholders. Memorandum from Ted W. Simon, PhD, June 23. Atlanta (GA): Office of Technical Services. Available from: http://risk.lsd.ornl.gov/homepage/ecoproc2.pdf. Accessed January 12, 2007.

[USEPA] US Environmental Protection Agency. 2001. Planning for ecological risk assessment: developing management objectives. Risk Assessment Forum, Washington (DC): USEPA-630-R-01-001A.

[USEPA] US Environmental Protection Agency. 2003. Generic ecological assessment endpoints (GEAE) for ecological risk assessment. Risk Assessment Forum. Washington (DC): USEPA/630/P-02/004F.

[USFWS] US Fish and Wildlife Service. 1980. Ecological services manuals, ESM 101, 102 and 103. Washington (DC): USFWS, Division of Ecological Services, United States Department of the Interior.

[USFWS] United States Fish and Wildlife Service 1973. Endangered Species Act of 1973: As Ammended throught the 108th Congress. US Department of Interior. Washington (DC). Available at http://www.fws.gov/endangered/esaall.pdf.

[USGS] US Geological Survey. 1997. Selected habitat suitability index model evaluations: information and technology report. Washington (DC): USGS, United States Department of the Interior. USGS/BRD/ITR-0005.

Vacher C, Bourguet D, Rousset F, Chevillon C, Hochberg ME. 2003. Modelling the spatial configuration of refuges for a sustainable control of pests: a case study of Bt cotton. J Evolut Biol 16:378–387.

Van DRG. 1994. Classical biological control of environmental pests. Florida Entomol. 77:20–33.

Van den Brink PJ, ter Braak CFJ. 1999. Principal response curves: analysis of time dependent multivariate responses of biological community to stress. Environ Toxicol Chem 18:138–148.

Van den Brink PJ, Hartgers EM, Fettweiss U, Crum SJH, Van Donk E, Brock TCM. 1997. Sensitivity of macrophyte-dominated freshwater microcosms to chronic levels of the herbicide linuron. I. Primary producers. Ecotox Environ Safe 38:13–24.

Van den Brink PJ, Van den Brink NW, Ter Braak CJH. 2003. Multivariate analysis of ecotoxicological data using ordination: demonstrations of utility on the basis of various examples. Austra J Ecotox 9:41–156.

Van Leeuwen CJ, Rijkeboer M, Niebeek G. 1986. Population dynamics of *Daphnia magna* as modified by chronic bromide stress. Hydrobiology 133:277–285.

Van Straalen NM, Hoffman AA. 2000. Review of experimental evidence for physiological costs of tolerance to toxicants. In: Kammenga J, Laskowski R, editors. Demography in ecotoxicology. West Sussex (UK): John Wiley & Sons Ltd.

Van Straalen NM, Kammenga JE. 1998. Assessment of ecotoxicity at the population level using demographic parameters. In Schuurmann G, Markets B, eds. Ecotoxicology. London (UK): John Wiley and Sons.

Van Straalen NM, Timmermans JTN. 2002. Genetic variation in toxicant-stressed populations: an evaluation of the genetic erosion hypothesis. Hum Ecol Risk Assess 8:983–1002.

Van Tienderen PH, De Haan AA, Van der Linden CG, Vosman B. 2002. Biodiversity assessment using markers for ecologically important traits. Trends Ecol Evol 17:577–582.

VanVeld P, Nacci, D. In press. Chemical tolerance: acclimations and adaptations to chemical stress. In: Di Giulio RT, Hinton DE, editors. The toxicology fishes. Washington (DC): Taylor and Francis Publishers.
Van Winkle W, Rose KA, Chambers RC. 1993. Individual-based approach to fish population dynamics: an overview. Trans Amer Fish Soc 122:397–403.
Varley GC, Gradwell GR, Hassel MP. 1973. Insect population ecology: An analytical approach. Oxford (UK): Blackwell Scientific Publications.
Varley GC, Gradwell GR. 1960. Key factors in population studies. J Anim Ecol 29:399–401.
Ver Hoef JM. 1996. Parametric empirical Bayes methods for ecological applications. Ecol Appl 6:1047–1055.
Verhulst PF. 1838. Notice sur la loi que la population suit dans son accroissement. Corr Math et Phys 10:113–121.
Wade PR. 2001. Bayesian methods in conservation biology. Conserv Biol 14:1308–1316.
Walker CH, Hopkin SP, Sibly RM, Peakall DB. 2001. Principles of ecotoxicology, 2nd ed. London (UK): Taylor & Francis.
Wallace AR. 1867. The Malay archipelago. Charles E Tuttle Co (September 2000). North Clarendon (VT).
Wallace AR. 1881. Island life. Prometheus books. repr. ed. (December 1997). Amherst (NY).
Walthall WK, Stark JD. 1997. Comparison of two population-level ecotoxicological endpoints: the intrinsic (r_m) and instantaneous (r_i) rates of increase. Environ Toxicol Chem 16:1068–1073.
Wang G, Liang XG, Wang FZ. 1999. The competitive dynamics of populations subject to an Allee effect. Ecol Model 124:183–192.
Waples RS. 2002. Definition and estimation of effective population size in the conservation of endangered species. In: Beissinger SR, McCoullough DR, editors. Population viability analysis. Chicago (IL): The University of Chicago Press.
Ware DM. 1984. Fitness of different reproductive strategies in teleost fishes. In: Fish reproduction. GW Potts, RJ Wootton, editors. London (UK): Academic Press Inc.
Ware DM, Tovey Hay D, McCarter B. 2000. Straying rates and stock structure of British Columbia herring. Fish Oceans Canada: Canadian Stock Assess Secretariat Res Document 2000/006.
Warren-Hicks W, Parkhurst BR, Baker SS. 1989. Ecological assessment of hazardous waste sites: a field and laboratory reference. Corvallis (OR): US Environmental Protection Agency (USEPA). Environmental Research Laboratory. USEPA/600/3-89-013.
Weeks AR, McKechnie SW, Hoffmann AA. 2002. Dissecting adaptive clinal variation: markers, inversions and size/stress associations in *Drosophila melanogaster* from a central field population. Ecol Lett 5:756–763.
Wennergren U, Stark J. 2000. Modeling long-term effects of pesticides on populations: beyond just counting dead animals. Ecol Appl 10:295–302.
Werner PA, Caswell H. 1977. Population growth rates and age versus stage distribution models for teasel (*Dipsacus sylvestris* Huds.). Ecology 58:1103–1111.
Westerberg L, Wennergren U. 2003. Predicting the spatial distribution of a population in a heterogeneous landscape. Ecol Model 166:53–65.
White GC. 1983. Numerical estimation of survival rates from band recovery and biotelemetry data. J Wildl Manage 47:716–728.
White GC, Anderson DR, Burnham KP, Otis DL. 1982. Capture-recapture and removal methods for sampling closed populations. Los Alamos (NM): Los Alamos National Laboratory (LANL). Report LA-8787-NERP.
White GC, Burnham KP. 1999. Program MARK: survival estimation from populations of marked animals. Bird Study Suppl 46:S120–S138.

White GC. 2000. Population viability analysis: data requirements and essential analyses. In: Boitani L, Fuller TK, editors. Research techniques in animal ecology. New York (NY): Columbia University Press. p. 288–331.

White TCR. 2001. Opposing paradigms: regulation or limitation of populations? Oikos 93:148–152.

Wickwire WT, Menzie CA, Burmistrov D, Hope BK. 2004. Incorporating spatial data into ecological risk assessments: the spatially explicit exposure module (SEEM) for ARAMS. In: Kapustka LA, Galbraith H, Luxon M, Biddinger GR, editors. ASTM International, West Conshohocken (PA): American Society for Testing and Materials (ASTM). STP 1458.

Wiens JA. 1977. On competition and variable environments. Amer Sci 65:590–597.

Wiens JA. 1989. Spatial scaling in ecology. Funct Ecol 3:385–397.

Wiens JA. 1996. Coping with variability in environmental impact assessment. In: Baird DJ, Maltby L, Greig-Smith PW, Douben PET, editors. Ecotoxicology: ecological dimensions. London (UK): Chapman & Hall. p. 55–70.

Wiens JA, Parker KR. 1995. Analyzing the effects of accidental environmental impacts: Approaches and assumptions. Ecol Appl 5:1069–1083.

Williams BK, Nichols JD, Conroy MJ. 2002. Analysis and management of animal populations. San Diego (CA): Academic Press, Inc.

Williams DW, Liebhold AM. 1995. Detection of delayed density dependence: effects of autocorrelation in an exogenous factor. Ecology 76:1005–1008.

Williams SE, Hero J-M. 1998. Rainforest frogs of the Australian Wet Tropics: guild classification and the ecological similarity of declining species. Proc R Soc Lond B 265:597–602.

Williamson M. 1972. Special topics in biology series. The analysis of biological populations. London (UK): Edward Arnold Limited.

Wilson, EO. 1999. *The Diversity of Life*. W. W. Norton & Co., New York, pp. 64 and 238.

Wintle BA, McCarthy MA, Volinsky CT, Kavanagh RP. 2003. The use of Bayesian model averaging to better represent uncertainty in ecological models. Conserv Biol 17:1579–1590.

With K, Gardner RH, Turner MG. 1997. Landscape connectivity and population distributions in heterogeneous landscapes. Oikos 78:151–169.

Wolda H, Dennis B, Taper ML. 1994. Density dependence, population persistence, and largely futile arguments—answers to Holyoak and Lawton (1993) and Hanski, Woiwod and Perry (1993). Oecology 98:229–234.

Wood SN. 1994. Obtaining birth and mortality patterns from structured population trajectories. Ecol Monogr 64:23–44.

Wood SN. 2001. Partially specified ecological models. Ecol Monogr 71:1–25.

Wood SN, Thomas MB. 1999. Super-sensitivity to structure in biological models. Proc R Soc Lond B 266:565–570.

Woodman JN, Cowling EB. 1987. Airborne chemicals and forest health. Environ Sci Technol 21:120–126.

Wu H, Li BL, Springer TA, Neill WH. 2000. Modelling animal movement as a persistent random walk in two dimensions: expected magnitude of net displacement. Ecol Model 132:115–124.

Wu J, Vankat JL, Barlas Y. 1993. Effects of patch connectivity and arrangement on animal metapopulation dynamics: a simulation study. Ecol Model 65:221–254.

Yen JL, Batterham P, Gelder B, McKenzie JA. 1996. Predicting resistance and managing susceptibility to cyromazine in the Australian sheep blowfly *Lucilia cuprina*. Aust J Exp Agr 36:413–420.

Young AG, Brown AHD, Murray BG, Thrall PH, Miller CH. 2000. Genetic erosion, restricted mating and reduced viability in fragmented populations of the endangered grassland herb *Rutidosis leptorrhynchoides*. In: Young AG, Clarke GM, editors. Genetics, demography and viability of fragmented populations. Cambridge (UK): Cambridge University Press.

Zheng J. 1996. Herring stock-recruitment relationships and recruitment patterns in the North Atlantic and Northeast Pacific oceans. Fish Res 26:257–277.

Appendix 1: Decision Context Scenarios

The purpose of this decision context scenario appendix is to provide examples of the regulatory context, the role of risk assessment in supporting management decisions, and the rationale for the use of population-level assessment methods in addressing management decisions. These scenarios are referred to in chapters throughout this book as a basis for specific examples and applications of population-level ERA. These examples are not intended to be all inclusive, but rather a subset of the decision contexts that might use population-level ERA. The four decision context scenarios presented herein include the following:

Appendix A1.1: Hazardous Waste Scenario
Appendix A1.2: European Water Framework Directive
Appendix A1.3: Consequences of Exceeding Water and Sediment Quality Standards
Appendix A1.4: Agricultural Pesticide Registration.

APPENDIX 1 (A1.1): HAZARDOUS WASTE SCENARIO

SCENARIO: HAZARDOUS WASTE SITE (HISTORIC RELEASES)

Regulatory Context

Hazardous waste sites exist around the world because of accidental spills and chemical handling procedures. These sites may be legacy sites (i.e., historic) or they may have occurred in present-day operations. How these sites were and are addressed and the regulatory context under which these releases were or are addressed vary greatly depending on numerous factors, such as these:

- local, national, and international laws, regulations, policies, narrative goals, and directives that govern an area where a spill or release occurs;
- whether these releases were accidental versus intentional;
- the scale of the release (e.g., 1 mL versus 1 tanker);
- the type of chemicals released (e.g., a chemical that is not very toxic and degrades rapidly versus one that is highly toxic and known to persist for decades);
- the time frame over which the release occurred (e.g., a single-pulse release versus a continuous discharge over several decades);
- the degree to which a chemical release could or can be contained and cleaned-up (e.g., a relatively immobile contaminant versus one that disperses rapidly and could not be contained);

- the environment to which the release occurred (e.g., a unique and highly valued natural ecosystem versus an active industrial area);
- the stakeholders involved or concerned about protecting the impacted environment; and
- the ecological entities identified for protection.

A hazardous waste site is a defined place in space and time with chemical contamination in soil, ground water, surface water, sediment, or biotic tissues. Examples of these hazardous waste sites include

- a fraction of an acre (could even be a small hole) where lead-based paint was previously dumped;
- an unlined 20-acre landfill used in the 1950s and 1960s, with known or unknown dischargers (potentially including persistent bioaccumulative compounds, such as polycyclic biphenyls);
- a government property or industrial facility, potentially 100 to10000 acres, with multiple areas ranging from contaminated by a range of chemicals specific to the operations of the area;
- a 55000-acre diverse area with terrestrial and aquatic ecosystem contaminated by the deposition of uncontrolled smelter emissions that may have occurred from the 1900s to the 1970s;
- a watershed where multiple chemicals from numerous sources have been released (e.g., lake, bay, river, estuary, wetland, and their associated landscapes);
- a train derailment, with chemical releases to soil and surface water features (recent or historic);
- an area impacted by acid mine drainage into a small stream, large river, or wetland;
- radiological releases over small or large spatial scales; and
- small or large areas that are contaminated from a chemical spill, where clean up has occurred (or has not), with chemical residues in the environment.

Various local, national, and international laws, regulations, policies, narrative goals, and directives influence the risk management context for hazardous waste site management. As described in Chapter 2, many laws, regulations, policies, narrative goals, and directives implicitly or explicitly identify populations as entities to be protected. Chemical residues that persist following an accidental release and chemical residues because of historic waste management practices are managed under a variety of regulatory programs, as described in greater detail in Chapter 2, such as the following.

United States of America

- Comprehensive Environmental Response, Compensation, and Liability Act and Superfund Amendments and Reauthorization Act
- Resource Conservation and Recover Act
- Endangered Species Act

- Clean Water Act
- Clean Air Act
- A broad range of state environmental regulatory programs

European Union

- European Union Directive 92/43 refers to the conservation of natural habitats and of wild fauna and flora aims (Article 2).
- European Union Directive 85/337 refers to specific sites and thus to both specific causal agents and receptors.

Canada

- Canadian Environmental Protection Act (CEPA)
- Province of British Columbia, Waste Management Act

Role of Risk Assessment in Supporting Management Decisions

ERAs are one of many considerations in the management decision-making process for hazardous waste sites. Information from an ERA is balanced with information such as risks to human health, costs associated with remedial efforts, and feasibility or effectiveness of remedial efforts. In particular, ERA results must be considered in the context of risks posed by remedial options, as the very habitat and receptors intended for protection with a management decision can be destroyed by a remedial action (e.g., excavation of contaminated sediments in a wetland would likely cause greater harm than low levels of residual chemicals in a wetland).

ERAs can be used to support management decisions at hazardous waste sites because they can be used to

- evaluate the likelihood that residual chemical concentrations in soil, ground water, surface water, sediment, and biotic tissues has caused adverse impacts to wildlife and their habitats (i.e., a retrospective evaluation of ecological risks);
- evaluate the likelihood that residual chemical concentrations in environmental media will cause adverse impacts to wildlife and their habitats in the future (i.e., a prospective evaluation of ecological risks);
- compare and prioritize sites that may require corrective action (i.e., relative risk ranking);
- estimate remedial action clean-up levels (if determined that remediation is necessary);
- compare net environmental benefits of multiple remedial alternatives; and
- evaluate the effectiveness of a previous remedial action.

Current Risk Assessment Approaches

Risk managers in the United States and Canada have a broad array of ERA approaches to currently use for managing risks, though one (ORDEQ 2001) explicitly identifies population-level directive for decision-making. More than 100 ERA guidance documents, procedural guidelines, or other resources — spanning just the last decade — have been published for use in the evaluation of ecological risks (see annotated compilations by Sorensen 1994; Sorensen and Margolin 1998). The majority of these follow the USEPA Framework for ERA (1992), the USEPA ERA Guidance for Superfund (1997), or the USEPA Guidelines for ERA (1998). Consistent with USEPA (1998), the overall ERA process generally consists of problem formulation, analysis, and risk characterization phases within the steps or tiers. A common theme and element of the array of ERA guidance documents and procedural guidelines is that the best ERAs are the ones that are appropriate for the specific risk management needs of the individual site.

ERA approaches generally involve multiple steps (e.g., USEPA 1997, 2000) or tiers (ORDEQ 2001; TCEQ 2001; ASTM 2002). Steps or tiers are intended to progress from conservative assessment of risks (using broad and general assumptions) to less conservative assessment of risks (using site specific information and assumptions, to the extent possible). Early steps or tiers involve chemical-specific hazard quotients (HQs), which is representative of potential impacts to oganisms.

As an alternative to chemical-specific HQs, several US states have adopted methodologies that consider a *de minimis* scale criterion, among other criteria, as part of the early steps and tiers in the ERA process (MADEP 1996; PADEP 1998; TCEQ 2001). Criteria that also must be met to use the *de minimis* scale criteria includes consideration of comparable habitat in the immediate vicinity of the hazardous site and reasonable expectation that pathways for threatened and endangered species (or otherwise valued habitats) are incomplete. These approaches suggest that wildlife populations will not be exposed to such small sites (PADEP 1998; TCEQ 2001) and that by focusing on habitat, management goals can be obtained (MADEP 1996).

The state of Oregon provides the only US ERA guidance document with an explicitly stated population-level directive, implemented in a Tier III ERA (ORDEQ 2001). Oregon's specific regulatory language states that an unacceptable risk occurs when there is a greater than 10% chance of more than 20% of the total local population receiving an exposure greater than the toxicity reference value for a hazardous substance.

Levels at Which Assessment Endpoints Are Defined

The risk management needs vary considerably between sites, but the overall management goals generally involve protection of populations, communities, and ecosystems (USEPA 1997, 1999, 2000). For example, relative to many hazardous waste sites, the USEPA states "Superfund's number 1 principle is to reduce ecological risks to levels that will result in the recovery and maintenance of healthy local populations and communities of biota" (USEPA 1999). Exception, however, is given to sensitive species and habitats, where management attention is given to individuals of threatened and endangered species and their habitats because these compromised populations are less capable of tolerating the loss of individuals.

Risk management goals and ERA assessment endpoints for hazardous waste sites (or any application of ERA for that matter) should be clearly stated (USEPA

2002). Example risk management goals and assessment endpoints for hazardous waste sites include (USEPA, 2002).

GOAL: Sustain small mammal populations
ASSESSMENT ENDPOINT: Survival and reproduction of small rodent species

GOAL: Sustain eastern bluebird populations
ASSESSMENT ENDPOINT: Eastern bluebird breeding success and site fidelity

GOAL: Viable, self-sustaining brook trout population that supports a sport fishery
ASSESSMENT ENDPOINT: Brook trout abundance, breeding success, fry survival, and adult return rates

Assessment and Analytical Methods

Assessment and analytical methods are selected based on the conceptual site model that depicts hazardous waste release mechanisms, potential ecological exposure pathways, and potential receptors, as illustrated in Figure A1.1 (pictorial) and Figure A1.2 (linear).

In early steps or tiers of an ERA, chemical concentrations in soil, water, and sediment are compared with media-specific ecotoxicological screening values (also called toxicological benchmarks, effects levels, and generic screening values) for the calculation of chemical-specific HQs. Short-term acute, subchronic, and chronic laboratory studies with surrogate species provide the basic information that serves as the foundation of many ecotoxicological screening values. Food web modeling, which can be used in early or mid-steps and tiers, uses dose-based no-observable-adverse-effects levels and lowest-observable-adverse effects levels, often derived from laboratory test species with estimates of similarities for wildlife species (e.g., Sample et al. 1996).

There are many examples of field-level studies at hazardous waste sites that have focused on population-level endpoints using empirical studies in addition to (or in lieu of) chemical-specific HQs (e.g., Fontenot et al. 1998, 2000; McGee et al. 1999; Baker et al. 2001; Boonstra and Bowman 2003; Meyer and Di Giulio 2003).

Indirect effects through food resource impacts and alteration of habitat can also be evaluated in addition to (or in lieu of) chemical-specific HQs (e.g., Kapustka et al. 2004).

Risk Characterization Methods and Risk Descriptors

Most often, risk is characterized using HQs generated by the comparison of media concentrations to ecotoxicologic screening values or estimated intake concentrations to dose-based toxicological benchmarks. Hazard quotients greater than 1 indicate that adverse impacts could have occurred in the past or may occur in the future. As

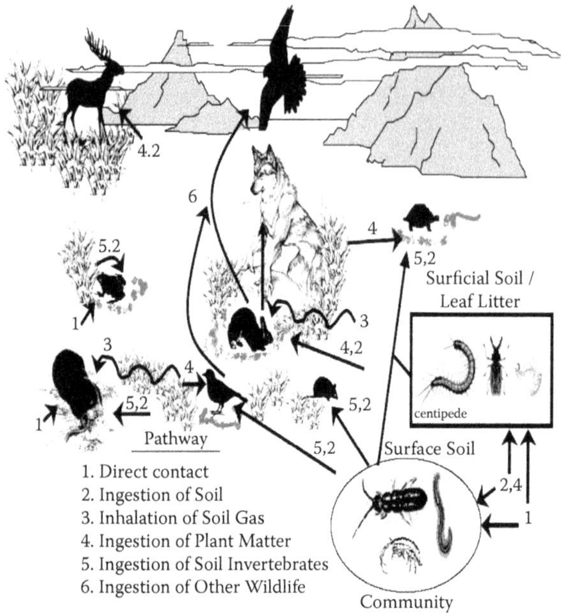

FIGURE A1.1 Pictorial conceptual site model.

the magnitude of HQ values increase, it is generally accepted that the likelihood for risk increases, but there is not a clearly defined relationship between HQs and risk.

Characterization and communication of risk using empirical or modeling studies is very specific to the nature of the studies, and should be based on a weight of evidence approach. For example, risks may be characterized by the change in clutch size or fledging survival from monitored nests (actual nests or nest boxes) compared with that expected for a similar species of birds. Similarly, small mammal trap and recapture results can be used to estimate age structure and density for a hazardous waste site compared to a reference location. Refer to Chapters 8 and 9 for more detailed information on empirical and modeling methods.

Rationale for Population-Level Assessment Methods

The use of population-level ERAs versus organism-level ERAs often depends on numerous factors, such as the risk management context, the size and ecological complexity of the site, the cost of the assessment, the cost of a clean-up decision (including the ecological cost of lost habitat from a clean-up), and the perceived value of the population-level information for risk management decision-making.

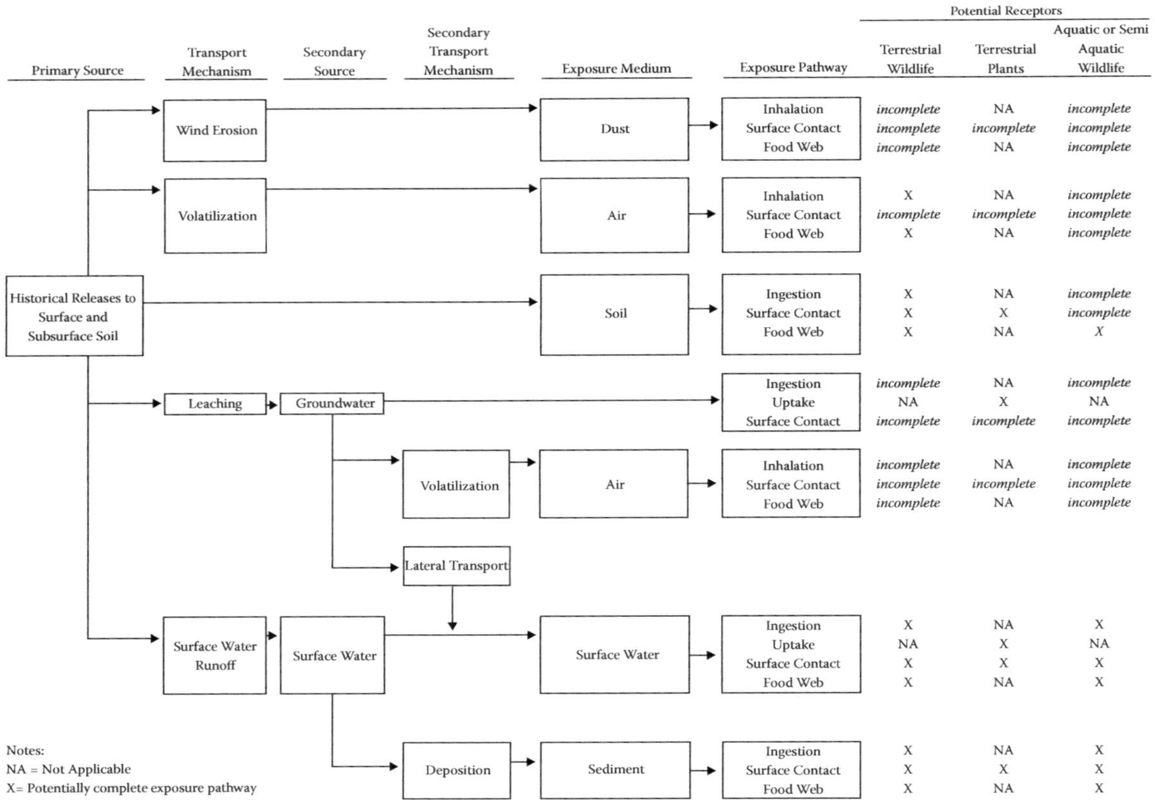

FIGURE A1.2 Schematic conceptual site model.

Chemical-specific HQs have become essentially synonymous with the initial steps (or tiers) of an ERA at chemical release sites — although less attention is given to actual ecological characteristics of a site until later in the ERA process (if at all). As a result, many ERAs are focused on potential impacts to organisms, whereas the risk management goals are focused toward populations. This is problematic because there is little consensus regarding how HQs can be used for risk management decision-making at the population, community, and ecosystem levels (although theoretical approaches that address issues of extrapolating from the organism to the population are discussed in scientific literature: e.g., Maltby et al. 2001; Landis 2002; Munns 2002). Furthermore, it has been argued that chemical-specific HQs often predict widespread mortality, while actual biological evidence of this effect cannot be found in the natural environment (Tannenbaum 2002, 2003; Tannenbaum et al. 2003).

Population-level ERAs are not appropriate for every hazardous waste site for a variety of reasons, such as the site is too small (i.e., populations are not exposed) or a threatened species is exposed (in which case the individual organism must be considered). However, where appropriate, populations and their attributes are responsive assessment endpoints for ERAs that can support most management decisions at hazardous waste sites. Population-level methods are being used at hazardous waste sites (as briefly described in this scenario, and described in greater detail in Chapters 8 and 9). Greater understanding and use of population-level ERA can be expected to increase the ecological relevance of decision-making, better address the intent of laws and regulations, allow better recognition of societal and stakeholder values, and provide managers with the breadth of information necessary to support practical, balanced, and implementable decisions. How a population is defined (i.e., the exposed population, local population, or the biological population) appears to remain the most significant challenge in the application of population-level ERA at hazardous waste sites.

APPENDIX 1 (A1.2): EUROPEAN WATER FRAMEWORK DIRECTIVE

SCENARIO: EUROPEAN WATER FRAMEWORK DIRECTIVE

Regulatory Context

The European Water Framework Directive (WFD; Directive 2000/60/EC) requires that surface waters (lakes, rivers, estuaries, coastal waters) within the jurisdiction of the European Union be classified on the basis of both ecological status and chemical status. Furthermore by a fixed deadline it requires that all controlled water bodies are in "good quality status" and this is specifically defined within the legislation. Surface water status is determined by the poorer of its ecological status and its chemical status (Article 2, paragraph 17).

Role of Risk Assessment in Supporting Management Decisions

Risk assessment is not mentioned specifically within the Water Framework Directive, but the normative definitions of ecological quality status specified within Annex 5 of the WFD include some explicit population-level characteristics (i.e., abundance, size and age structure) for fish. Most of the other criteria are in terms of community-level attributes that will depend implicitly on the way population dynamics respond to prevailing conditions.

Current Risk Assessment Approach

At the time of writing, the WFD is in the process of being implemented, and many of the details have yet to be clarified. Hence the current approach is under development. Nevertheless, to define ecological status, the approach will involve comparison of observations on defined "biological quality elements" at a particular site with those expected for that site as indicated by observations made on the same elements in near-pristine sites. To define chemical status, the approach will involve comparison of measured concentrations of specified chemicals with ecological quality standards following an approach similar to that outlined in Scenario A1.3.

Biological quality elements as defined by Annex 5 include

- phytoplankton: taxonomic composition, abundance, biomass, occurrence of blooms;
- macrophytes and phytobenthos: taxonomic composition, abundance, cover;
- benthic invertebrates: taxonomic composition, abundance, diversity; and
- fish: species composition, abundance, age structure.

Which and how many of these elements are used to define ecological quality status depends on the type of surface water body under consideration.

Rationale for Population-Level Assessment Methods

The WFD specifies that population-level endpoints (i.e., abundance and age structure) be measured, at least for fish. These endpoints might be enhanced by using population dynamics principles to extrapolate from these cross-sectional measures of populations taken at single points in time to future changes in abundance and population structure (Caswell 2001).

For the other taxonomic groups used to define ecological quality the focus of the WFD is more at the community level (e.g., diversity and total biomass). In this context, population models could be used to identify (and prioritize for monitoring) particularly sensitive communities on the basis of the lifecycle characteristics of their component species populations (Calow et al. 1997). Hence population-level assessments could be used to inform and refine the community assessments specified in WFD.

In that the WFD requires management back to "good quality status" in systems that are deemed to be less than "good" on the basis of either the biological elements or the chemical elements, it will be necessary to define remediation criteria. Once again, population models could be used to predict the relative potential and time scales for recovery of populations and communities (Barnthouse 2004).

APPENDIX 1 (A1.3): CONSEQUENCES OF EXCEEDING WATER AND SEDIMENT QUALITY STANDARDS

SCENARIO: ESTIMATING CONSEQUENCES OF EXCEEDING WATER AND SEDIMENT QUALITY STANDARDS

Regulatory Context

Several jurisdictions lay down a strategy for the establishment of harmonized quality standards for specified substances. For example, the European Water Framework Directive (WFD; Directive 2000/60/EC) is setting harmonized quality standards (QS) for approximately 30 priority substances identified in the context of the directive. The Oslo and Paris Commission (OSPAR) has developed ecotoxicological assessment criteria for chemical monitoring data from the Northeast Atlantic. Quality standards compose an essential component of aquatic monitoring programmes; the presumption being that QSs represent concentrations below which no harm to the aquatic environment is expected.

Role of Risk Assessment in Supporting Management Decisions

Chemical concentrations of substances for which standards have been defined observed in monitoring programs are compared with QSs. Those that exceed QSs trigger consideration of management action. The management can be in the form of controls on effluents or remediation of contaminated sediment.

Current Risk Assessment Approach

QSs are estimated from standard ecotoxicological endpoints with conservative uncertainty factors applied. This means that the QSs are based fundamentally on organism-level, rather than population-level responses. These observations are usually from a limited number of taxa and are often based on acute responses to high exposure concentrations. The worst-case endpoint is selected and divided by an uncertainty factor that takes into account extrapolation from simplified laboratory observations to the protection of more complex ecological systems. Alternatively the effects data may be compiled into frequency distributions from which percentiles (e.g., 10%, 50%) are used to define critical threshold levels (Long et al. 1995). The quality standards are listed in legal documents and are usually taken to be mandatory.

Either average or worst-case measured concentrations in water or sediment from monitoring stations are generally compared with the quality standards to provide an estimate of "risk" to aquatic ecosystems.

Rationale for Population-Level Assessment Methods

In defining water and sediment quality standards, it would be rare for the protection goal to be targeted at the organism level. Yet measurement endpoints used to establish environmental quality standards are almost exclusively in terms of organism-level effects. The extrapolation of the measurement endpoints to higher level effects is usually implicit and not very well defined. Therefore, there are at least 2 ways that the current approach could be refined by applying population-level assessment methods.

- The first refinement would be in the use of population-level, rather than organism-level, endpoints in the initial calculation of the environmental quality standards to make them more realistic ecologically (Forbes and Calow 1999).
- The second refinement would be to incorporate the spatiotemporal variability in exposure with respect to the distribution of populations of concern on relevant spatial and temporal scales. For example, the spatial distribution of exposure concentrations could be used to estimate the fraction of a particular population that is exposed above the QS. There could be very different consequences for population dynamics if there are a few "hot spots" of exposure versus a situation in which a large fraction of the water body slightly exceeded the QS.

APPENDIX 1 (A1.4): AGRICULTURAL PESTICIDE REGISTRATION

SCENARIO: AGRICULTURAL PESTICIDE REGISTRATION

Regulatory Context

Agricultural pesticides are any number of chemicals deliberately released into the environment to control pests that harm crops. Typical pesticide application methods (aerial spraying, seed-coating, foliar and granular application) can result in exposure of nontarget terrestrial and aquatic organisms to pesticide active ingredients. Consequently, authorities have established registration protocols for evaluating the risks associated with new pesticide use to nontarget plants and animals, and manage those risks by controlling pesticide use and application methods ("labeling"). Protocols also are in place for reregistration of existing pesticides already on the market.

The European Community (EC) has developed a legal framework for the regulation of pesticides in all member countries of the EC. The Commission of the European Communities, in collaboration with EC member countries, has responsibility for the registration of pesticide active ingredients (also referred to as active substances) for use in all EC member countries. Individual member countries, called member states, are responsible for the registration in their country of specific pesticide products containing active ingredients authorized for use by the Commission. This dual authority of the EC and its member states is granted by the Council of the European Community under Council Directive 91/414/EEC, adopted on 15 July 1991 and effective 25 July 1993. Standards and regulations for the classifications, labeling, and packaging of pesticides are set by Council Directive 67/548/EEC of 27 June 1967.

In the United States, pesticides are regulated under the Federal Insecticide, Fungicide and Rodenticide Act (FIFRA) and the Federal Food, Drug, and Cosmetic Act (FFDCA), both as amended by the Food Quality Protection Act (FQPA) of 1996. Under FIFRA, new pesticides must be registered by the USEPA before they can be sold or distributed. Use of existing pesticides is reviewed periodically to determine whether current labeling is sufficiently protective. The intent of FIFRA is to prevent "unreasonable adverse effects on the environment." FIFRA requires that certain data be provided by the registrant to support assessments of risk associated with pesticide use.

Role of Risk Assessment in Supporting Management Decisions

Relative to agricultural pesticide registration, the management goal of the USEPA is to protect the environment from any "unreasonable risk … taking into account the economic, social and environmental costs and benefits of the use of any pesticide." As summarized at http://www.epa.gov/pesticides/regulating/registering/index.htm, the process of registering a pesticide is a scientific, legal, and administrative procedure through which the USEPA examines the ingredients of the pesticide; the particular site or crop on which it is to be used; the amount, frequency, and timing of its use; and storage and disposal practices.

In evaluating a pesticide registration application, the USEPA assesses a wide variety of potential environmental and human health effects associated with use of the product. The producer of the pesticide (registrant) must provide data from tests done according to USEPA guidelines. These tests evaluate whether a pesticide has the potential to cause adverse effects on humans, wildlife, fish, and plants, including endangered species and nontarget organisms, as well as possible contamination of surface water or ground water from leaching, runoff, and spray drift. Using these data (augmented from other sources as necessary), ecological risk assessments are conducted to provide evaluations of environmental risks. Risk managers use the results of these assessments to determine whether intended use (labeling) is sufficiently protective of unreasonable adverse effects.

Current Risk Assessment Approach

The United States conducts a tiered assessment of new pesticides entering the market to evaluate the potential for adverse impacts of pesticide use on nontarget species. Consistent with USEPA (1998), each tier consists of problem formulation, analysis, and risk characterization phases. Information required to assess risks to nontarget organisms are derived from tests to determine pesticidal effects on birds, mammals, fish, terrestrial and aquatic invertebrates, and plants. These tests include short-term acute, subacute, reproduction, simulated field, and full field studies arrayed in a tiered system that progresses from basic laboratory tests to applied field tests. The results of each tier of tests are evaluated to determine the potential of the pesticide to cause harmful effects and to determine whether further testing is required. A purpose common to all data requirements is to help determine the need (and appropriate wording) for precautionary label statements to minimize the potential harm to nontarget organisms.

When the USEPA has reason to believe that the use of a pesticide may result in unreasonable adverse effects to people or the environment, a special review process is initiated. The goal of this process is to reduce the risks posed by the pesticide to an acceptable level while taking into consideration the benefits provided by the use

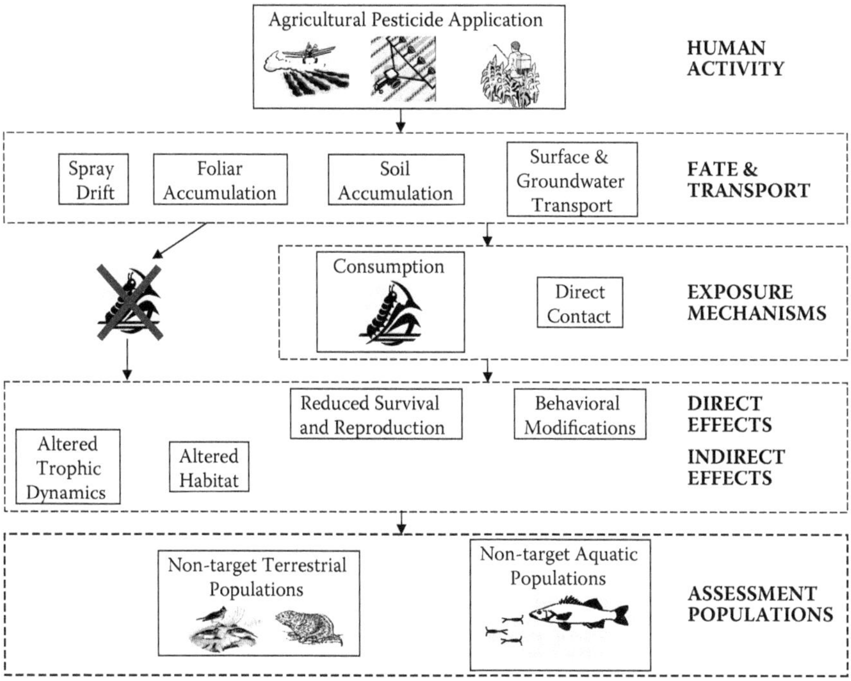

FIGURE A1.3 Simplified conceptual model.

of the pesticide. Such special reviews have been undertaken for a number of pesticides, including (granular) carbofuran, diazinon and toxaphene, because of suspected risks to wildlife.

Levels at Which Assessment Endpoints Are Defined

In the United States, early-tier assessments focus on direct acute and subchronic toxicological effects on individual organisms. Tests involving surrogate species (wildlife, aquatic life, and plants) provide measures of effect for assessment endpoints. Although explicit linkages are not made, such effects are assumed to provide insight to risks to populations and communities. In subsequent tiers, simulated or actual field data are used to examine acute and chronic adverse effects on captive or monitored fish and wildlife populations under natural or near-natural environments. Such assessments are required only when predictions as to possible adverse effects in less extensive studies cannot be made, or when the potential for harmful effects is high. The USEPA is developing methods to evaluate population and community-level risks using probabilistic exposure and effect modeling following the recommendations of the Ecological Committee on FIFRA Risk Assessment Methods (ECOFRAM 1999a, b).

A generic model linking pesticide application to nontarget populations is shown in Figure A1.3. Pesticides can be applied to crops using a number of different methods, ranging from aerial spraying to granular applications. The method of application determines in large part subsequent transport, degradation, and fate of active ingredients, resulting in their redistribution in terrestrial and aquatic systems. Nontarget organisms are exposed through a variety of mechanisms, including direct contact and consumption of contaminated prey, vegetation, and water. Exposure can result in direct effects on survival, reproduction, growth, and behavior. Additional indirect effects can be mediated through altered habitat and trophic dynamics. Ultimately, these effects can manifest at the population level in nontarget terrestrial and aquatic assessment populations.

Individual treated fields reside in landscapes consisting of multiple fields, hedgerows, and water bodies. Plot-level exposures to nontarget organisms are short-lived and often pulsed (depending on application practices). Although chronic effects can occur, acute mortality of directly exposed organisms is often the greatest concern. Because the landscape integrates exposures over many plots with different treatment regimes, landscape-scale exposures have a longer and patchier dynamic that often is seasonal in nature. Long-term exposures and chronic effects are more important at the landscape scale than at the plot scale.

Analytical Methods

In early (screening) tiers of the assessment, short-term acute and subchronic laboratory studies with surrogate species provide the basic information that serves to establish the acute toxicity of the active ingredient to test organisms and indicate whether further laboratory or field studies are needed. Toxicity information is summarized as standardized test statistics (endpoints). At this level of assessment, exposure by contact and food consumption is estimated using standardized assumptions for active ingredient application rates (and methods), persistence, mobility and degradation, exposure pathways, and fate. Single-event models provide rapid screens to help separate pesticides with low risk from ones requiring more refined assessments. Initially conservative assumptions can be replaced with site-specific information if available. If available or as needed, toxicity and exposure information can be augmented with field data.

Additional studies (i.e., avian, fish, and invertebrate reproduction, lifecycle studies, and plant field studies) may be required when earlier tier assessments suggest possible problems. Data from these studies are used to estimate the potential for chronic effects, taking into account measured or estimated residues in the environment, and to determine if additional field or laboratory data are necessary to evaluate risk further. Indirect effects through food resource impacts and alteration of habitat can also be evaluated at higher tiers. Environmental realism of exposure estimation is accomplished using simulation modeling and site-specific characteristics of transport and fate. Monitoring data and incidence reports (e.g., observed mortalities) can supplement these data and can be used to invoke special reviews to evaluate risks further.

Risk Characterization Methods

Risk is characterized using the risk quotient method, in which deterministically estimated environmental concentrations, based on maximum application rates, are compared with acute and chronic toxicity values.

USEPA is developing more refined risk assessment methods following the recommendations of ECOFRAM (1999a, b). These methods are intended to provide probabilistic risk assessments (at higher tiers) in terms of the magnitude and frequency of potential adverse effect that reflect environmental and biological variability in exposure and effect.

Risk Description and Communication Methods

At screening tiers, regulatory quotients (RQs) are compared with "levels of concern" that represent USEPA interpretive policy. Levels of concern address (LOCs) risk presumptive categories ranging from acute (potential for acute risk that may warrant regulatory action — RQ > 0.5 for mammals, birds, and aquatic life) to chronic risk (potential for chronic risk that may warrant regulatory action — chronic RC > 1). LOCs are communicated to decision-makers (risk managers) together with assumptions, uncertainties, and strengths and limitations of the assessment. Data describing exposure and effects collected under field conditions, if available, can be used to inform decision-makers of the veracity of assessment predictions.

Rationale for Population-Level Assessment Methods

Although specific reference to protecting populations is absent from FIFRA, the USEPA's Standard Evaluation Procedure for ecological risk assessment of pesticides considers adverse effects of pesticide use to include "reductions in populations of non-target organisms" explicitly in its definition of risk. Further, the impetus for EPA's ECOFRAM effort was to " … develop and validate risk assessment tools and processes that address increasing levels of biological organization (e.g., organisms, populations, communities, ecosystems), accounting for direct and indirect effects that pesticides may cause," and recommendations have been made for incorporating population-level assessment methods in higher tiers of pesticide risk assessments (ECOFRAM 1999a, b). No explicit assessment of risks to populations is made.

Appendix 2: Workshop Exercise: Application of 2 Modeling Techniques in a Theoretical Assessment for Agricultural Pesticide Registration

Chris Topping, Richard Sibly, Peter D. Delorme, Vibeke Moller, Alyce T. Fritz, Niels Elmegaard, and Wayne R. Munns, Jr.

An exercise was conducted during the workshop to illustrate how 3 of the modeling approaches described in Chapter 9 could be used to support pesticide registration decisions. The exercise involved a fictitious insecticide, developed to protect cereal crops in Danish agroecosystem landscapes, and ecological risks to populations of common skylark, *Alauda arvensis*. The common skylark is a ground-nesting species that feeds almost exclusively on insects during the breeding season, and thus would be maximally exposed to pesticides applied to field crops. The exercise employed both biologically structured and spatially explicit, individual-based models to project population-level effects from the mandatory premarket standard toxicity data assumed to be provided by the pesticide manufacturer. The general scenario assumed 80% of cereal fields growing winter wheat in an arbitrary Danish landscape to receive a single application of the pesticide in early to mid-May, with winter wheat representing 50% of the arable area. There were no other pesticide stressors present. However, other agricultural management stressors (e.g., soil cultivation, harvest) were to be simulated. Weather was assumed to be variable and typical of the landscape modelled. The regulatory context for the agricultural pesticide registration scenario is described in Appendix 1.

PESTICIDE TOXICITY DATA

An artificial toxicity data set was generated to reflect a hazard quotient of 20 for reproductive impacts, and a hazard quotient of 2 for direct mortality. To allow input into the population models, a comparison of the dietary toxicity to acute oral toxicity was conducted by converting dietary uptake to appropriate units (mg/kg diet to mg/kg bw or mg/individual) using allometric relationships for food consumption (USEPA 1993). Toxicity values were similar when converted; therefore, results from the oral toxicity tests were used. Concentration response curves were recalculated from the original acute oral toxicity units of mg/kg to mg/individual based on mean adult weights for bobwhite quail (Dunning 1993) and Odderskær et al. (1997) for skylark. This allowed a determination of the probability of effect (i.e., death) for a specified level of intake for an individual of average size. Pesticide concentrations in insects were determined using the field application rate and the nomogram of Hoerger and Kenaga (1972) as modified by Fletcher et al. (1994). Dietary requirements for skylarks were determined as a function of caloric requirements and energy content of food items. Based on this information, and the concentration in food and daily dietary intake, it was determined that for skylarks the probability of mortality as a result of acute oral toxicity was approximately 0.025 per application. Similarly, a reduction of approximately 12.5% in fecundity would occur in birds exposed to field concentrations.

MODEL DESCRIPTIONS

The first model was a generic, biologically structured model using population growth rate and mortality to predict the carrying capacity of the skylark population. This model was used to predict the direct impact of mortality on population size. The analysis assumed a uniform distribution of birds in the landscape, so that exposure was distributed pro-rata as a function of the area of landscape sprayed. All birds exposed were assumed to suffer increased mortality and decreased reproduction expected from field-rate exposure. Age of first reproduction and the interval between breeding seasons were both assumed to be 1 year. It was assumed further that

- survival of exposure to the pesticide was independent of survival of other mortality agents,
- reproductive performance was not depressed at first breeding or at end of life, and
- all juveniles had identical performance and so did all adults.

Population growth rate, r, at low density was estimated from the Euler-Lotka equation, which in this case took the form (Calow and Sibly 1990):

$$r = \log_e\left(\tfrac{1}{2} n S_j + S_a\right)$$

At higher densities it was assumed that population growth rate declined linearly with density, as in Sibly et al. (2000).

The spatially explicit model (SEM) system used in this exercise is ALMaSS (Topping et al. 2003a). This model takes ecological and behavioral information into account together with the spatiotemporal distribution of the stressor. The model is agent based; thus, individuals actively seek information from their environment to

TABLE A2.1
Scenario descriptions

Scenario	Acute toxicity	Reproductive depression	Indirect effects included*
Baseline	0	0	N/A
Acute toxicity only	2.5	0	Yes
Acute toxicity plus reproductive depression	2.5%	12.5%	Yes
Doubled effects of acute toxicity plus reproductive depression	5.0%	25.0%	Yes
Direct effects only	5.0%	12.5%	No

*Spatially explicit model only.

achieve their objectives of survival and reproduction. This complexity results in the population responding to local spatiotemporal changes in the distribution of the stressor, generating appropriate negative or positive feedback responses. The model is based on a comprehensive landscape model that incorporates topography, climate, and land management, providing locally based information to which the animal models respond.

A skylark model for ALMaSS already existed (Topping and Odderskær 2004) and was used for the simulations. This model was modified to respond to direct mortality resulting from pesticide application and to reductions in reproductive output. Direct mortality was implemented as a probability of mortality dependent on the dose of active ingredient received via diet. Decreased reproductive output was assumed for all female birds exposed based on the dose-response probability from the reproductive data set. This was implemented by determining the chance of each egg being infertile based on the probability of response from the dose-response data. In short, the SEM differed from the biologically structured model in that it incorporated spatiotemporal variation and nonequilibrium properties of the system, as well as by explicitly including individual variability.

RISK SCENARIOS

Several risk scenarios were modeled to examine the effect of varying adult mortality, reproductive effects, inclusion of indirect effects, and varying the simulated amount of arable landscape (Table A2.1). Under baseline conditions, none of the crop was treated with the pesticide, whereas all other scenarios assumed that 80% of winter wheat was treated. Scenarios modeled included effects of acute toxicity only (A, or effects of acute toxicity plus reproductive depression (reduced fecundity) (AR). A scenario in which effects were doubled (ARx2) was also investigated to allow for uncertainties in extrapolation. Indirect effects of insecticide application were investigated using the SEM, although 1 scenario considered direct effects only (ARd).

All scenarios were evaluated for a simulation period of 200 years from an initial standard starting set of conditions. The life-table under baseline conditions was

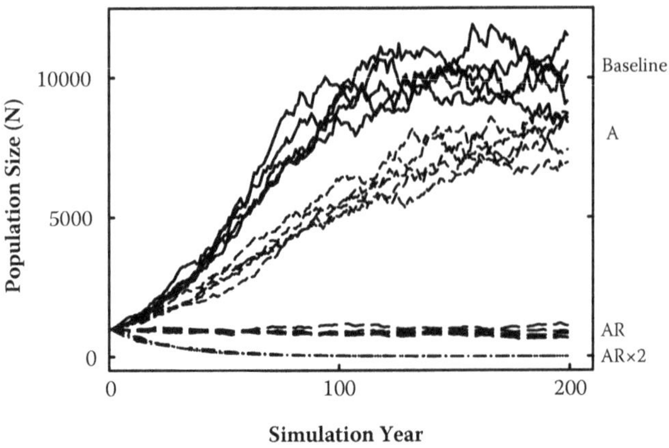

FIGURE A2.1 Skylark biologically structured model simulations.

assumed to be $S_j = 0.277$, $S_a = 0.5$, $n = 3.896$, giving a maximum possible value of $r = 0.0388$ in the scenario landscape. These values were specific to skylarks in the scenario landscape and were derived from the parameters used in the skylark model of Topping and Odderskær (2004). It was assumed that pesticide exposure reduced S_a by 0.025 (or 0.05 in the ARx2 scenario), and n by 12.5% (or 25% in the ARx2 scenario).

SIMULATION RESULTS

The biologically structured model predicted that, under baseline conditions, the population would increase roughly exponentially for the first 50 years, and reach an equilibrium abundance (carrying capacity) of 10000 birds soon after year 100 (Figure A2.1). Initial growth was slower in scenario A, and carrying capacity was reduced by 26% relative to Baseline. The skylark population declined throughout the simulation period in the AR simulation, but only very slightly. In the ARx2 simulation, the population declined rapidly and extinction occurred in 3 of the 5 replicates by year 200.

The SEM simulation results differed from those involving the biologically structured model in the degree of impact of the pesticide (Figure A2.2) temporally averaged population size was reduced by 25% relative to baseline abundance in scenario A, by 40% in scenario AR, and by 60% in scenario ARx2. Without indirect effects (ARd), there was only a slight reduction in temporally averaged population size compared with the baseline scenario, because of the population taking more than 100 years to reach the same carrying capacity as the baseline scenario. Population extinction occurred in 3 of the 10 replicate runs in ARx2, suggesting a high probability that the long-term population trajectory in this scenario is extinction.

The differences in population trajectories between the 2 models were due to a number of factors. A key difference between the models is that the SEM included compensating factors, such that at low densities, reproductive success rates for

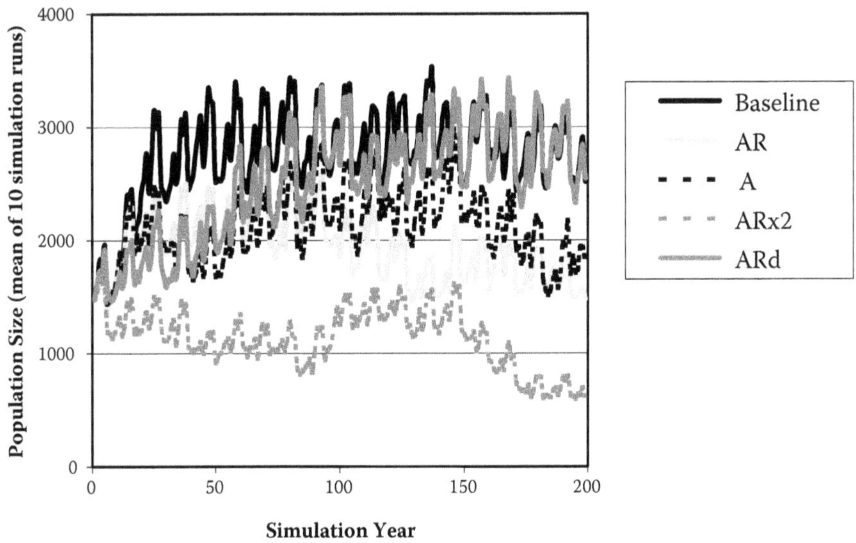

FIGURE A2.2 Skylark spatially explicit model simulations.

individuals were higher than at high densities when a large fraction of the population is nonbreeding. This resulted effectively in a much higher actual growth rate than the 4% assumed in the biologically structured model. In addition, the SEM incorporated birds living in optimal habitats (3% by area). These birds had a significantly higher reproductive rate because of good habitat quality, and thus the overall capability of the population to resist stress was improved by the surplus reproductive output of a small proportion of the birds. The contribution of immigrants from these habitats to the agricultural population was an emergent property of the SEM. To be reflected in the biologically structured model, this property would have to be parameterized explicitly in the generic model for every scenario. The models also differed in that the SEM assumed exposure of chicks only when they were in the fields at the time the pesticide was sprayed, whereas all chicks were exposed in the biologically structured model. The timing of realized impacts was also important. The SEM allowed the possibility of renesting (or remating and renesting if 1 of a pair has been killed) if the pesticide was applied early, and of having broods which were too early or too late to be affected by the pesticide. These possibilities were not reflected in the biologically structured model, and their effects were hard to predict in advance of running the SEM.

EXERCISE CONCLUSION

Clearly, not all aspects of a population-level ecological risk assessment were fully developed in this exercise. For example, there was little consideration of how birds would be exposed or of the standard procedures that normally would have been used in an assessment supporting pesticide registration. Further, modeling assumptions and uncertainties would need to be explored more fully before the assessments could be used in an actual decision-making context. Despite these limitations, we believe that this exercise provides an indication of what can be achieved and shows the potential utility of population modelling to support decision-making. Subsequent to this exercise, the use of these models in risk assessment was explored further by Topping et al. (2005) employing a similar comparative approach.

Appendix 3: Supplemental Reading

Supplemental citations are divided into the following 3 categories:

A3.1 Risk Assessment
A3.2 Ecotoxicology
A3.3 Population Ecology

A3.1 RISK ASSESSMENT

Adams SM, Bevelheimer MS, Greeley MS, Levine DA, Teh SJ. 1999. Ecological risk assessment in a large river-reservoir: 6. Bioindicators of fish population health. Environ Toxicol Chem 18:628–640.

Anderson DR, White GC, Burnham KP. 1995. Some specialized risk assessment methodologies for vertebrate populations. Environ Ecol Stat 2:91–115.

Ankley GT, DiToro DM, Hansen DJ, Berry WJ. 1996. Assessing the ecological risk of metals in sediments. Environ Toxicol Chem 15:2053–2055.

Babut M, Breuzin C. 2000. Relevance of the selection of pesticides for a river monitoring network with a simple risk assessment method. J Water Sci 13:363–378.

Barnthouse LW. 1992. The role of models in ecological risk assessment: a 1990's perspective. Environ Toxicol Chem 11:1751–1760.

Boughton D, Malvadkar U. 2002. Extinction risk in successional landscapes subject to catastrophic disturbances. Conserv Ecol 6:2. (Available from: http://www.consecol.org/vol6/iss2)

Briggs GB. 1996. Risk assessment policy for evaluating reproductive system toxicants and the impact of responses on sensitive populations. Toxicology 111:305–313.

Calow P, Forbes V. 2003. Does ecotoxicology inform ecological risk assessment? Environ Sci Technol 37:146A–151A.

Calow P. 1998. Ecological risk assessment: risk for what? How do we decide? Ecotoxicol Environ Safety 40:15–18.

Calow P, Sibly RM, Forbes VE. 1997. Risk assessment on the basis of simplified life-history scenarios. Environ Toxicol Chem 16:1983–1989.

Campbell PJ, Arnold DJS, Brock TCM, Grandy NJ, Heger W, Heimbach G, Maund SJ, Streloke M. Guidance document on higher tier aquatic risk assessment for pesticides (HARAP). Pensacola (FL): SETAC.

Caswell H. Prospective and retrospective perturbation analyses: their roles in conservation biology. Ecology 81:619–627.

Chapman PM, Wang FY, Janssen CR. 2003. Conducting ecological risk assessments of inorganic metals and metalloids: current status. Hum Ecol Risk Assess. 9:641–697.

Cook R, Suter G II. 1999. Ecological risk assessment in a large river-reservoir: 1. Introduction and background. Environ Toxicol Chem 18:581–588.

Crutchfield J, Ferson S. 2000. Predicting recovery of a fish population after heavy metal impacts. Environ Sci Policy 3:S183–S189.

Cura JJ. 1998. Ecological risk assessment. Water Environ Res 70:968–971.

Daniels RE, Allan JD. 1981. Life table evaluation of chronic exposure to a pesticide. Can J Fish Aquat Sci 38:485–494.

Detenbeck NE, Batterman SL, Brady VJ, Brazner JC, Snarski VM, Taylor DL, Thompson JA, Arthur JW. 2000. A test of watershed classification systems for ecological risk assessment. Environ Toxicol Chem 19:1174–1181.

Duke D, Taggart M. 2000. Uncertainty factors in screening ecological risk assessments. Environ Toxicol Chem 19:1668–1680.

Efroymson RA, Suter GW II. 2001. Ecological risk assessment framework for low-altitude aircraft overflights: II. Estimating effects on wildlife. Risk Anal 2:263–274.

Efroymson RA, Suter GW II, Rose WH. 2001. Ecological risk assessment framework for low-altitude aircraft overflights: I. Planning the analysis and estimating exposure. Risk Anal 21:251–262.

Emlen JM. 1989. Terrestrial population models for ecological risk assessment: a state-of-the-art review. Environ Toxicol Chem 8:831–842.

Fairbrother A. 2003. Lines of evidence in wildlife risk assessments. Hum Ecol Risk Assess 9:1475–1491.

Foley P. 1994. Predicting extinction times from environmental stochasticity and carrying capacity. Conserv Biol 8:124–137.

Forbes VE, Calow P, Sibly RM. 2001. Are current species extrapolation models a good basis for ecological risk assessment? Environ Toxicol Chem 20:442–447.

Forbes VE, Calow P. 2002. Extrapolation in ecological risk assessment: balancing pragmatism and precaution in chemical controls legislation. Bioscience 52:249–257.

Gagne JA. 1994. The future application of ecological models in environmental risk assessment. In: Kendall RJ. TE Lacher, editors. Wildlife Toxicology Population Model: Integrated Studies of Agroecosystems # 497–499. Pensacola (FL): SETAC Press.

Gentile JH, Harwell MA, Cropper W. 2001. Ecological conceptual models: a framework and case study on ecosystem management for south Florida sustainability. Sci Total Environ 274:231–253.

Ginzburg L, Ferson S. 1999. Population and ecosystem-level risk analysis under 316(b) of the Clean Water Act. Proceedings of the 1998 EPRI Clean Water Act section 316(b) technical workshop. Palo Alto (CA): Electric Power Research Institute.

Gleason TR, Munns WR, Nacci DE. 2000. Projecting population-level response of purple sea urchins to lead contamination for an estuarine ecological risk assessment. J Aquatic Ecosys Stress Recovery 7:177–185.

Goldwasser L, Ginzburg L, Ferson S. 2000. Variability and measurement error in extinction risk analysis: the northern spotted owl on the olympic peninsula. Quantitative methods for conservation biology. In: S Ferson, Hakoyama M, Iwasa HY, editors. Assessment of extinction risk from fluctuating population Size. J Theor Biol New York 2000 pp. 169–187.

Hakoyama H, IwasaY, Nakanishi J. 2000. Comparing risk factors for population extinction. J Theor Biol Vol. 204, Number 3, June 2000 pp. 327–336.

Hall LW, Giddings JM, Solomon KR. 1999. An ecological risk assessment for the use of irgarol 1051 as an algaecide for antifouling paints. Crit Rev Toxicol 29:367–437.

Hope BK, Peterson JA. 2000. A procedure for performing population-level. Ecological risk assessments. Environ Manage 25:281–289.

Janssen CR, DeSchamphelaere K, Heijerick D. 2000. Uncertainties in the environmental risk assessment of metals. Hum Ecol Risk Assess 6:1003–1018.

Jepson PC, Sherratt TN. 1996. The dimensions of space and time in the assessment of ecotoxicological risks. In: Baird DJ, Maltby L, Greig-Smith PW. Douben PET, editors. Ecotoxicology: ecological dimensions. London (UK): Chapman & Hall. p. 43–54.

Johnson AR. 2002. Landscape ecotoxicology and assessment of risk at multiple scales. Hum Ecol Risk Assess 8:127–146.

Jones D, Domotor S, Higley K, Kocher D, Bilyard G. 2003. Principles and issues in radiological ecological risk assessment. J Environ Radioact 66:19–39.

Jones D, Barnthouse L. 1999. Ecological risk assessment in a large river-reservoir: 3. Benthic invertebrates. Environ Toxicol Chem 18:599–609.

Klaine SJ, Cobb GP, Dickerson RL, Dixon KR, Kendall RJ, Smith EE, Solomon KR. 1996. An ecological risk assessment for the use of the biocide, *Dibromonitrilopropionamide* (DBNPA), in industrial cooling systems. Environ Toxicol Chem 15:21–30.

Kuhn A, Munns WR, Serbst J, Edwards P, Cantwell MG, Gleason T, Pelletier MC, Berry W. 2002. Evaluating the ecological significance of laboratory response data to predict population-level effects for the estuarine Amphipid ampelisca abdita. Environ Toxicol Chem 21:865–874.

Kuhn A, Munns WR, Poucher S, Champlin D, Lussier S. 2000. Prediction of population-level response from mysid toxicity test data using population modeling techniques. Environ Toxicol Chem 19:2364–2371.

Lackey RT. 1997. If ecological risk assessment is the answer, what is the question? Hum Ecol Risk Assess 3:921–928.

Lackey RT. 1997. Ecological risk assessment: use, abuse, and alternatives. Environ Manage 21:808–812.

Landis WG, Matthews RA, Matthews GB. 1996. The layered and historical nature of ecological systems and the risk assessment of pesticides. Environ Toxicol Chem 15:432–440.

Leeuwangh P, Brock TCM, Kersting K. 1994. An evaluation of four types of freshwater model ecosystem for assessing the hazard of pesticides. Hum Exp Toxicol 13:888–899.

Lowell RB, Culp JM, Dube MG. 2000. A weight-of-evidence approach for northern river risk assessment: integrating the effects of multiple stressors. Environ Toxicol Chem 19:1182–1190.

Maund SJ, Hamer MJ, Warinton JS, Kedwards TJ. 1998. Aquatic ecotoxicology of the pyrethroid insecticide lambda-cyhalothrin: considerations for higher-tier aquatic risk assessment. Pesticide Sci 54:408–417.

McClung G, Sayre PG. 1994. Ecological risk assessment case study: risk assessment for the release of recombinant Rhizobia at a small-scale agricultural field site. In: A review of ecological assessment case studies from a risk assessment perspective, Volume II. Washington, DC: US Environmental Protection Agency. (USEPA). EPA/630/R-94/003.

Menzie CA, Burmaster DE, Freshman DS. Callahan C. 1992. Assessment of methods for estimating ecological risk in the terrestrial component: a case study at the Baird and McGuire superfund site in Holbrook, Massachusetts. Environ Toxicol Chem 11:245–260.

Menzie CA, Freshman JS. 1997. An assessment of the risk assessment paradigm for ecological risk assessment. Hum Ecol Risk Assess 3:853–892.

Moraes R, Landis WG, Molander S. 2002. Regional risk assessment of a Brazilian Rain Forest Reserve. Hum Ecol Risk Assess 8:1779–1803.

Morton MG, Dickson K, Waller W, Acevedo M, Mayer F, Ablan M. 2000. Methodology for the evaluation of cumulative episodic exposure to chemical stressors in aquatic risk assessment. Environ Toxicol Chem 19:1213–1221.

Munkittrick KR, Dixon DG. 1989. A holistic approach to ecosystem health assessment using fish population characteristics. Hydrobiologia 188/189:123–135.

Munns WR, Walker HA, Paul JF, Gentile JH. 1996. Ocean disposal at the 106-mile deepwater dumpsite as an ecological risk assessment case study. J Mar Environ Engineering 3:279–297.

Munns WR. 2002. Axes of extrapolation in risk assessment. Hum Ecol Risk Assess 8:19–29.

Murata M, Masunaga S, Nakanishi J. 2003. Population-level ecological risk assessment of planar polychlorinated aromatic hydrocarbons in great cormorant (Phalacrocorax carbo) around Tokyo Bay, Japan. Environ Toxicol Chem 22:2508–2518.

Nacci DE, Serbst J, Gleason TR, Cayula S, Thursby G, Munns WR, RK Johnston. 2000. Biological responses of the sea urchin, Arbacia punctulata, to lead contamination for an estuarine ecological risk assessment. J Aquatic Ecosys Stress Recov 7:187–199.

Nakamaru M, Iwasa Y, Nakanishi J. 2002. Extinction risk to herring gull populations from DDT Exposure. Environ Toxicol Chem 21:195–202.

Nakamaru M, Iwasa Y, Nakanishi J. 2001. Ecological risk of DDT—a case study of biomagnification causing extinction risk of herring gull in Long Island. Environ Sci 14: 61–72.

Newman MC, Ownby DR, Mezin LCA, Powell DC, Christensen TRL, Lerberg SB, Anderson B. 2000. Applying species-sensitivity distributions in ecological risk assessment: assumptions of distribution type and sufficient numbers of species. Environ Toxicol Chem 19:508–55.

Osowksi SL, Swick JD, Carney GR, Pena HB, Danielson JE, Parrish DA. 2001. A watershed-based cumulative risk impact analysis: environmental vulnerability and impact criteria. Environ Mon Assess 66:159–185.

Pastorok RA, Bartell SM, Ferson S, Ginzburg LR. 2002. Ecological modeling in risk assessment: chemical effects on populations, ecosystems, and landscapes. Boca Raton (FL): CRC Press, Lewis Publishers.

Posthuma L, Traas TP, Suter GW, editors. 2002. Species sensitivity distributions in risk assessment, Boca Raton (FL): CRC Press.

Power M, Adams, SM, editors. 1997. Perspectives of the scientific community on the status of ecological risk assessment. Environ Manage 21:803–830.

Pulliam HR. 1994. Incorporating concepts from population and behavioral ecology into models of exposure to toxins and risk assessment. In: Kendall RJ, Lacher TE, editors. Wildlife toxicology population model. Boca Raton (FL): Lewis Publishers. p. 13–26.

Regan HM, Akcakaya HR, Ferson S, Root KV, Carroll S, Ginzburg LR. 2003. Treatments of uncertainty and variability in ecological risk assessment of single-species populations. Hum Ecol Risk Assess 9:889–906.

Ryti RT, Markweise J, Mirenda R, Soholt L. 2004. Preliminary remediation goals for terrestrial wildlife. Hum Ecol Risk Assess 10:1–14.

Saint-Jaques N, Harvey HH, Jackson DA. 1989. The use of white sucker (*Catostomus commersoni*) populations to assess the health of aquatic ecosystems exposed to low-level contaminant stress. Can J Fish Aquatic Sci 46:1455–1462.

Sears MK, Hellmich RL, Stanley-Horn DE, Oberhauser KS, Pleasants JM, Mattila HR, Siegfried BD, Dively GP. 2001. Impact of bt corn pollen on monarch butterfly populations: a risk assessment. Proc Natl Acad Sci USA 98:11937–11942.

Seiler FA, Alvarez JL. 1995. Toward a new risk assessment paradigm: variability, uncertainty, and errors. Technol J Franklin Inst 332A:221–235.

Sherratt TN, Roberts G, Williams P, Whitfield M, Biggs J, Shilabeer N, Maund SJ. 1999. A life-history approach to predicting the recovery of aquatic invertebrate populations after exposure to xenobiotic chemicals. Environ Toxicol Chem 18:2512–2518.

Sherratt TN, Jepson PC. 1993. A metapopulation approach to modeling the long-term impacts of pesticides on invertebrates. J Appl Ecol 30:696–705.

Sherry JP. 2003. The role of biomarkers in the health assessment of aquatic ecosystems. Aquatic Ecosys Health Manage 6:423–440.

Solomon KR, Sibley P. 2002. New concepts in ecological risk assessment: where do we go from here? Marine Pollut Bull 44:279–285.

Spencer M, Fisher NS, Wang WX, Ferson S. 2001. Temporal variability and ignorance in Monte Carlo contaminant bioaccumulation models: a case study with selenium in *Mytilus edulis*. Risk Anal 21:383–394.

Spencer M, Fisher NS, Wang WX. 1999. Exploring the effects of consumer-resource dynamics on bioaccumulation by aquatic herbivores. Environ Toxicol Chem 18:1582–1590.

Spencer M, Ginzburg LR, Goldstein RA. 1997. Community-level risk assessment, food chains and bioaccumulation. Environ Professional 19:90–97.

Spromberg JA, John BM, Landis WG. 2004. Metapopulation dynamics: indirect effects and multiple distinct outcomes in ecological risk assessment. In: Stark JD, Banks JE, Vargas RI, editors. How risky is risk assessment? The role that life history strategies play in susceptibility of species to stress. Proc Natl Acad Sci USA 101:732–736.

Suter GW II. 1999. A framework for assessment of ecological risks from multiple activities. Hum Ecol Risk Assess 5:397–413.

Suter GW II, Norton SB, Barnthouse LW. 2003. The evolution of frameworks for ecological risk assessment from the Red Book ancestor. Hum Ecol Risk Assess 9:1349–1360.

Suter GW II, Sample BE, DS Jones, Ashwood TL. 1994. Approach and strategy for performing ecological risk assessments for the U.S. Department of Energy's Oak Ridge Reservation: 1994 revision. ES/ER/TM-33/R1. Oak Ridge (TN): Oak Ridge National Laboratory.

Suter GW II, Barnthouse L. 1999. Ecological risk assessment in a large river-reservoir: 2. Fish community. Environ Toxicol Chem 18:589–598.

Suter GW II. 1999. Developing conceptual models for complex ecological risk assessments human ecology. Risk Assess 5:375–396.

Tanaka Y. 2003. Ecological risk assessment of pollutant chemicals: extinction risk based on population-level effects. Chemosphere 53:421–425

Tannenbaum LV, Bazar M, Hawkins MS, Cornaby BW, Ferguson EA, Carroll LC, Ryan PF. 2003. Rodent sperm analysis in field-based ecological risk assessment: pilot study at Ravenna Army Ammunition Plant, Ravenna, Ohio. Environ Pollut 123: 21–29.

Thacker JRM. Jepson PC. 1993. Pesticide risk assessment and non-target invertebrates: integrating population depletion, population recovery and experimental design. Bull Environ Contam Toxicol 51:523–531.

Theodorakis CW. 2001. Integration of genotoxic and population genetic endpoints in biomonitoring and risk assessment. Ecotoxicology 10:245–256.

Topping CJ, Odderskaer P. 2004. Modeling the influence of temporal and spatial factors on the assessment of impacts of pesticides on skylarks. Environ Toxicol Chem 23:509–520.

Traas TP, Stab JA, Kramer PRG, Cofino WP, Aldenberg T. 1996. Modeling and risk assessment of tributyltin accumulation in the food web of a shallow freshwater lake. Environ Sci Toxicol 30:1227–1237.

[USEPA] US Environmental Protection Agency. 1998. Guidelines for ecological risk assessment: risk assessment forum. Washington (DC): USEPA, EPA/630/R-95/002F.

Vanden Brink PJ, Hartgers EM, Gylstra R, Bransen F, Theo CM. 2002. Effects of a mixture of two insecticides in freshwater microcosms: II. Responses of plankton and ecological risk assessment. Ecotoxicology 11:181–197.

Walker R, Landis W, Brown P. 2001. Developing a regional ecological risk assessment: a case study of a Tasmanian agricultural catchment. Hum Ecol Risk Assess 7:402–417.

Wang G, Edge WD, Wolff J. 2001. Demographic uncertainty in ecological risk assessments. Ecol Model 136:95–102.

Weins J. 1996. Coping with variability in environmental impact assessment. In: Baird DJ, Maltby L, Greig-Smith PW, Douben PET, editors. Ecotoxicology: ecological dimensions. London (UK): Chapman & Hall. p. 55–70.

Wennergren U, Stark JD. 2000. Modeling long-term effects of pesticides on populations: beyond just counting dead animals. Ecol Appl 10:295–302.

Wennergren U, Stark JD. 1995. Can population effects of pesticides be predicted from demographic toxicological studies? J Econ Entomol 88:1089–1096.

Yoshinari T, Nakanishi J. 2000. Mean extinction time of populations under toxicant stress and ecological risk assessment. Environ Toxicol Chem 19:2856–2862.

A3.2 ECOTOXICOLOGY

Ares J. 2003. Time and space issues in ecotoxicology: population models, landscape pattern analysis, and long-range environmental chemistry. Environ Toxicol Chem 22:945–957.

Allan JD, Daniels RE. 1982. Life table evaluation of chronic exposure of *Eurytemora affinis* (Copepoda) to kepone. Mar Biol 66:176–184.

Baird DJ, Brock TCM, De Ruiter PC, Boxall ABA, Culp JM, Eldridge P, Hommen U, Jak RG, Kidd KA, DeWitt T. 2001. The food web approach in the environmental management of toxic substances. In: Baird DJ, Burton GA, editors. Ecological variability: separating natural from anthropogenic causes of ecosystem impairment. Pensacola (FL): SETAC. p. 83–122.

Baker RJ, Bickham AM, Bondarkov M, Gaschak SP, Matson CW, Rodgers BE, Wickliffe JK, Chesser RK. 2001. Consequences of polluted environments on population structure: the bank vole (*Clethrionomys Glareolus*) at Chernobyl. Ecotoxicology 10:211–216.

Banken JAO, Stark JD. 1998. Multiple routes of pesticide exposure and the risk of pesticides to biological controls: a study of neem and the devenspotted lady beetle (Coleoptera coccinellidae). J Econ Entomol 91:1–6.

Barata C, Baird DJ, Amat F, Soares AMVM. 2000. Comparing population response to contaminants between laboratory and field: An approach using daphnia magna ephippial egg banks. Funct Ecol 14:513–523.

Barata C, Baird DJ, Mitchell SE, Soares AMVM. 2002. Among- and within-population variability in tolerance to cadmium stress in natural populations of *Daphnia Magna*: implications for ecological risk assessment. Environ Toxicol Chem 21:1058–1064.

Barata C, Baird DJ, Soares AMVM. 2002. Demographic responses of a tropical Cladoceran to cadmium: effects of food supply and density. Ecol Appl 12:552–564.

Barron MG, Duvall SE, Barron KJ. 2004. Retrospective and current risks of mercury to panthers in the Florida Everglades. Ecotoxicology 13:223–229.

Benton MJ, Guttman SI. 1990. Effects of allozyme genotype on the survivorship of mayflies (Stenonem femoratum) exposed to copper. J North Am Benthol Soc 9:271–276.

Bergeron, JM, Crews D, McLachlan JA. 1994. PCBs as environmental estrogens: turtle sex determination as a biomarker for environmental contamination. Environ Health Perspect 102:780–781.

Beyer WN, Heinz GH. 2000. Implications of regulating environmental contaminants on the basis of wildlife populations and communities. Environ Toxicol Chem 19:1703–1704.

Blaustein AR, Romansic JM, Kiesecker JM, Hatch AC. 2003. Ultraviolet radiation, toxic chemicals and amphibian population declines. Divers Distrib 9:123–140.

Blockwell SJ, Maund SJ, Pascoe D. 1999. Effect of the organochlorine insecticide lindane on the population responses of the freshwater *Amphipod hyallella Azteca*. Environ Toxicol Chem 18:1264–1269.

Boone MD, Semlitsch RD. 2003. Interactions of bullfrog tadpole predators and an insecticide: predation release and facilitation. Oecology 137:610–616.

Brown AR, Riddle AM, Cunningham NL, Kedwards TJ, Shillabeer N, Hutchinson TH. 2003. Predicting the effects of endocrine disrupting chemicals on fish populations. Hum Ecol Risk Assess 9:761–788.

Cairns VW, Hodson PV, Nriagu JO, editors. 1984. Contaminants Effects In Fish. New York (NY): Wiley.

Chapman PM, Ho K, Munns WR, Solomon K, Weinstein MP. 2002. Issues in sediment toxicity and ecological risk assessment. Mar Pollut Bull 44:271–278.

Chaumot A, Charles S, Flammarion P, Auger P. 2003. Ecotoxicology and spatial modeling in population dynamics: an illustration with brown trout. Environ Toxicol Chem 22:958–969.

Chaumot A, Charles S, Flammarion P, Garric J, Auger P. 2002. Using aggregation methods to assess toxicant effects on population dynamics in spatial systems. Ecol Appl 12:1771–1784.

Crowe TP, Smith EL, Donkin P, Barnaby DL, Rowland SJ. 2004. Measurements of sub-lethal effects on individual organisms indicate community-level impacts of pollution. J Appl Ecol 41:114–123.

Davidson C, Shaffer HB, Jennings MR. 2001. Declines of the California red-legged rrog: spatial analysis of climate, UV-B, habitat and pesticides hypotheses. Ecol Appl 11:464–479.

Davis DD, Miller TJ. 2003. Population level effects of contaminated sediments on an estuarine fish, Fundulus heteroclitus. Am Fish Soc Annu Meeting 133:318–319.

DeCoen WM, Janssen CR. 2003. The missing biomarker link: relationship between effects on the cellular energy allocation biomarker of toxicant-stressed *Daphnia Magna* and corresponding population characteristics. Environ Toxicol Chem 22:1632–1641.

Downs TJ, Ambrose RF. 2001. Syntropic ecotoxicology: a heuristic model for understanding the vulnerability of ecological systems to stress. Ecosyst Health 7:266–283.

Duvall SE, MG Barron. 2000. A screening-level probabilistic ecological risk assessment of mercury in Florida Everglades food webs. Ecotoxicol Environ Safe 47:298–305.

Ferson S, Ginzburg LR, Goldstein RA. 1995. Inferring ecotoxicological risk from toxicity bioassays. Water Air Soil Pollut 90:71–82.

Flores JLG, Sarma SSS, Araiza MAF. 1999. Combined effects of chlorella density and methyl parathion concentration on the population growth of brachionus Calyciflorus (*Rotifera*). Bull Environ Contam Toxicol 62:769–775.

Forbes VE, Sibly RM, Calow P. 2001. Toxicant impacts on density-limited populations: a critical review of theory, practice, and results. Ecol Appl 11:1249–1257.

Fort DJ, Mclaughlin DW. 2003. Establishing cause-effect relationships for chemical stressors in amphibians: providing adequate data for the era. Multiple stressor effects in relation to declining amphibian populations. Am Soc Testing Mater Spec Tech Publ 1443:219–232.

Fox GA, Kennedy SW, Norstrom RJ, Wigfield DC. 1988. Porphyria in herring gulls: a biochemical response to chemical contamination of great lakes food chains. Environ Toxicol Chem 7:831–839.

Fry DM, Toone CK. 1981. DDT-induced feminization of gull embryos. Science 213:222–224.

Fuma S, Ishii N, Takeda NH, Miyamoto K, Yanagisawa K, Ichimasa Y, Saito M, Kawabata Z, Polikarpov GG. 2003. Ecological effects of various toxic agents on the aquatic microcosm in comparison with acute ionizing radiation. J Environ Radioact 67:1–14.

Gentile JH, Gentile SM, Hairston NG, Sullivan BK. 1982. The use of life-tables for evaluating the chronic toxicity of pollutants to Mysidopsis bahia. Hydrobiologia 93:179–187.

Gentile JH, Gentile SM, Hoffman G, Heltshe JF, Hairston N. 1983. The effects of a chronic mercury exposure on survival, reproduction and population dynamics of Mysidopsis bahia. Environ Toxicol Chem 2:61–68.

Gibbs PE, Bryan GW. 1986. Reproductive failure in populations of the dog-whelk, *Nucella lapillus*, caused by imposex induced by tributyltin from anti-fouling paints. J Mar Biol Assoc UK 66:767–777.

Gray LE, Kelce WR. 1996. Latent effects of pesticides and toxic substances on sexual differentiation of rodents. Toxicol Ind Health 12:515–531.

Gray MA, Metcalfe CD. 1997. Induction of testis-ova in Japanese medaka Oryzias latipes exposed to p-nonylphenol. Environ Toxicol Chem 16:1082–1086.

Guillette LJ, Crain DA, Rooney AA, Pickford DB. 1995. Organization versus activation: the role of endocrine-disrupting contaminants (EDCs) during embryonic development in wildlife. Environ Health Persp 103:157–164.

Guillette LJ, Gross TS, Masson TR, Matter JM, Percival HF, Woodward AR. 1994. Developmental anomalies of the gonad and abnormal sex hormone concentrations in juvenile alligators from contaminated and control lakes in Florida. Environ Health Persp 102:680–688.

Guckert JB. 1996. Toxicity assessment by community analysis. J Microbiol Meth 25:101–112.

Haimi J, Paavola S. 1998. Responses of two earthworm populations with different exposure histories to *Chlorophenol* contamination. Environ Toxicol Chem 17:1114–1117.

Hallam TG, Lassiter RR, Li J, McKinney W. 1990. Toxicant-induced mortality in models of daphnia populations. Environ Toxicol Chem 9:597–621.

Hitchcock DR, Mccutcheon SC, Smith MC. 2003. Using rotifer population demographic parameters to assess impacts of the degradation products from trinitrotoluene phytoremediation. Ecotoxicol Environ Safety 55:143–151.

Hooper HL, Sibly RM, Maund SJ, Hutchinson TH. 2003. The joint effects of larval density and ^{14}C-cypermethrin on the life history and population growth rate of the midge Chironomus riparius. J Appl Ecol 40:1049–1059.

Hunt JW, Anderson BS, Phillips BM, Tjeerdema RS, Taberski KM, Wilson CJ, Puckett HM, Stephenson M, Fairey R, Oakden J. 2001. A large-scale categorization of sites in San Francisco Bay, USA, based on the sediment quality triad, toxicity identification evaluations, and gradient studies. Environ Toxicol Chem 20:1252–1265.

Jak RG, Maas JL, Scholten MCT. 1998. Ecotoxicity of 3, 4-dichloroaniline in enclosed freshwater plankton communities at different nutrient levels. Ecotoxicology 7:49–60.

Jak RG, Ceulemans M, Scholten MCT, VanStraalen NM. 1998. Effects of Tributyltin on a coastal north sea plankton community in enclosures. Environ Toxicol Chem 17:1840–1847.

Jensen A, Forbes VE, Davis PE. 2001. Variation in cadmium uptake, feeding rate, and life-history effects in the gastropod Potamopyrgus antipodarum: linking toxicant effects on individuals to the population level. Environ Toxicol Chem 20:2503–2513.

Jobling S, Sheahan D, Osborne JA, Matthiessen P, Sumpter PJP. 1996. Inhibition of testicular growth in rainbow trout oncorhynchus mykiss exposed to estrogenic alkylphenolic chemicals. Environ Toxicol Chem 15:194–202.

Johnston RK, Munns WR, Tyler PL, Marajh-Whittemore P, Finkelstein K, Munney K, Short FT, Melville A, Hahn S. 2002. Weighing the evidence of ecological risk of chemical contamination in the estuarine environment adjacent to the Portsmouth Naval Shipyard, Kittery, Maine, USA. Environ Toxicol Chem 21:182–194.

Jones LA, Hajek RA. 1995. Effects of estrogenic chemicals on development. Environ Health Persp 103:63–67.

Kammenga JE, Laskowski R. 2000. Demography in ecotoxicology. Chichester (UK): John Wiley Inc.

Kammenga JE, VanKoert PHG, Koeman JH, Bakker J. 1997. Fitness consequences of toxic stress evaluated within the context of phenotypic plasticity. Ecol Appl 7:726–734.

Karasov WH, Meyer MW. 2000. Testing the role of contaminants in depressing avian numbers. Revista Chilena De Historia Natural 73:461–471.

Kareiva P, Stark J, Wennergren U. 1996. Using demography theory, community ecology and spatial models to illuminate ecotoxicology. In: Baird DJ, Maltby L, Greig-Smith PW, Douben PET, editors. Ecotoxicology: ecological dimensions. London (UK): Chapman & Hall. p. 13–23.

Kedwards TJ, Maund SJ, Chapman PF. 1999. Community level analysis of ecotoxicological field studies: II. Replicated-design studies. Environ Toxicol Chem 18:158–166.

Kim GW, Kwon T, Gearhiser M, Marschall E, Fisher S, Stein R. 2003. PCB biomagnification in the zebra mussel-round goby-smallmouth bass food chain in Lake Erie. Amer Fish Soc Annu Mtg 133:282–283.

Kramarz P, Stark JD. 2003. Population level effects of cadmium and the insecticide, imidacloprid to the parasitoid, Aphidius ervi, after exposure through its host, the Pea aphid, *acyrthosiphon pisum*. Biol Control 27:310–314.

Kuhn A, Poucher WR, Munns S, Champlin D, Lussier S. 2000. Prediction of population-level response from mysid toxicity test data using population modeling techniques. Environ Toxicol Chem 19:2364–2371.

Kuhn A, Munns WR, Champlin D, MicKinney R, Tagliabue J, Serbst J, Gleason T. 2001. Evaluation of the efficacy of extrapolation population modeling to predict the dynamics of *Americamysis bahia* populations in the laboratory. Environ Toxicol Chem 20:213–221.

Larno V, Laroche J, Launey S, Flammarion P, Devaux A. 2001. Responses of chub (*Leuciscus Cephalus*) populations to chemical stress, assessed by genetic markers, Dna damage and Cytochrome P4501a induction. Ecotoxicology 10:145–158.

Laroche J, Quiniou L, Juhel G, Auffret M, Moraga D. 2002. Genetic and physiological responses of flounder *Platichthys flesus* populations to chemical contamination in estuaries. Environ Toxicol Chem 21:2705–2712.

Larsen DK, Wagner I, Gustavson K, Forbes VE, Lund T. 2003. Long-term effect of sea-nine on natural coastal phytoplankton communities assessed by pollution induced community tolerance. Aquat Toxicol Amsterdam 62:35–44.

Lassiter RR, Hallam TG. 1990. Survival of the fattest: implications for acute effects of lipophilic chemicals on aquatic populations. Environ Toxicol Chem 9:585–596.

Lavie B, Nevo E. 1982. Heavy metal selection of phosphoglucose isomerase allozymes in marine gastropods. Mar Biol 71:17–22.

Lavie B, Nevo E, Zoller U. 1984. Differential ability of phosphoglucose isomerase allozyme genotypes of marine snails in nonionic detergent and crude oil-surfactant mixtures. Environ Res 35:270–276.

Linke-Gamenick I, Vismann B, Forbes VE. 2000. Effects of fluoranthene and ambient oxygen levels on survival and metabolism in three sibling species of capitella Polychaeta. Mar Ecol-Prog Ser 194:169–177.

Luna-Andrade A, Aguilar-Duran R, Nandini S, Sarma SSS. 2002. Combined effects of copper and microalgal (*Tetraselmis suecica*) concentrations on the population growth of *Brachionus plicatilis* Muller (Rotifera). Water Air Soil Pollut 141:143–153.

Maltby L, Clayton SA, Yu H, McLoughlin N, Wood RM, Yin D. 2000. Using single-species toxicity tests, community-level responses, and toxicity identification evaluations to investigate effluent impacts. Environ Toxicol Chem 19:151–157.

Maltby L. 1999. Studying stress: the importance of organism-level responses. Ecol Appl 9:431–440.

Marshall JS. 1962. The effects of continuous gamma radiation on the intrinsic rate of natural increase of Daphnia pulex. Ecology 43:598–607.

Maurer BA, Holt RD. 1996. Effects of chronic pesticide stress on wildlife populations in complex landscapes: processes at multiple scales. Environ Toxicol Chem 15:420–426.

Mayer FL, Deans CH, Smith AG. 1987. Inter-taxa correlations for toxicity to aquatic organisms. Gulf Breeze (FL): US Environmental Protection Agency, Office of Research and Development. EPA/600/X-87/332.

McGee BL, Fisher DJ, Yonkos LT, Ziegler GP, Turley S. 1999. Assessment of sediment contamination, acute toxicity and population viability of the estuarine amphipod Leptocheirus plumulosus in Baltimore Harbor, Maryland. Environ Toxicol Chem 18:2151–2160.

McTavish K, Stech H, Stay F. 1998. A modeling framework for exploring the population-level effects of endocrine disruptors. Environ Toxicol Chem 17:58–67.

Michael J. 2002. Impact of herbicides on the forest ecosystem, aquatic ecosystems and wildlife: the U.S. Experience. Rev Forestiere Fr (Nancy) 54:593–508.

Mineau P. 2002. Estimating the probability of bird mortality from pesticide sprays on the basis of the field study record. Environ Toxicol Chem 21:1497–1506.

Moe SJ, Stenseth NC, Smith RH. 2001. Effects of a toxicant on population growth rates: sublethal and delayed responses in blowfly populations. Funct Ecol 15:712–721.

Moller V, Forbes VE, Depledge MH. 1996. Population responses to acute and chronic cadmium exposure in sexual and asexual estuarine gastropods. Ecotoxicology 5:313–326.

Montuelle B, Balandras B, Volat B, Feray C. 2003. Effect of wastewater treatment plant discharges on the functional nitrifying communities in river sediments. Aquat Ecosys Health Manage 6:381–390.

Munns WR, Black DE, Gleason TR, Salomon K, Bengston D, Gutjahr-Gobell R. 1997. Evaluation of the effects of dioxin and PCBs on Fundulus heteroclitus populations using a modeling approach. Environ Toxicol Chem 16:1074–1081.

Paradis S, Wheatley B, Boswell-Purdy J, Belisle D, Cole M, Lickers H, Hayton A, Davies K. 1997. Mercury contamination through fish consumption: a model for predicting and preventing hazardous behavior on a community level. Water Air Soil Poll 97:147–158.

Pesch CE, Munns WR, Gutjahr-Gobell R. 1991. Effects of contaminated sediment on life history traits and population growth rate of Neanthes arenaceodentata (polychaeta: nereidae) in the laboratory. Environ Toxicol Chem 10:805–815.

Polgar TT, Summers JK, Cummins RA, Rose KA, Heimbuch DG. 1985. Investigation of relationships among pollutant loadings and fish stock levels in northeastern estuaries. Estuaries 8:125–135.

Posthuma L, Suter II GW, Traas TP, editors. 2002. Species sensitivity distributions for ecotoxicology. Boca Raton (FL): CRC Press.

Preston BL. 2002. Indirect effects in aquatic ecotoxicology: Implications for ecological risk assessment. Environ Manage 29:311–323.

Roex EWM, VanGestel CAM, VanWezel AP, VanStraalen NM. 2000. Ratios between acute aquatic toxicity and effects on population growth rates in relation to toxicant mode of action. Environ Toxicol Chem 19:685–693.

Roy LA, Armstrong JL, Sakamota K, Steinert S, Perkins E, Lomax DP, Johnson LL, Schlenk D. 2003. The relationships of biochemical endpoints to histopathology and population metrics in feral flatfish species collected near the municipal wastewater outfall of Orange County, California. Environ Toxicol Chem 22:1309–1317.

Salice CJ, Miller TJ. 2003. Population-level responses to long-term cadmium exposure in two strains of the freshwater gastropod Biomphalaria glabrata: results from a life-table response experiment. Environ Toxicol Chem 22:678–688.

Sarma SSS, Nandini S, Perez TR. 2001. Combined effects of mercury and algal food density on the population dynamics of Brachionus patulus (Rotifera). Bull Environ Contam Toxicol 67:841–847.

Selck H, Riemann B, Christoffersen K, Forbes VE, Gustavson K, Hansen BW, Jacobsen JA, Kusk OK, Petersen S. 2002. Comparing sensitivity of ecotoxicological effect end-points between laboratory and field. Ecotoxicol Environ Safe 52:97–112.

Shi D, Wen-Xiong W. 2004. Understanding the differences in Cd and Zn bioaccumulation and sub cellular storage among different populations of marine clams. Environ Sci Technol 38:449–456.

Sibly RM. 1996. Effects of pollutants on individual life histories and population growth rates. In: Newman MC, Jagoe CH, editors. Ecotoxicology. A hierarchical treatment. Boca Raton (FL): Lewis Publishers. p. 197–223.

Sibly RM, Newton I, Walker CH. 2000. Effects of dieldrin on population growth rates of sparrow hawks. 1963–1986. J Appl Ecol 37:540–546.

Sibley PKM, Chappel J, George TK, Solomon KR, Liber K. 2000. Integrating effects of stressors across levels of biological organization: examples using Organophosphorus insecticide mixtures in field-level exposures. J Aquat Ecosys Stress Recov 7:117–130.

Sierszen ME, Lozano SJ. 1998. Zooplankton population and community responses to the pesticide azinphos-methyl in freshwater littoral enclosures. Environ Toxicol Chem 17:907–1914.

Spurgeon DJ, Svendson C, Weeks JM, Hankard PK, Stubberud HE, Kammenga JE. 2003. Quantifying copper and cadmium impacts on intrinsic rate of population increase in the terrestrial *Oligochaete lumbricus rubellus*. Environ Toxicol Chem 22:1465–1472.

Stark JD. 2001. Population-level effects of the neem insecticide, neemix, on *Daphnia pulex*. J Environ Sci Health [B]: Pesti Food Contam Agric Wastes 36:457–465.

Stark JD, Banks JE. 2001. Selective pesticides: are they less hazardous to the environment? Bio Sci 51:980–982.

Stark JD, Banks JE. 2003. Population-level effects of pesticides and other toxicants on arthropods. Annu Rev Entomol 48:505–519.

Stark JD, Vargas RI. 2003. Demographic changes in daphnia pulex (leydig) after exposure to the insecticides spinosad and diazinon. Ecotoxicol Environ Safe 56:334–338.

Stark JD, Walthall WK. 2003. Agricultural adjuvants: acute mortality and effects on population growth rate of Daphnia pulex after chronic exposure. Environ Toxicol Chem 22:3056–3061.

Stark JD, Wennergren U. 1995. Can population effects of pesticides be predicted from demographic toxicological studies? J Econ Entomol 88:1089–1096.

Stark JD, Jepson PC, Mayer DF. 1995. Limitations to use of topical toxicity data for predictions of pesticide side effects in the field. J Econ Entomol 88:1081–1088.

Tanaka Y, Nakanishi J. 2001. Life history elasticity and the population-level effect of p-nonylphenol on *Daphnia galeata*. Ecol Res 16:41–48.

Tatara CP, Mulvey M, Newman MC. 2002. Genetic and demographic responses of mercury-exposed *mosquitofish* (*Gambusia holbrooki*) populations to chemical contamination in estuaries. Environ Toxicol Chem 21:2191–2197.

Tatara CP, Mulvey M, Newman MC. 1999. Genetic and demographic responses of *mosquitofish* (*Gambusia holbrooki*) populations exposed to mercury for multiple generations. Environ Toxicol Chem 18:2840–2845.

Theodorakis CW, Bickham JW, Lamb T, Medica PA, Lyne TB. 2001. Integration of genotoxicity and population genetic analyses in kangaroo rats (*Dipodomys merriami*) exposed to radionuclide contamination at the Nevada test site, USA. Environ Toxicol Chem 20:317–326.

Theodorakis CW, Shugart LR. 1997. Genetic ecotoxicology 2. Population genetic structure in *Mosquitofish* exposed in situ to *Radionuclides*. Ecotoxicology 6:335–354.

Thomas CFG, Hol EHA, Evert JW. 1990. Modeling the diffusion component of dispersal during the recovery of a population of *linyphiid* spiders from exposure to an insecticide. Funct Ecol 4:357–368.

Traas TR, Janse JH, Vanden Brink PJ, Brock TCM, Aldenberg T. 2004. Freshwater food web model for the combined effects of nutrients and insecticide stress and subsequent recovery. Environ Toxicol Chem 23:521–529.

Traas TP, Janse JH, Aldenberg T, Brock TCM. 1998. A food web model for fate, direct, and indirect effects of Dursban® 4E (Active ingredient *chlorpyrifos*) in freshwater microcosms. Aquat Ecol 32:179–190.

VanBeelen P, Fleuren-Kemila AK, Aldenberg T. 2000. The relation between extrapolated risk, expressed as potentially affected fraction, and community effects, expressed as pollution-induced community tolerance. Environ Toxicol Chem 20:1133–1140.

Vanden Brink PJ, VanWijngaarden RPA, Lucassen WGH, Brock TCM, Leeuwangh P. 1996. Effects of the insecticide Dursban® 4E (active ingredient chlorpyrifos) in outdoor experimental ditches: II. Invertebrate community responses and recovery. Environ Toxicol Chem 15:1143–1153.

Vander Hoeven N, Gerritsen AAM. 1997. Effects of Chlorpyrifos on individuals and populations of *Daphnia Pulex* in the laboratory and field. Environ Toxicol Chem 16:2438–2447.

Walker CH, Hopkin SP, Sibly RM, Peakall D. 2001. Principles of ecotoxicology. 2nd ed. London (UK): Taylor and Francis.

Walthall WK, Stark JD. 1997. Comparison of two population-level ecotoxicological endpoints: the intrinsic (R_m) and instantaneous (R_i) rates of increase. Environ Toxicol Chem 16:1068–1073.

Walthall WK, Stark JD. 1999. The acute and chronic toxicity of two xanthene dyes, fluorescein sodium salt and phloxine B, to Daphnia pulex (Leydig). Environ Pollut 104:207–215.

Watts MM, Pascoe D, Carroll K. 2002. Population responses of the freshwater amphipod *Gammarus pulex* (L.) to an environmental estrogen, 17 alpha-ethinylestradiol. Environ Toxicol Chem 21:445–450.

Wong DCL, Whittle D, Maltby L, Warren P. 2003. Multivariate analyses of invertebrate community responses to a C12–15AE-3S anionic surfactant in stream Mesocosms. Aquat Toxicol (Amsterdam) 62:105–117.

A3.3 POPULATION ECOLOGY

Akçakaya HR. 1991. A method for simulating demographic stochasticity. Ecol Mod 54:133–136.

Akçakaya HR. 2002. Estimating the variance of survival rates and fecundities. Anim Conserv 5:333–336.

Akçakaya HR, Radeloff VC, Mladenoff DJ, He HS. 2004. Integrating landscape and metapopulation modeling approaches: viability of the sharp-tailed grouse in a dynamic landscape. Conserv Biol 18:4.

Bezel VS, Pozolotina VN, Bel'skii, EA, Zhuikova TV. 2001. Variation in population parameters: adaptation to toxic environmental factors. Russ J Ecol 32:413–419.

Biek R, Funk WC, Maxell BA, Mills LS. 2002. What is missing in amphibian decline research: Insight from ecological sensitivity analysis. Conserv Biol 16:728–734.

Brook BW, O'Grady JJ, Chapman AP, Burgman MA, Akcakaya HR, Frankham R. 2000. Predictive accuracy of population viability analysis in conservation biology. Nature 404:385–387.

Calow P, Forbes VE. 1998. How do physiological responses to stress translate into ecological and evolutionary processes? Comp Biochem Physiol A 120:11–16.
Caswell H. 1996. Demography meets ecotoxicology: untangling the population level effects of toxic substances. In: Newman MC, Jagoe CH, editors. Ecotoxicology: a hierarchical treatment. Boca Raton (FL): Lewis, Inc. p. 255–292.
Chaumot A, Charles S, Flammarion P, Auger P. 2003. Do migratory or demographic disruptions rule the population impact of pollution in spatial networks? Theor Popul Biol 64:473–480.
Cole LC. 1954. The population consequences of life history phenomena. Q Rev Biol 29:103–137.
Conway G. 2000. Genetically modified crops: Risks and promise. Conserv Ecol 4:2.
Emlen JM, Pikitch EK. 1989. Animal population dynamics: Identification of critical components. Ecol Model 44:253–273.
Ferson S. 1994. Using fuzzy arithmetic in Monte Carlo simulation of fishery populations. In: Kruse G, editor. Management strategies for exploited fish populations, proceedings of the international symposium on management strategies for exploited fish populations, 1992, Alaska Sea Grant College Program, AK-SG-93-02. p. 595–608.
Ferson S, Burgman M. 1995. Correlations, dependency bounds and extinction risks. Biol Conserv 73:101–105.
Ferson S, Ginzburg L, Silvers A. 1989. Extreme event risk analysis for age-structured populations. Ecol Model 47:175–187.
Forbes VE, Sibly RM, Linke-Gamenick I. 2003. Joint effects of a toxicant and population density on population dynamics: An experimental study using *Capitella* sp. I (*Polychaeta*). Ecol Appl 13:1094–1103. Issue 2.
Fuji T, Hope BK, Foxwell J, Stirling BA. 2000. Derivation of population-level ecological remedial action objectives: Tualitin River case study. Hum Ecol Risk Assess 6:1075–1086.
Ginzburg LR, Ferson S. 1992. Assessing the effect of compensation on the risk of population decline and extinction. Estuarine research in the 1980s. In: Lavett Smith C, editor. The Hudson River environmental society seventh symposium on Hudson River ecology. Albany (NY): State University of New York Press. p. 392–403.
Ginzburg LR, Slobodkin LB, Johnson K, Bindman AG. 1982. Quasi-extinction probabilities as a measure of impact on population growth. Risk Anal. 21:171–181.
Glaser D, Connolly JP. 2000. The use of ecotoxicology and population models in natural remediation. Natural remediation of environmental contaminants: its role in ecological risk assessment and risk management. General Publications Series. Pensacola (FL): SETAC. p. 121–157.
Hallam TG, Lassiter RR, Li J, Suarez LA. 1990. Modeling individuals employing an integrated energy response: application to daphnia. Ecology 71:938–954.
Hammers-Wirtz M, Ratte HT. 2000. Offspring fitness in *Daphnia*: is the *Daphnia* reproduction test appropriate for extrapolating effects on the population level? Environ Toxicol Chem 19:1856–1866.
Hansen F, Forbes VE, Forbes TL. 1999. Using elasticity analysis of demographic models to link toxicant effects on individuals to the population level: an example. Funct Ecol 13:877–879.
Holloway GJ, Sibly RM, Povey SR. 1990. Evolution in toxin-stressed environments. Funct Ecol 4:289–294.
Hooper HL, Sibly RM, Hutchinson TH, Maund SJ. 2003. The influence of larval density, food availability and habitat longevity on the life history and population growth rate of the midge Chironomus riparius. Oikos 102:515–524.

Iwasa Y, Hakoyama H. 1998. Extinction rate of a population with both demographic and environmental stochasticity. Theor Popul Biol 53:1–15.

Jolly GM. 1965. Explicit estimates from capture-recapture data with both death and immigration-stochastic model. Biometrika 52:225–247.

Kalmus H, Smith CAB. 1960. Evolutionary origin of sexual differentiation and the sex-ratio. Nature 186:1004–1006.

Karels A, Markkula E, Oikari A. 2001. Reproductive, biochemical, physiological, and population responses in perch (*Perca fluviatilis* L.) and roach (*Rutilus rutilus* l.) downstream of two elemental chlorine-free pulp and paper mills. Environ Toxicol Chem 20:1517–1527.

Kuhn A, Munns WR, Serbst JR, Edwards P, Cantwell MG, Gleason T, Pelletier M, Berry W. 2002. Evaluating the ecological significance of laboratory response data to predict population-level effects for the estuarine amphipod *Ampelisca abdita*. Environ Toxicol Chem 21:865–874.

Kuhn A, Munns WR, Champlin D, McKinney R, Tagliabue M, Serbst J, Gleason T. 2001. Evaluation of the efficacy of extrapolation population modeling to predict the dynamics of *Americamysis bahia* populations in the laboratory. Environ Toxicol Chem 20:213–221.

Li DM, Guo ZW. 2000. Some aspects of ecological modeling developments in China. Ecol Model 132:3–10.

Liess M. 2002. Population response to toxicants is altered by intraspecific interaction. Environ Toxicol Chem 21:138–142.

Liess M, Schulz R. 1999. Linking insecticide contamination and population response in an agricultural stream. Environ Toxicol Chem 18:1948–1955.

Lingjaerde OC, Stenseth NC, Kristoffersen AB, Smith RH, Moe SJ, Read JM, Daniels S, Simkiss K. 2001. Exploring the density-dependent structure of blowfly populations by nonparametric additive modeling. Ecology 82:2645–2658.

Linke-Gamenick I, Forbes VE, Sibly RM. 1999. Density-dependent effects of a toxicant on life-history traits and population dynamics of a *Capitellid polychaete*. Mar Ecol-Prog Ser 184:139–148.

Matson CW, Rodgers BE, Chesser RK, Baker RJ. 2000. Genetic diversity of *Clethrionomys glareolus* populations from highly contaminated sites in the Chernobyl region, Ukraine. Environ Toxicol Chem 19:2130–2135.

McCarthy MA, Burgman MA, Ferson S. 1995. The use of logistic regression in sensitivity analysis of population viability models. Biol Conserv 73:93–100.

Meyer JN, Di Giuliuo RT. 2003. Heritable adaptation and fitness costs in killifish (Fundulus Beteroclitus) inhabiting a polluted estuary. Ecol Appl 13:490–503.

Moe SJ, Stenseth NC, Smith RH. 2002. Density-dependent compensation in blowfly populations gives indirectly positive effects of a toxicant. Ecology 83:1597–1603.

Nakamaru M, Iwasa Y, Nakanishi J. 2003. Extinction risk of a bird population, using the stochastic differential equation model. Chemosphere 53:377–387.

Nevo E, Noy R, Lavie B, Bieles A, Muchtar S. 1986. Genetic diversity and resistance to marine pollution. Biol J Linn Soc 29:139–144.

Nisbet RM, Muller EB, Lika K, Kooijman S. 2000. From molecules to ecosystems through dynamic energy budget models. J Anim Ecol 69:913–926.

Peterson G, Cunningham S. 2000. The risks and benefits of genetically modified crops: a multidisciplinary perspective. Conserv Ecol 4:13.

Roark SA, Andrews JF, Guttman SI. 2001. Population genetic structure of the western *Mosquitofish, Gambusia Affinis*, in a highly channelized portion of the San Antonio River in San Antonio, TX. Ecotoxicology 10:223–227.

Salminen J, Haimi J. 2001. Life history and spatial distribution of the Enchytraeid worm Cognetta sphagnetorum (Oligochaeta) in metal-polluted soil: below-ground sink-source population dynamics? Environ Toxicol Chem 20:1993–1999.

Samson FB, Perez-Trejo F, Salwasser H, Ruggiero LF, Shaffer ML. 1985. On determining and managing minimum population size. Wildl Soc Bull 14:425–433.

Schlueter MA, Guttman SI, Oris JT, Bailer JA. 1997. Differential survival of fathead minnows, *Pimephales Promelas*, as affected by copper exposure, prior population stress, and *Allozyme Genotypes*. Environ Toxicol Chem 16:939–947.

Shaffer ML, Samson FB. 1985. Population size and extinction: a note on determining critical population sizes. Am Nat 125:144–152.

Shaw JL, Manning JP. 1996. Evaluating macroinvertebrate population and community level effects in outdoor microcosms: use of *in situ* bioassays and multivariate analysis. Environ Toxicol Chem 15:608–617.

Sinclair ARE. 1996. Mammalian populations: fluctuation, regulation, life history theory and their implications for conservation. In: Floyd RB, Sheppard AW, DeBarro PJ editors. Frontiers of population ecology. Victoria (Australia): Commonwealth Scientific and Industrial Research Organisation (CSIRO). p. 127–154.

Snell TW, Serra M. 2000. Using probability of extinction to evaluate the ecological significance of toxicant effects. Environ Toxicol Chem 19:2357–2363.

Stacy PB, Taper M. 1992. Environmental variation and the persistence of small populations. Ecol Appl 2:18–29.

Stark JD, Banken JAO. 1999. Importance of population structure at the time of toxicant exposure. Ecotoxicol Environ Safety 42:282–287.

Stark JD, Tanigoshi L, Bounfour M, Antonelli A. 1997. Reproductive potential: Its influence on the susceptibility of a species to pesticides. Ecotoxicol Environ Safe 37:273–279.

Staton JL, Schizas NV, Chandler GT, Coull BC, Quattro JM. 2001. Ecotoxicology and population genetics: the emergence of 'Phylogeographic and evolutionary ecotoxicology.' Ecotoxicology 10:217–222.

Tanaka Y, Nakanishi J. 2001. Life history elasticity and the population-level effect of p-nonylphenol on Daphnia galeata. Ecol Res 16:41–48.

Tanaka Y, Nakanishi J. 2001. Model selection and parameterization of the concentration-response functions for population-level effects. Environ Toxicol Chem 20:1857–1865.

Walthall WK, Stark JD. 1997. A comparison of acute mortality and population growth rate as endpoints of toxicological effects. Ecotoxicol Environ Safe 37:45–52.

Wang G, Edge W, Wolff J. 2001. Rainfall and guthion 2S interactions affect gray-tailed vole demography. Ecol Appl 11:928–933.

Wu JG, David JL. 2002. A spatially explicit hierarchical approach to modeling complex ecological systems: theory and applications. Ecol Model 153:7–26.

Index

A

B

C

D

E

F

G

H

I

K

L

M

P

Q

R

S

Y

Z

Other titles from the Society of Environmental Toxicology and Chemistry (SETAC)

Mercury Cycling in a Wetland-Dominated Ecosystem: A Multidisciplinary Study
O'Driscoll, Rencz, Lean
2005

Atrazine in North American Surface Waters: A Probabilistic Aquatic Ecological Risk Assessment
Giddings, editor
2005

Effects of Pesticides in the Field
Liess, Brown, Dohmen, Duquesne, Hart, Heimbach, Kreuger, Lagadic, Maund, Reinert, Streloke, Tarazona
2005

Human Pharmaceuticals: Assessing the Impacts on Aquatic Ecosystems
Williams, editor
2005

Toxicity of Dietborne Metals to Aquatic Organisms
Meyer, Adams, Brix, Luoma, Stubblefield, Wood, editors
2005

Toxicity Reduction and Toxicity Identification Evaluations for Effluents, Ambient Waters, and Other Aqueous Media
Norberg-King, Ausley, Burton, Goodfellow, Miller, Waller, editors
2005

Use of Sediment Quality Guidelines and Related Tools for the Assessment of Contaminated Sediments
Wenning, Batley, Ingersoll, Moore, editors
2005

Life-Cycle Assessment of Metals
Dubreuil, editor
2005

Working Environment in Life-Cycle Assessment
Poulsen and Jensen, editors
2005

Life-Cycle Management
Hunkeler, Saur, Rebitzer, Finkbeiner, Schmidt, Jensen, Stranddorf, Christiansen
2004

Scenarios in Life-Cycle Assessment
Rebitzer and Ekvall, editors

Ecological Assessment of Aquatic Resources: Linking Science to Decision-Making
Barbour, Norton, Preston, Thornton, editors
2004

Life-Cycle Assessment and SETAC: 1991–1999
15 LCA publications on CD-ROM
2003

Amphibian Decline: An Integrated Analysis of Multiple Stressor Effects
Greg Linder, Sherry K. Krest, Donald W. Sparling
2003

Metals in Aquatic Systems:
A Review of Exposure, Bioaccumulation, and Toxicity Models
Paquin, Farley, Santore, Kavvadas, Mooney, Winfield, Wu, Di Toro
2003

Silver: Environmental Transport, Fate, Effects, and Models:
Papers from Environmental Toxicology and Chemistry, 1983 to 2002
Gorusch, Kramer, La Point
2003

Code of Life-Cycle Inventory Practice
de Beaufort-Langeveld, Bretz, van Hoof, Hischier, Jean, Tanner, Huijbregts, editors
2003

Contaminated Soils: From Soil–Chemical Interactions to Ecosystem Management
Lanno, editor
2003

Environmental Impacts of Pulp and Paper Waste Streams
Stuthridge, van den Heuvel, Marvin, Slade, Gifford, editors
2003

Life-Cycle Assessment in Building and Construction
Kotaji, Edwards, Shuurmans, editors
2003

Porewater Toxicity Testing: Biological, Chemical, and Ecological Considerations
Carr and Nipper, editors
2003

Reevaluation of the State of the Science for Water-Quality Criteria Development
Reiley, Stubblefield, Adams, Di Toro, Erickson, Hodson, Keating Jr, editors
2003

Community-Level Aquatic System Studies—Interpretation Criteria (CLASSIC)
Giddings, Brock, Heger, Heimbach, Maund, Norman, Ratte, Schäfers, Streloke, editors
2002

SETAC

A Professional Society for Environmental Scientists and Engineers and Related Disciplines Concerned with Environmental Quality

The Society of Environmental Toxicology and Chemistry (SETAC), with offices currently in North America and Europe, is a nonprofit, professional society established to provide a forum for individuals and institutions engaged in the study of environmental problems, management and regulation of natural resources, education, research and development, and manufacturing and distribution.

Specific goals of the society are:

- Promote research, education, and training in the environmental sciences.
- Promote the systematic application of all relevant scientific disciplines to the evaluation of chemical hazards.
- Participate in the scientific interpretation of issues concerned with hazard assessment and risk analysis.
- Support the development of ecologically acceptable practices and principles.
- Provide a forum (meetings and publications) for communication among professionals in government, business, academia, and other segments of society involved in the use, protection, and management of our environment.

These goals are pursued through the conduct of numerous activities, which include:

- Hold annual meetings with study and workshop sessions, platform and poster papers, and achievement and merit awards.
- Sponsor a monthly scientific journal, a newsletter, and special technical publications.
- Provide funds for education and training through the SETAC Scholarship/Fellowship Program.
- Organize and sponsor chapters to provide a forum for the presentation of scientific data and for the interchange and study of information about local concerns.
- Provide advice and counsel to technical and nontechnical persons through a number of standing and ad hoc committees.

SETAC membership currently is composed of more than 5000 individuals from government, academia, business, and public-interest groups with technical backgrounds in chemistry, toxicology, biology, ecology, atmospheric sciences, health sciences, earth sciences, and engineering.

If you have training in these or related disciplines and are engaged in the study, use, or management of environmental resources, SETAC can fulfill your professional affiliation needs.

All members receive a newsletter highlighting environmental topics and SETAC activities, and reduced fees for the Annual Meeting and SETAC special publications.

All members except Students and Senior Active Members receive monthly issues of *Environmental Toxicology and Chemistry (ET&C)* and *Integrated Environmental Assessment and Management (IEAM)*, peer-reviewed journals of the Society. Student and Senior Active Members may subscribe to the journal. Members may hold office and, with the Emeritus Members, constitute the voting membership.

If you desire further information, contact the appropriate SETAC Office.

1010 North 12th Avenue
Pensacola, Florida 32501-3367 USA
T 850 469 1500 F 850 469 9778
E setac@setac.org

Avenue de la Toison d'Or 67
B-1060 Brussels, Belgium
T 32 2 772 72 81 F 32 2 770 53 86
E setac@setaceu.org

www.setac.org
Environmental Quality Through Science®

For Product Safety Concerns and Information please contact our EU
representative GPSR@taylorandfrancis.com
Taylor & Francis Verlag GmbH, Kaufingerstraße 24, 80331 München, Germany

www.ingramcontent.com/pod-product-compliance
Ingram Content Group UK Ltd.
Pitfield, Milton Keynes, MK11 3LW, UK
UKHW020022200926
475702UK00014B/2009

* 9 7 8 0 3 6 7 4 5 2 9 2 6 *